# Optical and Spectroscopic Properties of Rare-Earth-Doped Crystals

# Optical and Spectroscopic Properties of Rare-Earth-Doped Crystals

Editors

**Alessandra Toncelli**
**Željka Antić**

MDPI • Basel • Beijing • Wuhan • Barcelona • Belgrade • Manchester • Tokyo • Cluj • Tianjin

*Editors*
Alessandra Toncelli
University of Pisa
Italy

Željka Antić
University of Belgrade
Serbia

*Editorial Office*
MDPI
St. Alban-Anlage 66
4052 Basel, Switzerland

This is a reprint of articles from the Special Issue published online in the open access journal *Crystals* (ISSN 2073-4352) (available at: https://www.mdpi.com/journal/crystals/special_issues/RE_crystals).

For citation purposes, cite each article independently as indicated on the article page online and as indicated below:

LastName, A.A.; LastName, B.B.; LastName, C.C. Article Title. *Journal Name* **Year**, *Volume Number*, Page Range.

**ISBN 978-3-0365-7382-3 (Hbk)**
**ISBN 978-3-0365-7383-0 (PDF)**

© 2023 by the authors. Articles in this book are Open Access and distributed under the Creative Commons Attribution (CC BY) license, which allows users to download, copy and build upon published articles, as long as the author and publisher are properly credited, which ensures maximum dissemination and a wider impact of our publications.

The book as a whole is distributed by MDPI under the terms and conditions of the Creative Commons license CC BY-NC-ND.

# Contents

**About the Editors** . . . . . . . . . . . . . . . . . . . . . . . . . . . . . . . . . . . . . . . . . . . . . . . . . . . . . . . . . . . . . . . . vii

**Preface to "Optical and Spectroscopic Properties of Rare-Earth-Doped Crystals"** . . . . . . . . ix

**Aleksandar Ćirić, Tamara Gavrilović and Miroslav D. Dramićanin**
Luminescence Intensity Ratio Thermometry with $Er^{3+}$: Performance Overview
Reprinted from: *Crystals* 2021, 11, 189, doi:10.3390/cryst11020189 . . . . . . . . . . . . . . . . . . . 1

**Madeleine Fellner, Alberto Soppelsa and Alessandro Lauria**
Heat-Induced Transformation of Luminescent, Size Tuneable, Anisotropic $Eu:Lu(OH)_2Cl$ Microparticles to Micro-Structurally Controlled $Eu:Lu_2O_3$ Microplatelets
Reprinted from: *Crystals* 2021, 11, 992, doi:10.3390/cryst11080992 . . . . . . . . . . . . . . . . . . . 21

**Xavier H. Guichard, Francesco Bernasconi and Alessandro Lauria**
Charge Compensation in Europium-Doped Hafnia Nanoparticles: Solvothermal Synthesis and Colloidal Dispersion
Reprinted from: *Crystals* 2021, 11, 1042, doi:10.3390/cryst11091042 . . . . . . . . . . . . . . . . . . 29

**Soung-Soo Yi and Jae-Yong Jung**
Calcium Tungstate Doped with Rare Earth Ions Synthesized at Low Temperatures for Photoactive Composite and Anti-Counterfeiting Applications
Reprinted from: *Crystals* 2021, 11, 1214, doi:10.3390/cryst11101214 . . . . . . . . . . . . . . . . . . 39

**Kory Burns, Paris C. Reuel, Fernando Guerrero, Eric Lang, Ping Lu, Assel Aitkaliyeva, Khalid Hattar and Timothy J. Boyle**
Thermal Stability and Radiation Tolerance of Lanthanide-Doped Cerium Oxide Nanocubes
Reprinted from: *Crystals* 2021, 11, 1369, doi:10.3390/cryst11111369 . . . . . . . . . . . . . . . . . . 49

**Neeraj Panwar, Kuldeep Singh, Komal Kanwar, Yugandhar Bitla, Surendra Kumar and Venkata Sreenivas Puli**
Low-Temperature Magnetic and Magnetocaloric Properties of Manganese-Substituted $Gd_{0.5}Er_{0.5}CrO_3$ Orthochromites
Reprinted from: *Crystals* 2022, 12, 263, doi:10.3390/cryst12020263 . . . . . . . . . . . . . . . . . . . 69

**Milica Sekulić, Tatjana Dramićanin, AleksandarĆirić, Ljubica Đačanin Far, Miroslav D. Dramićanin and Vesna Đorđević**
Photoluminescence of the $Eu^{3+}$-Activated $Y_xLu_{1-x}NbO_4$ (x = 0, 0.25, 0.5, 0.75, 1) Solid-Solution Phosphors
Reprinted from: *Crystals* 2022, 12, 427, doi:10.3390/cryst12030427 . . . . . . . . . . . . . . . . . . . 83

**Chao Lv, Hong-Xin Yin, Yan-Long Liu, Xu-Xin Chen, Ming-He Sun and Hong-Liang Zhao**
Preparation of Cerium Oxide via Microwave Heating: Research on Effect of Temperature Field on Particles
Reprinted from: *Crystals* 2022, 12, 843, doi:10.3390/cryst12060843 . . . . . . . . . . . . . . . . . . . 95

**Soung-Soo Yi and Jae-Yong Jung**
Room Temperature Synthesis of Various Color Emission Rare-Earth Doped Strontium Tungstate Phosphors Applicable to Fingerprint Identification
Reprinted from: *Crystals* 2022, 12, 915, doi:10.3390/cryst12070915 . . . . . . . . . . . . . . . . . . . 107

**Weili Wang, Shihai Miao, Dongxun Chen and Yanjie Liang**
Rapid Aqueous-Phase Synthesis and Photoluminescence Properties of $K_{0.3}Bi_{0.7}F_{2.4}:Ln^{3+}$ (Ln = Eu, Tb, Pr, Nd, Sm, Dy) Nanocrystalline Particles
Reprinted from: *Crystals* 2022, 12, 963, doi:10.3390/cryst12070963 . . . . . . . . . . . . . . . . . . . 117

**Fulvia Gennari, Milica Sekulić, Tanja Barudžija, Željka Antić, Miroslav D. Dramićanin and Alessandra Toncelli**
Infrared Photoluminescence of Nd-Doped Sesquioxide and Fluoride Nanocrystals: A Comparative Study
Reprinted from: *Crystals* **2022**, *12*, 1071, doi:10.3390/cryst12081071 . . . . . . . . . . . . . . . . . **127**

**Yazhao Wang, Zhonghua Zhu, Shengdi Ta, Zeyu Cheng, Peng Zhang, Ninghan Zeng, et al.**
Optical Properties of Yttria-Stabilized Zirconia Single-Crystals Doped with Terbium Oxide
Reprinted from: *Crystals* **2022**, *12*, 1081, doi:10.3390/cryst12081081 . . . . . . . . . . . . . . . . . **139**

**Gábor Mandula, Zsolt Kis, Krisztián Lengyel, László Kovács and Éva Tichy-Rács**
Saturation Spectroscopic Studies on $Yb^{3+}$ and $Er^{3+}$ Ions in $Li_6Y(BO_3)_3$ Single Crystals
Reprinted from: *Crystals* **2022**, *12*, 1151, doi:10.3390/cryst12081151 . . . . . . . . . . . . . . . . . **153**

# About the Editors

**Alessandra Toncelli**

Prof. Alessandra Toncelli (ORCID No. 0000-0003-4400-8808) obtained her PhD in Physics in 1998 and has been the Associate Professor at the University of Pisa since 2017. Her research interests comprise the growth, development and characterization of optical materials and the devices used in the whole optical range, from UV-VIS to mid-infrared and THz regions. During her career, she characterized and optimized many materials, mainly rare-earth-doped crystals, aimed to be used in many different applications, such as photonics, scintillation, quantum information technologies, biomedicine, etc.

She has served in the organizing committee of many international conferences among the most renowned in the area (such as CLEO/Europe and the Europhoton Conference on Solid-State and Fiber Coherent Light Sources) and has been involved in many national and international projects.

In these research areas, she has published more than 190 articles in peer-reviewed international journals and holds two patents. She currently holds an h index of 51 in Scopus and 49 in ISI WoS.

**Željka Antić**

Dr. Željka Antić (ORCID No. 0000-0002-7990-2001), obtained her PhD in Chemistry and Chemical Technology from the University of Belgrade in 2010. She then spent three years as a postdoctoral fellow at the Department of Chemical and Materials Engineering at the University of Alberta, Edmonton, Canada. Currently, Dr. Antić is employed at the Vinča Institute of Nuclear Sciences—National Institute of the Republic of Serbia as a Research Professor. Dr. Antić is working in the Group for Optical Materials and Spectroscopy—OMAS (www.omasgroup.org)—and is the Head of the Laboratory for the Synthesis of Materials and Nanomaterials in the Center of Excellence for Photoconversion (www.converse-civ.org). Her research focuses on novel and innovative routes for luminescent nanomaterials/materials/thin-film synthesis, characterization, development and application. She is a co-founder of the International Conference on the Physics of Optical Materials and Devices—ICOM (www.icomonline.org)—and a member of the Organizing Committee of the International Conference on Phosphor Thermometry—ICPT (www.icpt-phosphor.com)—and the Serbian Conference on Materials Application and Technology—SCOM (www.razvojnauke.org/#SCOM_Conference). Dr. Antić was involved in numerous national and international projects, including the European Union's Horizon 2020 FET Open program project "Nanoparticles-based 2D thermal bioimaging technologies" (http://nanotbtech.web.ua.pt/) and NATO's Science for Peace and Security Multiyear project "The Optical Nose Grid for Large Indoor Area Explosives' Vapours Monitoring" (https://orion-sps.org/).

# Preface to "Optical and Spectroscopic Properties of Rare-Earth-Doped Crystals"

Photonics applications based on rare-earth (RE)-doped crystals are developing in many different fields, such as photovoltaic, laser technology, optical data storage, sensing, bioimaging, diagnosis and therapy. RE-doped inorganic bulk materials have long been known to cause luminescence emissions that are spectrally distributed throughout the whole optical range, from the ultraviolet (UV) to the mid-infrared region, with unique features which have made these materials very important, especially for laser applications. Moreover, when grown in nanometric size, these materials exhibit peculiar behaviors for their efficiency, lifetimes, energy transfer processes, interaction with the environment, etc., which have stimulated new research devoted, on the one hand, to the physical understanding of these phenomena, and on the other hand, to the development of many new applications. This book is dedicated to studying RE-doped bulk and nanocrystalline materials from a scientific point of view and to presenting their possible applications in any field.

This book is intended to serve as a unique multidisciplinary forum covering all aspects of science, technology and applications of RE-doped crystals, starting from the growth techniques with specific attention to the emission features of these materials and their applications.

**Alessandra Toncelli and Željka Antić**
*Editors*

Article

# Luminescence Intensity Ratio Thermometry with $Er^{3+}$: Performance Overview

Aleksandar Ćirić *, Tamara Gavrilović and Miroslav D. Dramićanin *

Vinča Institute of Nuclear Sciences—National Institute of the Republic of Serbia, University of Belgrade, P.O. Box 522, 11001 Belgrade, Serbia; tashichica@gmail.com
* Correspondence: aleksandar.ciric@ff.bg.ac.rs (A.Ć.); dramican@gmail.com (M.D.D.); Tel.: +381-63-7261078 (A.Ć.); +381-62-503669 (M.D.D.)

**Abstract:** The figures of merit of luminescence intensity ratio (LIR) thermometry for $Er^{3+}$ in 40 different crystals and glasses have been calculated and compared. For calculations, the relevant data has been collected from the literature while the missing data were derived from available absorption and emission spectra. The calculated parameters include Judd–Ofelt parameters, refractive indexes, Slater integrals, spin–orbit coupling parameters, reduced matrix elements (RMEs), energy differences between emitting levels used for LIR, absolute, and relative sensitivities. We found a slight variation of RMEs between hosts because of variations in values of Slater integrals and spin–orbit coupling parameters, and we calculated their average values over 40 hosts. The calculations showed that crystals perform better than glasses in $Er^{3+}$-based thermometry, and we identified hosts that have large values of both absolute and relative sensitivity.

**Keywords:** luminescence thermometry; phosphors; $Er^{3+}$; Judd–Ofelt; Slater integrals

---

**Citation:** Ćirić, A.; Gavrilović, T.; Dramićanin, M.D. Luminescence Intensity Ratio Thermometry with $Er^{3+}$: Performance Overview. *Crystals* **2021**, *11*, 189. https://doi.org/10.3390/cryst11020189

**Academic Editor:** Alessandro Chiasera

Received: 3 February 2021
Accepted: 10 February 2021
Published: 14 February 2021

**Publisher's Note:** MDPI stays neutral with regard to jurisdictional claims in published maps and institutional affiliations.

**Copyright:** © 2021 by the authors. Licensee MDPI, Basel, Switzerland. This article is an open access article distributed under the terms and conditions of the Creative Commons Attribution (CC BY) license (https://creativecommons.org/licenses/by/4.0/).

## 1. Introduction

The measurements of temperature, one of seven fundamental physical quantities, can be classified according to the nature of contact between the measurement object and instrument to invasive (where there is direct contact, e.g., thermocouples, thermistors), semi-invasive (where measuring object is altered in a way to enable contactless measurements), and non-invasive (where the temperature is estimated remotely, e.g., optical pyrometers) [1]. The first type necessarily perturbs the temperature of measurement objects which limits its use in microscopic objects. In addition, such approaches are difficult to implement on moving objects or in harsh environments, for example, in high-intensity electromagnetic fields, radioactive, or chemically challenging surroundings. Thus, the current market of thermometers, accounting for more than 80% of all sensors [2], demands methods that allow for remote or microscopic measurements. Among many perspective optical semi-invasive techniques, luminescence thermometry which uses thermographic phosphors has drawn the largest attention [3,4]. The thermographic phosphor probe can be incorporated within the measured object or on its surface, on macroscale to nanoscale sizes, or can be mounted on the surface of the fiber-optic cables and bring to proximity of measuring objects. Luminescent thermometry has found a range of valuable applications, from engineering to biomedical [5], and, currently, it is a widely researched topic with an exponentially increasing number of published research papers [6].

Presently, many types of materials are used for the construction of thermometry probes. These include rare-earth and transition metal activated phosphors, semiconductor quantum dots, organic dyes, and metal-organic complexes, carbon dots, and luminescent polymers. Among the rare-earth crystals are by far the most exploited type [5], usually exploited in the so-called luminescence intensity ratio (LIR, sometimes called fluorescence intensity ratio (FIR) or labeled as Δ) temperature read-out scheme that is based on the temperature-dependent intensity ratio of emissions from the thermally coupled excited levels of rare-earth ions.

The $Er^{3+}$ is considered to be workhorse in LIR-based luminescence thermometry. LIR read-out is obtained with both downshifting and up-conversion emissions, the latter most frequently sensitized by $Yb^{3+}$. The popularity of $Er^{3+}$ in luminescence thermometry is a consequence of its efficient emission and ~700 cm$^{-1}$ energy difference between its thermally coupled $^4S_{3/2}$ and $^2H_{11/2}$ levels. Such energy difference is ideal for thermometry in the physiological range of temperatures (30–50 °C) since it provides the maximal sensitivity of measurement in this range (~ 1.1%K$^{-1}$ at 303 K). To increase sensitivity, in recent times, the emission from the $Er^{3+}$ $^4F_{7/2}$ level is used for LIR considering the higher energy difference between $^4F_{7/2}$ and $^4S_{3/2}$ levels compared to the energy difference between $^2H_{11/2}$ and $^4S_{3/2}$ levels [7]. This $Er^{3+}$ LIR variant has perspective applications at high temperatures and greatly widens the sensor's operating temperature range.

Since $Er^{3+}$ can activate many hosts there are many investigations of luminescence thermometry using emissions of this ion. These studies are lengthy and cumbersome involving material synthesis and characterizations, measurements of emission spectra at various temperatures, complex data fitting and analysis, and evaluation of thermometric performance. Therefore, to alleviate this problem, the theoretical prediction of $Er^{3+}$ thermometric performance in different hosts may be useful as a guide for the host selection.

Our previous research has demonstrated that LIR and figures of merit of luminescence thermometry can be predicted by a theoretical model that involves the famous Judd–Ofelt (JO) theory with the high matching to experimental data [6,8]. JO theory explains and predicts the intensities of the trivalent rare-earth ions' ($RE^{3+}$) f-f electronic transitions, and its parameters include all phenomenological mechanisms responsible for the line strengths observed in both absorption and emission spectra.

Here, we aimed to perform the theoretical analysis of $Er^{3+}$ emissions involved in the LIR thermometry, linking the temperature sensing performance of materials (LIR absolute and relative sensitivities) with their composition and structural properties. For this extensive analysis we have selected 40 different materials in crystal and glass form, and collected the relevant data from the literature. The refractive index values were calculated from Sellmeier's equation (where available). The energy level positions are given for each host. For 16 hosts, the reduced matrix elements (RMEs) are recalculated from the Slater integrals and spin–orbit (s–o) parameters. The relation between material properties, JO parameters, Slater integrals, and s–o parameters is provided.

## 2. Methods

### 2.1. Luminescence Intensity Ratio Method

Out of all the luminescence temperature read-out methods the LIR has been the most frequently investigated since it is simple, requires inexpensive equipment, and, most importantly, is ratiometric and self-referenced [7]. The temperature read-out is not affected by fluctuations in excitation power. The most interesting materials for LIR thermometry are trivalent lanthanides, as they have sharp emission lines and plenty to choose from, from ultraviolet (UV) to near-infrared (NIR) spectral regions. LIR is defined as the ratio of two emissions that varies with temperature [9]. In the case of two energy levels separated by < 2000 cm$^{-1}$, it is said that they are thermally coupled, as the higher energy level can be effectively populated by the thermal energy. The ratio of populations of optical centers in the energetically higher (H) and lower (L) levels is then proportional to the Boltzmann distribution: $N_H/N_L \sim \exp(-\Delta E/kT)$, where $\Delta E$ is the energy difference between the levels of interest, k = 0.695 cm$^{-1}$ K$^{-1}$ is the Boltzmann constant, and T is the temperature. As the intensities are proportional to the population, the equation for LIR for the Boltzmann thermometer is:

$$LIR(T) = \frac{I_H(T)}{I_L(T)} = B \times \exp\left(-\frac{\Delta E}{kT}\right) \quad (1)$$

where B is the temperature invariant parameter that depends only on the host.

The thermometer's performance is estimated by the absolute ($S_a$) and relative ($S_r$) sensitivities, and temperature resolution ($\Delta T$) given by [10]:

$$S_a = \left|\frac{\partial LIR}{\partial T}\right|, \quad S_r = \frac{S_a}{LIR} \cdot 100\%, \quad \Delta T = \frac{\sigma_a}{S_a} = \frac{\sigma_r}{S_r} \qquad (2)$$

where $\sigma_a$ and $\sigma_r$ are the absolute and relative uncertainties in measurement of LIR, presented as standard deviations.

For the temperature dependence of $LIR$ given by Equation (1), sensitivities have the following form:

$$S_a = \frac{\Delta E}{kT^2} B \exp\left(-\frac{\Delta E}{kT}\right), \quad S_r = \frac{\Delta E}{kT^2} \cdot 100\% \qquad (3)$$

The absolute sensitivity reaches maximum at $T = \Delta E / 2k$ with the value of [8]:

$$S_{amax} = \frac{4Bk}{e^2 \Delta E} \qquad (4)$$

where $e = 2.718$ is a number (the natural logarithm base).

The ideal situation for LIR is the Boltzmann luminescence thermometer since it is easily calibrated with the well-known and simple theory. According to Equation (3), the relative sensitivity only depends on the value of energy difference between thermalized energy levels. The choice of levels with the energy gap larger than 2000 cm$^{-1}$ may result in the thermalization loss at low temperatures, and even around room temperatures, while the small energy gap gives small relative sensitivities. One should consider that for achieving the Boltzmann's thermal equilibrium some other conditions must be fulfilled besides the suitable energy difference between the levels, as recently demonstrated by Geitenbeek et al. [11] and Suta et al. [12]. Furthermore, considering the adjacent energy levels of trivalent rare-earth used for thermometry, the largest energy gap is in the Eu$^{3+}$ (between $^5D_1$ and $^5D_0$ levels) and is approximately 1750 cm$^{-1}$.

Thus, the current research of LIR of a single emission center is aimed at increasing the relative sensitivity without the loss of thermalization (and deviation from the Boltzmann distribution). One recently demonstrated solution is the inclusion of the third, non-adjacent level, with higher energy, which is thermalized with the second level. If the first and second levels are thermalized, and second and third levels are thermalized, then the ratios of emission intensities of the first and third levels will follow the Boltzmann distribution, even if their separation is greater than stated above, see Figure 1 for the case of Er$^{3+}$. The conventional LIR of Er$^{3+}$ is equal to the ratio of emissions from $^2H_{11/2}$ (~523 nm) and $^4S_{3/2}$ (~542 nm) levels, which are separated by ~700 cm$^{-1}$, thus giving the relative sensitivity of ~1.1% K$^{-1}$. By observing intensities to the ground level from $^4F_{7/2}$ (~485 nm) with the $^4S_{3/2}$, it is evident that their relative change is much larger than with $^2H_{11/2}$. This larger energy difference, according to Equation (3), ultimately results in more than a three-fold increase in relative sensitivity.

## 2.2. Judd–Ofelt Theory and Its Relevance for Luminescence Thermometry

The electronic configuration of trivalent erbium is that of Xenon plus the 11 electrons in the 4f shell, i.e., Er$^{3+}$ has [Xe]4f$^{11}$ electronic configuration. With only 3 electrons missing from the completely filled 4f shell, Er$^{3+}$ shares the same LS terms and LSJ levels as the Nd$^{3+}$ who has 3 electrons in the 4f shell. The transitions from one to another level are followed by the reception or release of energy. The probability for such phenomena for a given set of initial and final levels is given by the wavefunctions and the appropriate moment operator. The exchange of energy in the intra-configurational 4f transitions with the highest intensity is of induced electric dipole and magnetic dipole types [13]. What was puzzling only half a century ago was the origin of these "electric dipole" interactions, as they were clearly and strictly forbidden by the Parity selection rule, also known as the Laporte rule. The solution to this problem came in 1962 in the papers simultaneously published by Judd [14] and Ofelt [15], to what is latter known as the Judd–Ofelt theory (JO). For the sake of brevity, it will not be explained here, and the reader is instead referred to the excellent references [16–18]; however, we will touch on several basics that are the most relevant for the present research.

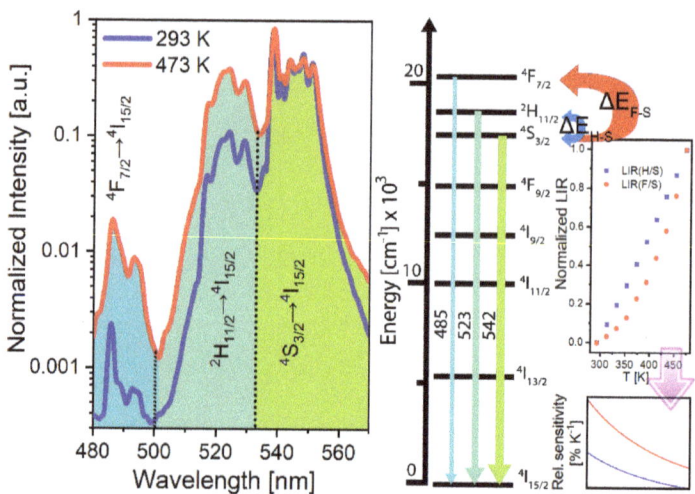

**Figure 1.** Emission spectra of YF$_3$:Er$^{3+}$ recorded at 293 K and 473 K, the energy level diagram depicting the emissions $^4F_{7/2}$, $^2H_{11/2}$, $^4S_{3/2} \rightarrow {}^4I_{15/2}$ of Er$^{3+}$, the energy difference between $^2H_{11/2}$ and $^4F_{7/2}$ levels from $^4S_{3/2}$ level (blue and red arrows, respectively), the normalized LIRs of $^2H_{11/2} \rightarrow {}^4I_{15/2}/{}^4S_{3/2} \rightarrow {}^4I_{15/2}$ and $^4F_{7/2} \rightarrow {}^4I_{15/2}/{}^4S_{3/2} \rightarrow {}^4I_{15/2}$ and the corresponding relative sensitivities on the given temperature range.

In RE$^{3+}$ (trivalent rare-earth) ions in general, the electrostatic ($H_e$) and s–o ($H_{so}$) interactions between 4f electrons are dominant and are of the approximately same magnitude, thus the Hamiltonian can be given approximately as $H = H_e + H_{so}$ [17].

The electrostatic Hamiltonian can be reduced to the electron-electron repulsion form [17], which can be split into its radial ($F^k$) and angular ($f_k$) parts:

$$H_e = \frac{1}{4\pi\varepsilon_0} \sum_{i<j}^{11} \frac{e^2}{r_{ij}} = \sum_k F^k f_k \quad (5)$$

The radial parameters are the Slater integrals given by [19]:

$$F^k(4f) = \frac{e^2}{4\pi\varepsilon_0} \iint_0^\infty \frac{r_<^k}{r_>^{k+1}} R_{4f}^2(r_i) R_{4f}^2(r_j) dr_i dr_j \quad (6)$$

where $r_>$ is greater and $r_<$ is smaller than $r_i$ and $r_j$, and $R$ are the radial parts of the wavefunction. The Slater integrals can be evaluated by the Hartree-Fock method; however, it does not provide accurate results, and it is best to obtain them semi-empirically by adjusting them to the experimentally observed energies of 4f levels [20].

$H_{so}$ mixes all states that have the same $J$ quantum number, and it is proportional to the s–o coupling parameter, $\zeta$, which is further proportional to the number of electrons within the 4f shell. Er$^{3+}$ has a relatively high value of $\zeta$ in comparison with other trivalent lanthanide ions, providing a large mixing of states [21]. In this intermediate coupling approximation scheme, the wavefunctions are expressed as a linear combination of all other states in the configuration with the same $J$ quantum number [17,18,22,23]:

$$\left|4f^{11}SLJ\right\rangle' = \sum_i c_i \left|4f^{11}S'L'J\right\rangle, \quad \sum_i c_i^2 = 1 \tag{7}$$

As the 4f electrons are shielded by the outer higher-energy electrons the crystal field (CF) introduces only a perturbation to the Hamiltonian [16]. Nevertheless, that perturbation weakens the already mentioned Laporte (parity) selection rule that forbids the ED transitions within the configuration. The 4f–4f transitions of electric dipole (ED) type become allowed and are known as the induced ED [24]. The radiative transition probability for such spontaneous emission is then equal to [25]:

$$A_{SLJ \to S'L'J'} = \frac{64\pi^4 \tilde{\nu}_{SLJ \to S'L'J'}^3}{3h(2J+1)} (\chi_{ED} D_{ED} + \chi_{MD} D_{MD}) \tag{8}$$

or for the purely induced ED emission (MD is an abbreviation for the magnetic dipole):

$$A_{SLJ \to S'L'J'} = \frac{64\pi^4 \tilde{\nu}_{SLJ \to S'L'J'}^3}{3h(2J+1)} \chi_{ED} D_{ED} \tag{9}$$

where $h = 6.626 \times 10^{-27}$ erg·s is the Planck's constant. $X$ is the local field correction, $\tilde{\nu}_{SLJ \to S'L'J'}$ is the emission barycenter energy, and $D$ is the dipole strength given in esu$^2$ cm$^2$ units. The emission barycenter is [26]:

$$\tilde{\nu}_{SLJ \to S'L'J'} = \frac{\int \nu_{SLJ \to S'L'J'} i_{SLJ \to S'L'J'}(\nu) d\nu}{\int i_{SLJ \to S'L'J'}(\nu) d\nu} \tag{10}$$

where $i$ is the intensity at a given energy. The local field correction for ED emission is given by [27]:

$$\chi_{ED} = \frac{n(n^2+2)^2}{9} \tag{11}$$

where $n$ is the refractive index that should be given at the wavelength of the barycenter of the emission. It can be calculated from the Sellmeier's equation for a given material, which is given in the form [28,29]:

$$n(\lambda) = \sqrt{1 + \sum_{i=1}^{3} \frac{B_i \lambda^2}{\lambda^2 - C_i}} \tag{12}$$

In the JO scheme, the ED strength is given by [26]:

$$D_{ED}^{SLJ \to S'L'J'} = e^2 \sum_{\lambda=2,4,6} \Omega_\lambda U_{SLJ \to S'L'J'}^\lambda \tag{13}$$

where $e = 4.803 \times 10^{10}$ esu, $U_{SLJ \to S'L'J'}^\lambda$ is the abbreviation for squared RMEs $|\langle 4f^{11}SLJ \| U^\lambda \| 4f^{11}S'L'J' \rangle|^2$, which in turn can be calculated from the Slater integrals and the s–o coupling parameter. $\Omega_\lambda$ are the JO intensity parameters, obtained semi-empirically or by the ab initio calculations (from the crystal-field parameters).

The integrated emission intensity for the transition $SLJ \to S'L'J'$ is given by [30,31]:

$$I_{SLJ \to S'L'J'} = \int i_{SLJ \to S'L'J'}(\tilde{\nu}) d\tilde{\nu} = h\tilde{\nu}_{SLJ \to S'L'J'} N_{SLJ} A_{SLJ \to S'L'J'} \tag{14}$$

or without the $h\nu$ if the spectrum is recorded in counts instead of power units [32].

LIR of two emissions from the thermally coupled levels is then given by:

$$LIR(T) = \frac{I_H}{I_L} = \frac{\nu_H N_H A_H}{\nu_L N_L A_L} \quad (15)$$

where $I_{H/L}$ are the integrated intensities from the higher and lower level, respectively (without $\nu_H/\nu_L$ if recorded in counts).

According to the Boltzmann distribution, the optical center population is given by:

$$\frac{N_H}{N_L} = \frac{g_H}{g_L} \exp\left(-\frac{\Delta E}{kT}\right) \quad (16)$$

where $g = 2J + 1$ are the degeneracies of the selected levels.

Equation (15) can be rewritten as Equation (1) where B is the temperature invariant parameter that is given by:

$$B = \frac{\nu_H g_H A_H}{\nu_L g_L A_L} \quad (17)$$

or if the intensities are recorded in counts instead of power units, without the $\nu_H/\nu_L$. As we have demonstrated in our previous article [8], by inserting Equation (8) into Equation (17), the LIR, the absolute sensitivity (and everything related to it) and temperature resolution can be predicted by JO parameters, as the B parameter can be obtained from:

$$B = \left(\frac{\tilde{\nu}_H}{\tilde{\nu}_L}\right)^4 \frac{\chi^H_{ED} D^H_{ED} + \chi^H_{MD} D^H_{MD}}{\chi^L_{ED} D^L_{ED} + \chi^L_{MD} D^L_{MD}} \quad (18)$$

or in the case of the pure ED transitions:

$$B = \left(\frac{\tilde{\nu}_H}{\tilde{\nu}_L}\right)^4 \frac{\chi^H_{ED} D^H_{ED}}{\chi^L_{ED} D^L_{ED}} \quad (19)$$

For the case of spectra recorded in counts, $\nu_H/\nu_L$ should be to the power of 3.

The shielding of 4f electrons by electrons from outer orbitals ensures that the $RE^{3+}$ spectra are featured by sharp peaks whose energies are almost host-independent. This is reflected in the almost host invariant reduced matrix elements. However, as the Slater parameters deviate significantly in $Er^{3+}$, using such approximation may introduce significant errors. For the analysis of this type, it is more accurate to use the reduced matrix elements that are calculated from Slater integrals and s–o coupling parameters, which are calculated semi-empirically from the positions of the energy levels. Analogously, the small variations in energy level positions may provide significant variations in energy level differences, and thus large deviations in absolute and relative sensitivities. Finally, the small differences in refractive index become enormous when they propagate in the local field correction coefficient and, thus, it is of utmost importance to use accurate values. In this study, the observed levels are energetically very close, thus, it is a good approximation to consider the refractive index as wavelength-independent; however, the exact method is always preferred.

### 3. Results and Discussion

#### 3.1. Calculations of $Er^{3+}$ Radiative Properties in Different Hosts

For the study, we have selected 40 different hosts doped with $Er^{3+}$ (Table 1), from the literature that contained the most complete set of data needed for the analysis presented in this paper. As the JO parametrization is traditionally performed semi-empirically from the absorption spectrum, powders and non-transparent materials are not included in this analysis.

Table 1. A collection of 40 Er$^{3+}$ doped hosts used in this study. Form: C–Crystal, G–Glass.

| No. | Host | Abbreviation | Name | Er$^{3+}$ Concentration | Form | Ref. |
|---|---|---|---|---|---|---|
| 1 * | β-NaGdF$_4$ | | Sodium Gadolinium Fluoride | 1% | C | [33] |
| 2 | LaF$_3$ | | Lanthanum fluoride | 0.05% | C | [34,35] |
| 3 | Y$_3$Al$_5$O$_{12}$ | YAG | Yttrium Aluminium Garnet | 1.2 at% | C | [36] |
| 4 | LaCl$_3$ | | Lanthanum Chloride | 1% | C | [17] |
| 5 | La$_2$O$_2$S | | Lanthanum oxysulfide | 1 mol% | C | [37,38] |
| 6 | Y$_2$O$_2$S | | Yttrium oxysulfide | 1 mol% | C | [37,38] |
| 7 | Sb$_2$O$_3$-35P$_2$O$_5$-5MgO-AgCl | SPMEA | Antimony Phosphate | 0.17 at% | G | [39] |
| 8 | YAlO$_3$ | | Yttrium Orthoaluminate | 1.5 at% | C | [40] |
| 9 | KCaF$_3$ | | Kalium Calcium Fluoride | 1.62 at% | C | [41] |
| 10 | Y$_2$O$_3$ | | Yttrium oxide | 1% | C | [42,43] |
| 11 | YVO$_4$ | | Yttrium Vanadate | 2.5% | C | [44] |
| 12 | PbF$_2$-GaF$_3$-(Zn,Mn)F$_2$ | PbZnGaLaF | Lead-Based fluoride | 0.6% | G | [45] |
| 13 * | CaYAlO$_4$ | CYAO | Yttrium Calcium Aluminate | 0.5 at% | C | [46] |
| 14 | Gd$_3$Ga$_5$O$_{12}$ | GGG | Gadolinium Gallium Garnet | 1.2 at% | C | [36] |
| 15 | Y$_3$Sc$_2$Ga$_3$O$_{12}$ | YSGG | Yttrium Scandium Gallium Garnet | 1.2 at% | C | [36] |
| 16 | PbO·H$_3$BO$_3$·TiO$_2$·AlF$_3$ | LBTAF | Lead Borate Titanate Aluminium Fluoride | 4.8 at% | G | [47] |
| 17 | Li$_2$CO$_3$·H$_3$BO$_3$ | LiBO | Lithium Borate | 1 mol% | G | [48] |
| 18 | Li$_2$CO$_3$·H$_3$BO$_3$·MgCO$_3$ | MgLiBO | Magnesium Lithium Borate | 1 mol% | G | [48] |
| 19 | Li$_2$CO$_3$·H$_3$BO$_3$·MgCO$_3$ | CaLiBO | Calcium Lithium Borate | 1 mol% | G | [48] |
| 20 | Li$_2$CO$_3$·H$_3$BO$_3$·SrCO$_3$ | SrLiBO | Strontium Lithium Borate | 1 mol% | G | [48] |
| 21 | Li$_2$CO$_3$·H$_3$BO$_3$·BaCO$_3$ | BaLiBO | Barium Lithium Borate | 1 mol% | G | [48] |
| 22 | ZrO$_2$·YO$_{1.5}$ | YSZ | Yttria stabilized Zirconia | 0.7 at% | C | [49] |
| 23 | LiNbO$_3$ | | Lithium Niobate | $1.5 \times 10^{19}$ cm$^{-3}$ | C | [50] |
| 24 | PbO-PbF$_2$ | | Oxyfluoride | 1.35 wt% | G | [51] |
| 25 | ZrF$_4$·BaF$_2$·LaF$_3$·AlF$_3$ | ZBLA | Fluorozirconate | 0.5% | G | [52] |
| 26 | ZnO·Al$_2$O$_3$·Bi$_2$O$_3$·B$_2$O$_3$ | ZnAlBiB | Zinc Alumino Bismuth Borate | 0.5 mol% | G | [53] |
| 27 | NaPO$_3$·TeO$_2$·AlF$_3$·LiF | LiTFP | Lithium Fluorophosphate | 0.12 at% | G | [54] |
| 28 | NaPO$_3$·TeO$_2$·AlF$_3$·NaF | NaTFP | Sodium Fluorophosphate | 0.12 at% | G | [54] |
| 29 | NaPO$_3$·TeO$_2$·AlF$_3$·KF | KTFP | Kalium Fluorophosphate | 0.12 at% | G | [54] |
| 30 | Kigre patented | | Phosphate | 1.51 wt% | G | [55] |
| 31 | TeO$_2$·PbF$_2$·AlF$_3$ | | Fluoro tellurite | 0.625 at% | G | [56] |
| 32 | SrGdGa$_3$O$_7$ | | Strontium Gadolinium Gallium Garnet | $4.2 \times 10^{21}$ cm$^{-3}$ | C | [57] |
| 33 | YPO$_4$ | | Yttrium Phosphate | 0.6 at% | C | [58] |
| 34 | Y$_2$SiO$_5$ | YSO | Yttrium Orthosilicate | 2 mol%$^b$ | C | [59] |
| 35 | BaGd$_2$(MoO$_4$)$_4$ | BGM | Barium Gadolinium Molybdate | 1.4 at% | C | [60] |
| 36 | NaY(MoO$_4$)$_2$ | NYM | Sodium Yttrium Molybdate | 1.25 at% | C | [61] |
| 37 | Gd$_2$(MoO$_4$)$_3$ | | Gadolinium Molybdate | 1% | C | [62] |
| 38 | LiLa(MoO$_4$)$_2$ | | Lithium Lanthanum Molybdate | 0.55 at% | C | [63] |
| 39 * | LiLa(WO$_4$)$_2$ | | Lithium Lanthanum Tungstanate | 0.65% | C | [64] |
| 40 | KY(WO$_4$)$_2$ | | Kalium Yttrium Tungstanate | 0.5% | C | [65] |

* Co-doped with Yb$^{3+}$.

In the 3rd column, Table 2 gives the energies of the $^4S_{3/2}$, $^2H_{11/2}$, and $^4F_{7/2}$ levels used for the two LIRs that will be theoretically investigated. As stated in the introduction, this is important in the estimation of the thermometric figures of merit, and it is linked to the Slater integrals and s–o parameters. The table also includes the Slater integrals and s–o coupling parameters of the 16 out of the 40 hosts, and the JO intensity parameters for all the hosts, taken from references Table 1.

**Table 2.** Energies of $^4S_{3/2}$, $^2H_{11/2}$ and $^4F_{7/2}$ levels in hosts listed in Table 1, their Slater integrals and spin-orbit (s–o) coupling parameters, and JO intensity parameters as reported in the corresponding references.

| No. | Host | Energy [cm$^{-1}$] | | | Slater Integrals and s–o [cm$^{-1}$] | | | | JO Parameters × 10$^{20}$ [cm$^2$] | | |
|---|---|---|---|---|---|---|---|---|---|---|---|
| | | $^4S_{3/2}$ | $^2H_{11/2}$ | $^4F_{7/2}$ | $F_2$ | $F_4$ | $F_6$ | $\zeta$ | $\Omega_2$ | $\Omega_4$ | $\Omega_6$ |
| 1 | β–NaGdF$_4$ | 18459 | 19186 | 20483 | 429.4 | 68.7 | 7.1 | 2403 | 4.97 | 1.16 | 2.03 |
| 2 | LaF$_3$ | 18353 | 19118 | 20412 | 435.7 | 67.2 | 7.4 | 2351 | 3.9 | 1.0 | 2.3 |
| 3 | YAG | 18166 | 18967 | 20344 | 433.2 | 65.0 | 6.5 | 2345 | 0.74 | 0.92 | 0.70 |
| 4 | LaCl$_3$ | 18383 | 19068 | 20414 | 433.2 | 66.9 | 7.3 | 2386 | 5.45 | 2.08 | 0.69 |
| 5 | La$_2$O$_2$S | 18236 | 18930 | 20320 | 445.8 | 66.6 | 7.5 | 2395 | 4.32 | 2.32 | 1.17 |
| 6 | Y$_2$O$_2$S | 18236 | 18930 | 20320 | 444.0 | 68.1 | 7.4 | 2391 | 2.71 | 2.10 | 1.85 |
| 7 [d] | SPMEA | 18182 | 19175 | 20576 | NA | NA | NA | NA | 7.89 | 3.27 | 1.07 |
| 8 | YAlO$_3$ | 18350 | 19150 | 20300 | NA | NA | NA | NA | 0.95 | 0.58 | 0.55 |
| 9 [d] | KCaF$_3$ | 18450 | 19305 | 20576 | NA | NA | NA | NA | 0.74 | 0.87 | 0.57 |
| 10 [b] | Y$_2$O$_3$ | 18071 | 18931 | 20266 | 429.6 | 65.0 | 7.1 | 2383 | 4.59 | 1.21 | 0.48 |
| 11 [b] | YVO$_4$ | 18209 | 19059 | 20371 | 440.8 | 66.8 | 7.3 | 2381 | 13.45 | 2.23 | 1.67 |
| 12 | PbZnGaLaF | 18552 | 19193 | 20618 | 444.0 | 64.6 | 6.9 | 2395 | 1.54 | 1.13 | 1.19 |
| 13 | CYAO | 18298 | 19040 | 20454 | NA | NA | NA | NA | 3.78 | 2.52 | 1.91 |
| 14 [c] | GGG | 18450 | 19168 | 20387 | NA | NA | NA | NA | 0.70 | 0.37 | 0.86 |
| 15 [c] | YSGG | 18433 | 19127 | 20345 | NA | NA | NA | NA | 0.92 | 0.48 | 0.87 |
| 16 [a] | LBTAF | 18382 | 19194 | 20450 | 433.9 | 67.0 | 6.7 | 2386 | 5.89 | 1.10 | 1.47 |
| 17 [a] | LiBO | 18413 | 19205 | 20488 | 437.8 | 65.1 | 6.8 | 2377 | 3.24 | 0.92 | 0.82 |
| 18 [a] | MgLiBO | 18389 | 19168 | 20462 | 437.8 | 64.5 | 6.8 | 2373 | 1.33 | 0.39 | 0.62 |
| 19 [a] | CaLiBO | 18413 | 19205 | 20488 | 438.2 | 64.9 | 6.8 | 2376 | 3.68 | 0.76 | 1.52 |
| 20 [a] | SrLiBO | 18413 | 19205 | 20488 | 438.6 | 64.7 | 6.8 | 2381 | 2.53 | 0.39 | 1.10 |
| 21 [a] | BaLiBO | 18403 | 19205 | 20488 | 438.2 | 64.9 | 6.8 | 2382 | 1.80 | 0.28 | 0.90 |
| 22 [c] | YSZ | 18416 | 19342 | 20534 | NA | NA | NA | NA | 1.50 | 0.50 | 0.22 |
| 23 [c] | LiNbO$_3$ | 18248 | 19047 | 20492 | NA | NA | NA | NA | 7.29 | 2.24 | 1.27 |
| 24 [a] | PbO-PbF$_2$ | 18501 | 19297 | 20601 | 437.8 | 66.6 | 7.3 | 2461 | 3.22 | 1.34 | 0.61 |
| 25 | ZBLA | 18450 | 19193 | 20534 | NA | NA | NA | NA | 2.54 | 1.39 | 0.97 |
| 26 [c] | ZnAlBiB | 18484 | 19193 | 20491 | NA | NA | NA | NA | 2.10 | 1.53 | 1.43 |
| 27 | LiTFP | 18344 | 19189 | 20486 | NA | NA | NA | NA | 4.70 | 1.21 | 1.30 |
| 28 | NaTFP | 18377 | 19226 | 20486 | NA | NA | NA | NA | 5.92 | 1.07 | 1.44 |
| 29 | KTFP | 18377 | 19226 | 20486 | NA | NA | NA | NA | 5.09 | 0.69 | 1.45 |
| 30 | Kigre patented | 18350 | 19150 | 20300 | NA | NA | NA | NA | 6.28 | 1.03 | 1.39 |
| 31 [c] | TeO$_2$·PbF$_2$·AlF$_3$ | 18332 | 18957 | 20434 | NA | NA | NA | NA | 5.52 | 2.07 | 1.00 |
| 32 | SrGdGa$_3$O$_7$ | 18135 | 19131 | 20411 | NA | NA | NA | NA | 2.46 | 1.24 | 0.51 |
| 33 | YPO$_4$ | 18348 | 19083 | 20449 | 435.7 | 67.2 | 7.4 | 2351 | 3.02 | 3.07 | 2.58 |
| 34 [e] | YSO | 18348 | 19083 | 20449 | NA | NA | NA | NA | 1.29 | 0.29 | 2.78 |
| 35 [c] | BGM | 18348 | 19083 | 20408 | NA | NA | NA | NA | 12.33 | 1.96 | 0.96 |
| 36 [c] | NYM | 18148 | 19157 | 20449 | NA | NA | NA | NA | 13.34 | 1.69 | 2.29 |
| 37 | Gd$_2$(MoO$_4$)$_3$ | 18348 | 19157 | 20449 | NA | NA | NA | NA | 11.74 | 8.16 | 3.98 |
| 38 | LiLa(MoO$_4$)$_2$ | 18348 | 19157 | 20449 | NA | NA | NA | NA | 8.07 | 1.06 | 0.83 |
| 39 | LiLa(WO$_4$)$_2$ | 18348 | 19047 | 20345 | NA | NA | NA | NA | 9.03 | 2.02 | 0.59 |
| 40 | KY(WO$_4$)$_2$ | 18420 | 19190 | 20450 | NA | NA | NA | NA | 7.08 | 2.30 | 1.01 |

[a] Slater integrals calculated from $F^{2,4,6}$ parameters by Equation (15) in Ref. [17]. [b] Slater integrals calculated from Racah parameters by Equation (17) in Ref. [17]. [c] Slater integrals and spin–orbit coupling parameter not provided, the RME values the authors used are by Carnall in Ref. [66], or by [d] Weber in Ref. [35]. [e] Energy levels are not given in the literature, values in the table are provided approximately.

Figure 2 presents the variation of Slater integrals and s–o coupling parameters in those 16 hosts. Although there are no large differences in parameters between the crystals and glasses, there are certain trends that may be observed by the type of compound. Deviations in parameters' values from host to host can be large, so the use of Carnall or Weber tables [22,35] for Er$^{3+}$ RMEs can introduce large errors in the later calculations.

Figure 3 presents the JO parameters as given in Table 2. Glass hosts have smaller values of JO parameters than crystals, on average. When crystals are analyzed, the largest values of $\Omega_2$ parameter are found in tungstates and molybdates, while the smallest values are in garnets, phosphates, silicates, and oxysulfides. $\Omega_6$ are expectedly higher in fluorides, phosphates, and silicates. In glasses, borate glasses have lower $\Omega_2$, while phosphate glasses have higher $\Omega_2$. $\Omega_6$ is on average higher in phosphate glasses. No clear correlation could be given for the $\Omega_4$ parameter in crystals or glasses.

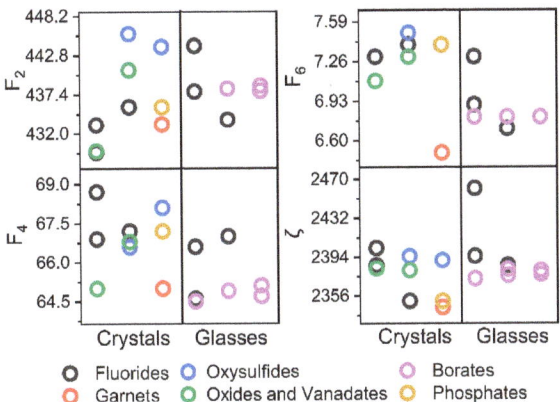

Figure 2. Slater integrals and s–o coupling parameters as listed in Table 2.

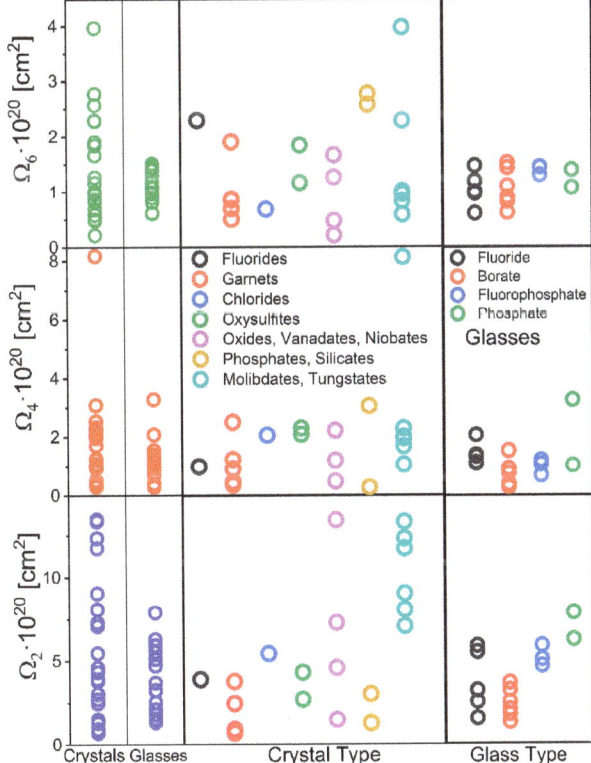

Figure 3. Judd–Ofelt parameters in different hosts, as given in Table 2.

The squared RMEs for each transition investigated for LIR are given in Table 3. This list can be used beyond the scope of this paper for accurate calculations of JO parameters.

The deviations from the average RME values are given in Figure 4, and they are large for the $^2H_{11/2} \rightarrow {}^4I_{15/2}$ transition. Thus, the use of Carnall's or Weber's values [22,35] might introduce significant errors in the JO parameters estimation, as the RMEs were calculated for the LaF$_3$ and YAlO$_3$, respectively. The average RMEs values calculated from Table 3 are given in Table 4, together with the deviations from the values by Carnall and Weber. The refractive index values taken from the corresponding references are also listed in Table 3. If Sellmeier's equation is given, the refractive index is calculated at the wavelength of the emission. From the refractive index value, the local field correction is calculated according to Equation (11). The induced ED strengths (the last column of Table 3) are calculated for each transition and for each using JO parameters from Table 2 and local field corrections and RMEs from Table 3.

**Table 3.** Squared RMEs for hosts in Table 1, recalculated by RELIC software [17] from Slater integrals and s–o coupling parameters in Table 2, refractive index values, local corrections for emission, and induced electric dipole strengths. Note: if Slater integrals were not provided in Table 2, the squared RMEs will be given from the tables by Carnall [66], unless indicated that the authors used tables by Weber [35].

| No. | Initial Level | Squared RME U$^2$ | Squared RME U$^4$ | Squared RME U$^6$ | n | $\chi_{ED}$ (Emission) | Ref. for n | D [esu$^2$ cm$^2$] × 10$^{40}$ |
|---|---|---|---|---|---|---|---|---|
|  | $^4S_{3/2}$ | 0 | 0 | 0.2216 | 1.499 | 3.00 |  | 10.38 |
| 1 [b] | $^2H_{11/2}$ | 0.7247 | 0.4159 | 0.0925 | 1.499 | 3.00 | [33] | 98.55 |
|  | $^4F_{7/2}$ | 0 | 0.1461 | 0.6298 | 1.499 | 3.00 |  | 33.40 |
|  | $^4S_{3/2}$ | 0 | 0 | 0.2275 | 1.516 | 3.11 |  | 12.07 |
| 2 | $^2H_{11/2}$ | 0.7141 | 0.4112 | 0.0867 | 1.518 | 3.12 | [67] | 78.33 |
|  | $^4F_{7/2}$ | 0 | 0.1473 | 0.6285 | 1.522 | 3.15 |  | 36.75 |
|  | $^4S_{3/2}$ | 0 | 0 | 0.2134 | 1.836 | 5.88 |  | 3.45 |
| 3 | $^2H_{11/2}$ | 0.5816 | 0.3350 | 0.0756 | 1.838 | 5.91 | [36] | 18.26 |
|  | $^4F_{7/2}$ | 0 | 0.1465 | 0.6192 | 1.842 | 5.95 |  | 13.11 |
|  | $^4S_{3/2}$ | 0 | 0 | 0.2226 | 1.7 | 4.52 |  | 3.54 |
| 4 [b] | $^2H_{11/2}$ | 0.7205 | 0.4152 | 0.0911 | 1.7 | 4.52 | [68] | 111.96 |
|  | $^4F_{7/2}$ | 0 | 0.1467 | 0.6274 | 1.7 | 4.52 |  | 17.03 |
|  | $^4S_{3/2}$ | 0 | 0 | 0.2240 | 2.2 | 11.44 |  | 6.05 |
| 5 [b] | $^2H_{11/2}$ | 0.6872 | 0.3971 | 0.0849 | 2.2 | 11.44 | [69] | 92.03 |
|  | $^4F_{7/2}$ | 0 | 0.1474 | 0.6247 | 2.2 | 11.44 |  | 24.75 |
|  | $^4S_{3/2}$ | 0 | 0 | 0.2257 | 2.2 | 11.44 |  | 9.63 |
| 6 [b] | $^2H_{11/2}$ | 0.6891 | 0.3968 | 0.0841 | 2.2 | 11.44 | [69] | 65.89 |
|  | $^4F_{7/2}$ | 0 | 0.1472 | 0.6272 | 2.2 | 11.44 |  | 33.90 |
|  | $^4S_{3/2}$ | 0 | 0 | 0.2211 | 2.35 | 14.70 |  | 5.64 |
| 7 [c] | $^2H_{11/2}$ | 0.7125 | 0.4123 | 0.0925 | 2.35 | 14.70 | [39] | 161.57 |
|  | $^4F_{7/2}$ | 0 | 0.1468 | 0.6266 | 2.35 | 14.70 |  | 26.55 |
|  | $^4S_{3/2}$ | 0 | 0 | 0.2211 | 1.946 | 7.24 |  | 2.81 |
| 8 [a] | $^2H_{11/2}$ | 0.7125 | 0.4123 | 0.0925 | 1.948 | 7.27 | [40,41] | 22.30 |
|  | $^4F_{7/2}$ | 0 | 0.1468 | 0.6266 | 1.953 | 7.34 |  | 9.91 |
|  | $^4S_{3/2}$ | 0 | 0 | 0.2285 | 1.402 | 2.45 |  | 3.00 |
| 9 | $^2H_{11/2}$ | 0.7056 | 0.4109 | 0.0870 | 1.404 | 2.46 | [35] | 21.44 |
|  | $^4F_{7/2}$ | 0 | 0.1467 | 0.6273 | 1.406 | 2.47 |  | 11.19 |
|  | $^4S_{3/2}$ | 0 | 0 | 0.2171 | 1.938 | 7.13 |  | 2.40 |
| 10 | $^2H_{11/2}$ | 0.6964 | 0.4022 | 0.0912 | 1.942 | 7.19 | [70] | 85.98 |
|  | $^4F_{7/2}$ | 0 | 0.1464 | 0.6238 | 1.948 | 7.27 |  | 10.99 |
|  | $^4S_{3/2}$ | 0 | 0 | 0.2231 | 2.017 | 8.25 |  | 8.59 |
| 11 | $^2H_{11/2}$ | 0.6796 | 0.3919 | 0.0843 | 2.023 | 8.34 | [71] | 234.27 |
|  | $^4F_{7/2}$ | 0 | 0.1471 | 0.6253 | 2.036 | 8.54 |  | 31.66 |
|  | $^4S_{3/2}$ | 0 | 0 | 0.2182 | 1.611 | 3.78 |  | 5.99 |
| 12 [b] | $^2H_{11/2}$ | 0.6547 | 0.3795 | 0.0838 | 1.611 | 3.78 | [45] | 35.45 |
|  | $^4F_{7/2}$ | 0 | 0.1471 | 0.6200 | 1.611 | 3.78 |  | 20.85 |

Table 3. Cont.

| No. | Initial Level | Squared RME | | | n | χED (Emission) | Ref. for n | D [esu² cm²] × 10⁴⁰ |
|---|---|---|---|---|---|---|---|---|
| | | $U^2$ | $U^4$ | $U^6$ | | | | |
| | $^4S_{3/2}$ | 0 | 0 | 0.2285 | 1.85 | 6.04 | | 10.07 |
| 13 bd | $^2H_{11/2}$ | 0.7056 | 0.4109 | 0.0870 | 1.85 | 6.04 | [72] | 89.25 |
| | $^4F_{7/2}$ | 0 | 0.1467 | 0.6273 | 1.85 | 6.04 | | 36.17 |
| | $^4S_{3/2}$ | 0 | 0 | 0.2211 | 1.987 | 7.81 | | 4.39 |
| 14 ac | $^2H_{11/2}$ | 0.7125 | 0.4123 | 0.0925 | 1.982 | 7.74 | [36] | 16.86 |
| | $^4F_{7/2}$ | 0 | 0.1468 | 0.6266 | 1.998 | 7.97 | | 13.68 |
| | $^4S_{3/2}$ | 0 | 0 | 0.2211 | 1.944 | 7.21 | | 4.44 |
| 15 ac | $^2H_{11/2}$ | 0.7125 | 0.4123 | 0.0925 | 1.948 | 7.27 | [36] | 21.54 |
| | $^4F_{7/2}$ | 0 | 0.1468 | 0.6266 | 1.954 | 7.35 | | 14.20 |
| | $^4S_{3/2}$ | 0 | 0 | 0.2157 | 1.564 | 3.44 | | 7.31 |
| 16 b | $^2H_{11/2}$ | 0.6296 | 0.3618 | 0.0815 | 1.564 | 3.44 | [47] | 97.49 |
| | $^4F_{7/2}$ | 0 | 0.1463 | 0.6235 | 1.564 | 3.44 | | 24.86 |
| | $^4S_{3/2}$ | 0 | 0 | 0.2145 | 1.478 | 2.88 | | 4.06 |
| 17 b | $^2H_{11/2}$ | 0.6115 | 0.3531 | 0.0795 | 1.478 | 2.88 | [48] | 54.70 |
| | $^4F_{7/2}$ | 0 | 0.1466 | 0.6196 | 1.478 | 2.88 | | 14.83 |
| | $^4S_{3/2}$ | 0 | 0 | 0.2138 | 1.476 | 2.86 | | 3.06 |
| 18 b | $^2H_{11/2}$ | 0.6068 | 0.3507 | 0.0792 | 1.476 | 2.86 | [48] | 22.91 |
| | $^4F_{7/2}$ | 0 | 0.1467 | 0.6185 | 1.476 | 2.86 | | 10.17 |
| | $^4S_{3/2}$ | 0 | 0 | 0.2143 | 1.480 | 2.89 | | 7.51 |
| 19 b | $^2H_{11/2}$ | 0.6081 | 0.3512 | 0.0791 | 1.480 | 2.89 | [48] | 60.55 |
| | $^4F_{7/2}$ | 0 | 0.1467 | 0.6191 | 1.480 | 2.89 | | 24.28 |
| | $^4S_{3/2}$ | 0 | 0 | 0.2135 | 1.479 | 2.88 | | 5.42 |
| 20 b | $^2H_{11/2}$ | 0.6067 | 0.3507 | 0.0794 | 1.479 | 2.88 | [48] | 40.58 |
| | $^4F_{7/2}$ | 0 | 0.1466 | 0.6184 | 1.479 | 2.88 | | 17.01 |
| | $^4S_{3/2}$ | 0 | 0 | 0.2137 | 1.481 | 2.89 | | 4.44 |
| 21 b | $^2H_{11/2}$ | 0.6100 | 0.3524 | 0.0798 | 1.481 | 2.89 | [48] | 29.26 |
| | $^4F_{7/2}$ | 0 | 0.1466 | 0.6190 | 1.481 | 2.89 | | 13.80 |
| | $^4S_{3/2}$ | 0 | 0 | 0.2211 | 2.167 | 10.80 | | 1.12 |
| 22 c | $^2H_{11/2}$ | 0.7125 | 0.4123 | 0.0925 | 2.172 | 10.89 | [73] | 29.88 |
| | $^4F_{7/2}$ | 0 | 0.1468 | 0.6266 | 2.180 | 11.04 | | 4.87 |
| | $^4S_{3/2}$ | 0 | 0 | 0.2211 | 2.316 | 13.95 | | 6.48 |
| 23 c | $^2H_{11/2}$ | 0.7125 | 0.4123 | 0.0925 | 2.331 | 14.31 | [74] | 143.84 |
| | $^4F_{7/2}$ | 0 | 0.1468 | 0.6266 | 2.349 | 14.75 | | 25.94 |
| | $^4S_{3/2}$ | 0 | 0 | 0.2156 | 1.779 | 5.27 | | 3.03 |
| 24 b | $^2H_{11/2}$ | 0.7176 | 0.4147 | 0.0959 | 1.779 | 5.27 | [51] | 67.47 |
| | $^4F_{7/2}$ | 0 | 0.1461 | 0.6243 | 1.779 | 5.27 | | 13.30 |
| | $^4S_{3/2}$ | 0 | 0 | 0.2285 | 1.518 | 3.12 | | 5.11 |
| 25 d | $^2H_{11/2}$ | 0.7056 | 0.4109 | 0.0870 | 1.519 | 3.13 | [52] | 56.47 |
| | $^4F_{7/2}$ | 0 | 0.1467 | 0.6273 | 1.520 | 3.14 | | 18.74 |
| | $^4S_{3/2}$ | 0 | 0 | 0.2211 | 1.819 | 5.70 | | 7.29 |
| 26 b | $^2H_{11/2}$ | 0.7125 | 0.4123 | 0.0925 | 1.819 | 5.70 | [53] | 52.12 |
| | $^4F_{7/2}$ | 0 | 0.1468 | 0.6266 | 1.819 | 5.70 | | 25.85 |
| | $^4S_{3/2}$ | 0 | 0 | 0.2285 | 1.584 | 3.58 | | 6.85 |
| 27 bd | $^2H_{11/2}$ | 0.7056 | 0.4109 | 0.0870 | 1.584 | 3.58 | [75] | 90.58 |
| | $^4F_{7/2}$ | 0 | 0.1467 | 0.6273 | 1.584 | 3.58 | | 22.91 |
| | $^4S_{3/2}$ | 0 | 0 | 0.2285 | 1.587 | 3.60 | | 7.59 |
| 28 bd | $^2H_{11/2}$ | 0.7056 | 0.4109 | 0.0870 | 1.587 | 3.60 | [75] | 109.39 |
| | $^4F_{7/2}$ | 0 | 0.1467 | 0.6273 | 1.587 | 3.60 | | 24.46 |
| | $^4S_{3/2}$ | 0 | 0 | 0.2285 | 1.588 | 3.61 | | 7.64 |
| 29 bd | $^2H_{11/2}$ | 0.7056 | 0.4109 | 0.0870 | 1.588 | 3.61 | [75] | 92.30 |
| | $^4F_{7/2}$ | 0 | 0.1467 | 0.6273 | 1.588 | 3.61 | | 23.32 |
| | $^4S_{3/2}$ | 0 | 0 | 0.2211 | 1.581 | 3.56 | | 7.09 |
| 30 a | $^2H_{11/2}$ | 0.7125 | 0.4125 | 0.0925 | 1.587 | 3.60 | [55] | 115.99 |
| | $^4F_{7/2}$ | 0 | 0.1469 | 0.6266 | 1.599 | 3.69 | | 23.58 |

Table 3. Cont.

| No. | Initial Level | Squared RME | | | n | $\chi_{ED}$ (Emission) | Ref. for n | D [esu² cm²] × 10⁴⁰ |
|---|---|---|---|---|---|---|---|---|
| | | U² | U⁴ | U⁶ | | | | |
| 31 [bc] | $^4S_{3/2}$ | 0 | 0 | 0.2211 | 2.116 | 9.86 | | 5.10 |
| | $^2H_{11/2}$ | 0.7125 | 0.4123 | 0.0925 | 2.116 | 9.86 | [56] | 112.55 |
| | $^4F_{7/2}$ | 0 | 0.1468 | 0.6266 | 2.116 | 9.86 | | 21.46 |
| 32 [bd] | $^4S_{3/2}$ | 0 | 0 | 0.2285 | 1.831 | 5.83 | | 2.69 |
| | $^2H_{11/2}$ | 0.7056 | 0.4109 | 0.0870 | 1.831 | 5.83 | [57] | 52.82 |
| | $^4F_{7/2}$ | 0 | 0.1467 | 0.6273 | 1.831 | 5.83 | | 11.58 |
| 33 [b] | $^4S_{3/2}$ | 0 | 0 | 0.2275 | 1.77 | 5.18 | | 13.54 |
| | $^2H_{11/2}$ | 0.7141 | 0.4112 | 0.0866 | 1.77 | 5.18 | [58] | 84.03 |
| | $^4F_{7/2}$ | 0 | 0.1473 | 0.6285 | 1.77 | 5.18 | | 47.84 |
| 34 [bd] | $^4S_{3/2}$ | 0 | 0 | 0.2285 | 1.8 | 5.49 | | 14.65 |
| | $^2H_{11/2}$ | 0.7056 | 0.4109 | 0.0870 | 1.8 | 5.49 | [76] | 29.33 |
| | $^4F_{7/2}$ | 0 | 0.1467 | 0.6273 | 1.8 | 5.49 | | 41.21 |
| 35 [bc] | $^4S_{3/2}$ | 0 | 0 | 0.2211 | 2.02 | 8.30 | | 4.90 |
| | $^2H_{11/2}$ | 0.7125 | 0.4123 | 0.0925 | 2.02 | 8.30 | [60] | 223.35 |
| | $^4F_{7/2}$ | 0 | 0.1468 | 0.6266 | 2.02 | 8.30 | | 20.51 |
| 36 [c] | $^4S_{3/2}$ | 0 | 0 | 0.2211 | 2.01 | 8.15 | | 11.68 |
| | $^2H_{11/2}$ | 0.7125 | 0.4123 | 0.0925 | 2.00 | 8.00 | [61] | 240.22 |
| | $^4F_{7/2}$ | 0 | 0.1468 | 0.6266 | 2.00 | 8.00 | | 38.82 |
| 37 [ad] | $^4S_{3/2}$ | 0 | 0 | 0.2211 | 2.16 | 10.66 | | 20.30 |
| | $^2H_{11/2}$ | 0.7125 | 0.4125 | 0.0925 | 2.16 | 10.66 | [55,77] | 279.11 |
| | $^4F_{7/2}$ | 0 | 0.1469 | 0.6266 | 2.16 | 10.66 | | 85.18 |
| 38 [d] | $^4S_{3/2}$ | 0 | 0 | 0.2285 | 2.05 | 8.76 | | 4.38 |
| | $^2H_{11/2}$ | 0.7056 | 0.4109 | 0.0870 | 2.05 | 8.76 | [63] | 143.07 |
| | $^4F_{7/2}$ | 0 | 0.1467 | 0.6273 | 2.05 | 8.76 | | 15.60 |
| 39 [bd] | $^4S_{3/2}$ | 0 | 0 | 0.2285 | 2.0 | 8.00 | | 3.11 |
| | $^2H_{11/2}$ | 0.7056 | 0.4109 | 0.0870 | 2.0 | 8.00 | [78] | 167.32 |
| | $^4F_{7/2}$ | 0 | 0.1467 | 0.6273 | 2.0 | 8.00 | | 15.37 |
| 40 [bd] | $^4S_{3/2}$ | 0 | 0 | 0.2285 | 2.0 | 8.00 | | 5.32 |
| | $^2H_{11/2}$ | 0.7056 | 0.4109 | 0.0870 | 2.0 | 8.00 | [79] | 139.07 |
| | $^4F_{7/2}$ | 0 | 0.1467 | 0.6273 | 2.0 | 8.00 | | 22.40 |

[a] RME values not calculated by RELIC software, but given in the corresponding reference. [b] Refractive Index values approx. wavelength-independent. [c] RME values from Carnall [66], [d] from Weber [35].

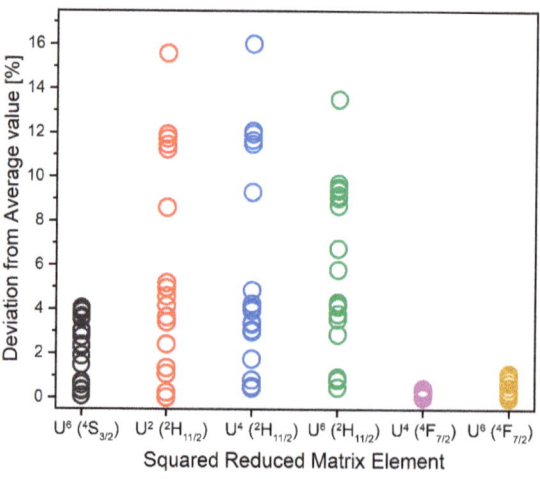

**Figure 4.** Deviation of the RMEs listed in Table 3 from their average values.

**Table 4.** Average RME values estimated from squared RMEs listed in Table 3. Deviations of average values from squared RMEs reported by Carnall (C) and Weber (W), in percentage.

| Initial Level | $U^2$ | $U^4$ | $U^6$ | $U^2$(C) | $U^4$(C) | $U^6$(C) | $U^2$(W) | $U^4$(W) | $U^6$(W) |
|---|---|---|---|---|---|---|---|---|---|
| $^4S_{3/2}$ | 0 | 0 | 0.2224 | 0 | 0 | 0.6 | 0 | 0 | 2.7 |
| $^2H_{11/2}$ | 0.6889 | 0.3989 | 0.0874 | 3.3 | 3.3 | 5.5 | 2.4 | 2.9 | 0.5 |
| $^4F_{7/2}$ | 0 | 0.1468 | 0.6254 | 0 | 0 | 0.2 | 0 | 0.1 | 0.3 |

### 3.2. Calculations of LIR Parameters

For this theoretical analysis, two $Er^{3+}$-based LIRs are considered, the traditional LIR that uses the temperature-dependent ratio of emissions from $^4S_{3/2}$ and $^2H_{11}$ levels, and the relatively novel concept that uses the temperature-dependent ratio emissions from $^4S_{3/2}$ and $^4F_{7/2}$ levels. Table 5 provides the energy differences between $^4S_{3/2}$ and $^2H_{11}$ and $^4S_{3/2}$ and $^4F_{7/2}$ that are used to calculate the room-temperature-relative sensitivities for each host using Equation (2). The temperature invariant B parameters are calculated from the data in Table 3 using Equation (19) (version for spectra recorded in counts). Then, using Equations (2)–(4) and calculated B values, it was possible to derive the LIR's absolute sensitivity, the maximal absolute sensitivity value, and the temperature at which maximal absolute sensitivity occurs.

The relation between relative and absolute sensitivities of traditional LIR (that uses $Er^{3+}$ emissions from $^2H_{11/2}$ and $^4S_{3/2}$ levels) for different hosts is presented in Figure 5a–c. As a rule of thumb, the higher the sensitivity value the better is the performance of thermometry. From Figure 5a, one can see that glasses tend to perform slightly weaker than crystals, on average. Figure 5b compares the LIR performance of different crystals. Fluorides', garnets', phosphates', and silicates' performances are worse than for other hosts. The best results are obtained with simple oxides, vanadates, niobates, molybdates, and tungstates. Figure 5c illustrates the performances of only glass hosts. Even the number of hosts in this set is rather small, it is possible to observe that $Er^{3+}$ activated borate glasses perform worse than other glasses. Fluorophosphate glasses show high relative sensitivities, but somewhat small absolute sensitivities. The best combination of sensitivities is achieved in PbO-PbF$_2$ glass. Similar conclusions can be drawn for the novel LIR type (that uses $Er^{3+}$ emissions from $^4F_{7/2}$ and $^4S_{3/2}$ levels), Figure 5d–f. Among different glasses, tellurite-fluoride glasses show the best performance. For crystals, the situation is almost equivalent to that of traditional LIR.

Figure 6a–c gives the relation between relative sensitivity and absolute sensitivity at the temperature at which the absolute sensitivity has its maximum for the traditional LIR, while Figure 6d–f show the same relationship for the novel type LIR. Analogous conclusions can be drawn as in the previous analysis (Figure 5). Among glass hosts, tellurite-fluoride, tungstate, and molybdate glasses show the best performances. Among crystals, the performance trend is almost the same, but the NaY(MoO$_4$)$_2$ shows the worst performance at elevated temperatures. The best overall performer is LiLa(WO$_4$)$_2$.

As a limit of the study, we must note that the values of the energy levels, Slater integrals and s–o parameters, refractive index values, and JO parameters are taken from literature, so one cannot estimate the level of their accuracy. The extreme outliers are to be taken with caution.

**Table 5.** Calculated luminescence thermometry parameters: energy gaps ($\Delta E$) from Er$^{3+}$ $^4S_{3/2}$ level to $^2H_{11/2}$ and $^4F_{7/2}$, relative temperature sensitivities ($S_r$) for LIRs between selected levels, $B$ LIR parameters, absolute sensitivities at room temperature ($S_a$), maximum sensitivity value ($S_{amax}$), temperatures at which maximum absolute sensitivity occurs ($T(S_{amax})$), and relative sensitivities at $T(S_{amax})$ ($S_r(T(S_{amax}))$).

| No. | Higher Level | $\Delta E$ | $S_r$ (300 K) [% K$^{-1}$] | $B$ | $S_a$ (300 K) [K$^{-1}$] | max($S_a$) [K$^{-1}$] | T(max($S_a$)) [K] (°C) | $S_r$ (T(max($S_a$))) [% K$^{-1}$] |
|---|---|---|---|---|---|---|---|---|
| 1 | $^2H_{11/2}$ | 727 | 1.16 | 10.66 | 0.003792 | 0.0055 | 523 (250) | 0.38 |
|   | $^4F_{7/2}$ | 2024 | 3.24 | 4.40 | 0.000009 | 0.0008 | 1456 (1183) | 0.14 |
| 2 | $^2H_{11/2}$ | 765 | 1.22 | 7.37 | 0.002297 | 0.0036 | 550 (277) | 0.36 |
|   | $^4F_{7/2}$ | 2059 | 3.29 | 4.24 | 0.000007 | 0.0008 | 1481 (1208) | 0.14 |
| 3 | $^2H_{11/2}$ | 801 | 1.28 | 6.05 | 0.001663 | 0.0028 | 576 (303) | 0.35 |
|   | $^4F_{7/2}$ | 2178 | 3.48 | 5.40 | 0.000005 | 0.0009 | 1567 (1294) | 0.13 |
| 4 | $^2H_{11/2}$ | 685 | 1.10 | 35.26 | 0.014453 | 0.0194 | 493 (220) | 0.41 |
|   | $^4F_{7/2}$ | 2031 | 3.25 | 6.58 | 0.000013 | 0.0012 | 1461 (1188) | 0.14 |
| 5 | $^2H_{11/2}$ | 694 | 1.10 | 17.03 | 0.006772 | 0.0092 | 499 (226) | 0.40 |
|   | $^4F_{7/2}$ | 2084 | 3.30 | 5.66 | 0.000009 | 0.0010 | 1499 (1226) | 0.13 |
| 6 | $^2H_{11/2}$ | 694 | 1.10 | 7.65 | 0.003043 | 0.0042 | 499 (226) | 0.40 |
|   | $^4F_{7/2}$ | 2084 | 3.30 | 4.87 | 0.000007 | 0.0009 | 1499 (1226) | 0.13 |
| 7 | $^2H_{11/2}$ | 993 | 1.59 | 33.60 | 0.004557 | 0.0128 | 714 (441) | 0.28 |
|   | $^4F_{7/2}$ | 2394 | 3.83 | 6.82 | 0.000003 | 0.0011 | 1722 (1449) | 0.12 |
| 8 | $^2H_{11/2}$ | 800 | 1.28 | 9.07 | 0.002501 | 0.0043 | 576 (303) | 0.35 |
|   | $^4F_{7/2}$ | 1950 | 3.12 | 4.85 | 0.000013 | 0.0009 | 1403 (1130) | 0.14 |
| 9 | $^2H_{11/2}$ | 855 | 1.37 | 8.21 | 0.001858 | 0.0036 | 615 (342) | 0.33 |
|   | $^4F_{7/2}$ | 2126 | 3.40 | 5.21 | 0.000007 | 0.0009 | 1529 (1256) | 0.13 |
| 10 | $^2H_{11/2}$ | 860 | 1.37 | 41.42 | 0.009209 | 0.0181 | 619 (346) | 0.32 |
|   | $^4F_{7/2}$ | 2195 | 3.51 | 6.57 | 0.000006 | 0.0011 | 1579 (1306) | 0.13 |
| 11 | $^2H_{11/2}$ | 850 | 1.36 | 31.60 | 0.007284 | 0.0140 | 612 (339) | 0.33 |
|   | $^4F_{7/2}$ | 2162 | 3.46 | 5.34 | 0.000006 | 0.0009 | 1555 (1282) | 0.13 |
| 12 | $^2H_{11/2}$ | 641 | 1.02 | 6.55 | 0.003104 | 0.0038 | 461 (188) | 0.43 |
|   | $^4F_{7/2}$ | 2066 | 3.30 | 4.78 | 0.000008 | 0.0009 | 1486 (1213) | 0.13 |
| 13 | $^2H_{11/2}$ | 742 | 1.19 | 9.99 | 0.003373 | 0.0051 | 534 (261) | 0.37 |
|   | $^4F_{7/2}$ | 2156 | 3.45 | 5.02 | 0.000006 | 0.0009 | 1551 (1278) | 0.13 |
| 14 | $^2H_{11/2}$ | 718 | 1.15 | 4.27 | 0.001566 | 0.0022 | 517 (244) | 0.39 |
|   | $^4F_{7/2}$ | 1937 | 3.10 | 4.29 | 0.000012 | 0.0008 | 1394 (1121) | 0.14 |
| 15 | $^2H_{11/2}$ | 694 | 1.11 | 5.46 | 0.002173 | 0.0030 | 499 (226) | 0.40 |
|   | $^4F_{7/2}$ | 1912 | 3.06 | 4.38 | 0.000014 | 0.0009 | 1376 (1103) | 0.15 |
| 16 | $^2H_{11/2}$ | 812 | 1.30 | 15.17 | 0.004009 | 0.0070 | 584 (311) | 0.34 |
|   | $^4F_{7/2}$ | 2068 | 3.31 | 4.68 | 0.000008 | 0.0009 | 1488 (1215) | 0.13 |
| 17 | $^2H_{11/2}$ | 792 | 1.27 | 15.30 | 0.004339 | 0.0073 | 570 (297) | 0.35 |
|   | $^4F_{7/2}$ | 2075 | 3.32 | 5.04 | 0.000008 | 0.0009 | 1493 (1220) | 0.13 |
| 18 | $^2H_{11/2}$ | 779 | 1.25 | 8.48 | 0.002519 | 0.0041 | 560 (287) | 0.36 |
|   | $^4F_{7/2}$ | 2073 | 3.31 | 4.58 | 0.000007 | 0.0008 | 1491 (1218) | 0.13 |
| 19 | $^2H_{11/2}$ | 792 | 1.27 | 9.14 | 0.002594 | 0.0043 | 570 (297) | 0.35 |
|   | $^4F_{7/2}$ | 2075 | 3.32 | 4.45 | 0.000007 | 0.0008 | 1493 (1220) | 0.13 |
| 20 | $^2H_{11/2}$ | 792 | 1.27 | 8.50 | 0.002411 | 0.0040 | 570 (297) | 0.35 |
|   | $^4F_{7/2}$ | 2075 | 3.32 | 4.33 | 0.000007 | 0.0008 | 1493 (1220) | 0.13 |
| 21 | $^2H_{11/2}$ | 802 | 1.28 | 7.50 | 0.002052 | 0.0035 | 577 (304) | 0.35 |
|   | $^4F_{7/2}$ | 2085 | 3.33 | 4.29 | 0.000006 | 0.0008 | 1500 (1227) | 0.13 |
| 22 | $^2H_{11/2}$ | 926 | 1.48 | 31.12 | 0.005428 | 0.0126 | 666 (393) | 0.30 |
|   | $^4F_{7/2}$ | 2118 | 3.39 | 6.16 | 0.000008 | 0.0011 | 1524 (1251) | 0.13 |
| 23 | $^2H_{11/2}$ | 799 | 1.28 | 25.90 | 0.007167 | 0.0122 | 575 (302) | 0.35 |
|   | $^4F_{7/2}$ | 2244 | 3.59 | 6.00 | 0.000005 | 0.0010 | 1614 (1341) | 0.12 |
| 24 | $^2H_{11/2}$ | 796 | 1.27 | 25.24 | 0.007058 | 0.0119 | 573 (300) | 0.35 |
|   | $^4F_{7/2}$ | 2100 | 3.36 | 6.05 | 0.000009 | 0.0011 | 1511 (1238) | 0.13 |
| 25 | $^2H_{11/2}$ | 743 | 1.19 | 12.46 | 0.004194 | 0.0063 | 535 (262) | 0.37 |
|   | $^4F_{7/2}$ | 2084 | 3.33 | 5.07 | 0.000008 | 0.0009 | 1499 (1226) | 0.13 |
| 26 | $^2H_{11/2}$ | 709 | 1.13 | 8.00 | 0.003025 | 0.0042 | 510 (237) | 0.39 |
|   | $^4F_{7/2}$ | 2007 | 3.21 | 4.83 | 0.000010 | 0.0009 | 1444 (1171) | 0.14 |

Table 5. Cont.

| No. | Higher Level | ΔE | $S_r$ (300 K) [% K$^{-1}$] | B | $S_a$ (300 K) [K$^{-1}$] | max($S_a$) [K$^{-1}$] | T(max($S_a$)) [K] (°C) | $S_r$ (T(max($S_a$))) [% K$^{-1}$] |
|---|---|---|---|---|---|---|---|---|
| 27 | $^2H_{11/2}$ | 845 | 1.35 | 15.13 | 0.003551 | 0.0067 | 608 (335) | 0.33 |
|    | $^4F_{7/2}$  | 2142 | 3.42 | 4.66 | 0.000006 | 0.0008 | 1541 (1268) | 0.13 |
| 28 | $^2H_{11/2}$ | 849 | 1.36 | 16.50 | 0.003818 | 0.0073 | 611 (338) | 0.33 |
|    | $^4F_{7/2}$  | 2109 | 3.37 | 4.46 | 0.000006 | 0.0008 | 1517 (1244) | 0.13 |
| 29 | $^2H_{11/2}$ | 849 | 1.36 | 13.83 | 0.003199 | 0.0061 | 611 (338) | 0.33 |
|    | $^4F_{7/2}$  | 2109 | 3.37 | 4.23 | 0.000006 | 0.0008 | 1517 (1244) | 0.13 |
| 30 | $^2H_{11/2}$ | 800 | 1.28 | 18.82 | 0.005190 | 0.0089 | 576 (303) | 0.35 |
|    | $^4F_{7/2}$  | 1950 | 3.12 | 4.67 | 0.000013 | 0.0009 | 1403 (1130) | 0.14 |
| 31 | $^2H_{11/2}$ | 625 | 1.00 | 24.40 | 0.012168 | 0.0147 | 450 (177) | 0.44 |
|    | $^4F_{7/2}$  | 2102 | 3.36 | 5.83 | 0.000008 | 0.0010 | 1512 (1239) | 0.13 |
| 32 | $^2H_{11/2}$ | 996 | 1.59 | 23.07 | 0.003093 | 0.0087 | 717 (444) | 0.28 |
|    | $^4F_{7/2}$  | 2276 | 3.64 | 6.14 | 0.000004 | 0.0010 | 1637 (1364) | 0.12 |
| 33 | $^2H_{11/2}$ | 735 | 1.18 | 6.98 | 0.002416 | 0.0036 | 529 (256) | 0.38 |
|    | $^4F_{7/2}$  | 2101 | 3.36 | 4.89 | 0.000007 | 0.0009 | 1512 (1239) | 0.13 |
| 34 | $^2H_{11/2}$ | 735 | 1.18 | 2.25 | 0.000779 | 0.0012 | 529 (256) | 0.38 |
|    | $^4F_{7/2}$  | 2101 | 3.36 | 3.89 | 0.000005 | 0.0007 | 1512 (1239) | 0.13 |
| 35 | $^2H_{11/2}$ | 735 | 1.18 | 51.32 | 0.017757 | 0.0263 | 529 (256) | 0.38 |
|    | $^4F_{7/2}$  | 2060 | 3.29 | 5.77 | 0.000010 | 0.0011 | 1482 (1209) | 0.13 |
| 36 | $^2H_{11/2}$ | 1009 | 1.61 | 23.75 | 0.003032 | 0.0089 | 726 (453) | 0.28 |
|    | $^4F_{7/2}$  | 2301 | 3.68 | 4.67 | 0.000003 | 0.0008 | 1655 (1382) | 0.12 |
| 37 | $^2H_{11/2}$ | 809 | 1.29 | 15.65 | 0.004179 | 0.0073 | 582 (309) | 0.34 |
|    | $^4F_{7/2}$  | 2101 | 3.36 | 5.81 | 0.000008 | 0.0010 | 1512 (1239) | 0.13 |
| 38 | $^2H_{11/2}$ | 809 | 1.29 | 37.22 | 0.009940 | 0.0173 | 582 (309) | 0.34 |
|    | $^4F_{7/2}$  | 2101 | 3.36 | 4.94 | 0.000007 | 0.0009 | 1512 (1239) | 0.13 |
| 39 | $^2H_{11/2}$ | 699 | 1.12 | 60.18 | 0.023537 | 0.0324 | 503 (230) | 0.40 |
|    | $^4F_{7/2}$  | 1997 | 3.19 | 6.74 | 0.000015 | 0.0013 | 1437 (1164) | 0.14 |
| 40 | $^2H_{11/2}$ | 770 | 1.23 | 29.54 | 0.009052 | 0.0144 | 554 (281) | 0.36 |
|    | $^4F_{7/2}$  | 2030 | 3.25 | 5.76 | 0.000011 | 0.0011 | 1460 (1187) | 0.14 |

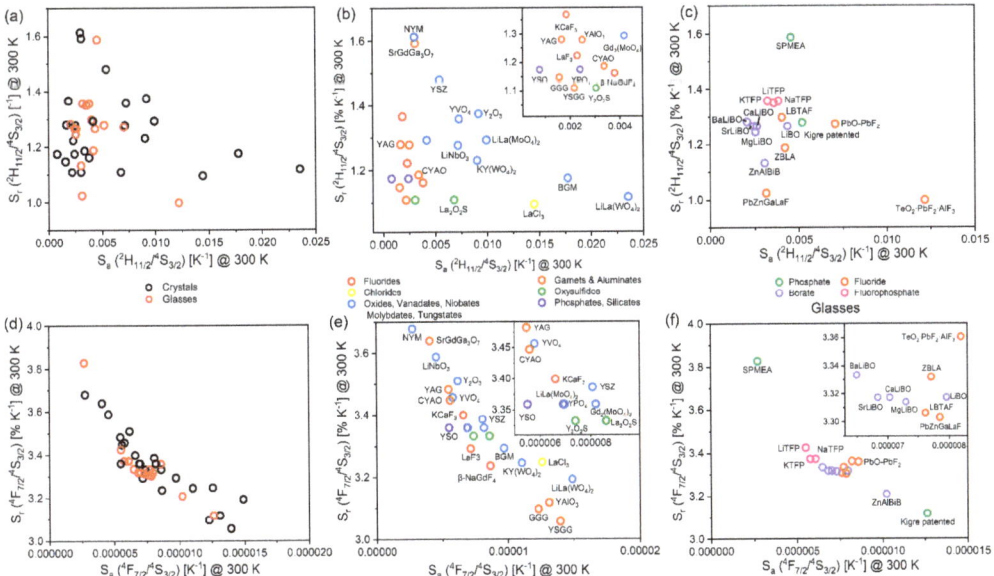

**Figure 5.** Relative sensitivities vs. absolute sensitivities at 300 K. (a–c) for LIR by $^2H_{11/2}$ higher level, (d–f) by $^4F_{7/2}$. (a,d) comparison of crystals and glasses, (b,e) between crystal types, (c,f) between different glasses.

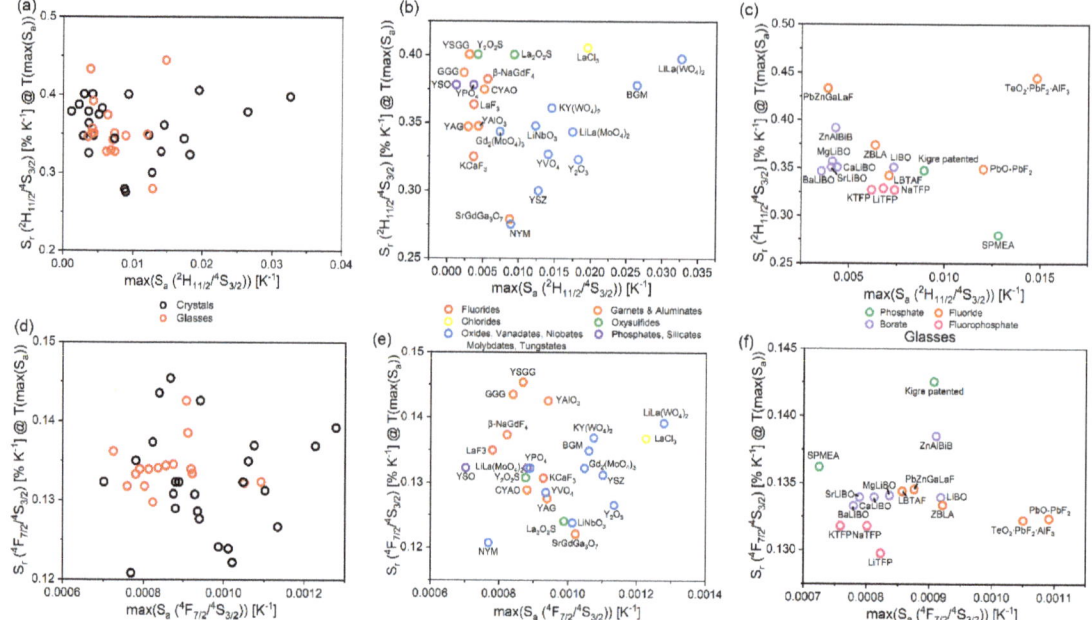

**Figure 6.** Relative sensitivities vs. absolute sensitivities at temperatures with maximum absolute sensitivities. (a–c) for LIR by $^2H_{11/2}$ higher level, (d–f) by $^4F_{7/2}$. (a,d) comparison of crystals and classes, (b,e) between crystal types, (c,f) between different glasses.

## 4. Conclusions

The conventional thermometric characterizations are lengthy, complicated, and expensive. Given that there is an infinite number of possible hosts and doping concentrations of luminescent activators, the guidelines in selecting the appropriate material are important, and they can be provided by the Judd–Ofelt thermometric model which predicts thermometric figures of merit from its 3 intensity parameters.

$Er^{3+}$ deserves special attention in luminescence thermometry. It features LIR between $^2H_{11/2}$ and $^4S_{3/2}$ levels with energy separation of ~700 cm$^{-1}$, and a recently introduced LIR between $^4F_{7/2}$ and $^4S_{3/2}$ levels, whose higher energy separation allows for up to 3× larger relative sensitivity. The performances of 40 various crystals and glasses were predicted by the Judd–Ofelt thermometric model, and guidelines were set to aid the search for the best phosphor for LIR thermometry.

It was demonstrated that the Slater integrals and s–o coupling parameters significantly vary from host to host so that their values should not be adopted from other hosts. Consequently, for $Er^{3+}$, the squared reduced matrix elements also significantly vary between hosts (especially for the $^2H_{11/2} \rightarrow {}^4I_{15/2}$ transition). Therefore, RMEs from frequently used Carnall or Weber tables should be replaced by the average RMEs for the three transitions that are used in these LIR read-out schemes, if the exact RMEs cannot be obtained. This will allow for the improved precision in the prediction of thermometric sensor performances, as well as for the improved Judd–Ofelt parametrization of $Er^{3+}$ doped compounds.

**Author Contributions:** Conceptualization, A.Ć. and M.D.D.; methodology, A.Ć. and M.D.D.; validation, A.Ć.; formal analysis, A.Ć.; investigation, A.Ć. and T.G.; resources, A.Ć. and T.G.; data curation, A.Ć.; writing—original draft preparation, A.Ć. and M.D.D.; writing—review and editing, A.Ć. and M.D.D.; visualization, A.Ć.; supervision, M.D.D.; project administration, M.D.D.; funding acquisition, M.D.D. All authors have read and agreed to the published version of the manuscript.

**Funding:** This research was supported by the NATO Science for Peace and Security Programme under award id. [G5751] and by the Ministry of Education, Science, and Technological Development of the Republic of Serbia.

**Institutional Review Board Statement:** Not applicable.

**Informed Consent Statement:** Not applicable.

**Data Availability Statement:** Data are available from Aleksandar Ćirić upon reasonable request.

**Conflicts of Interest:** The authors declare no conflict of interest.

# References

1. Childs, P.R.N.; Greenwood, J.R.; Long, C.A. Review of temperature measurement. *Rev. Sci. Instrum.* **2000**, *71*, 2959–2978. [CrossRef]
2. Gschwend, P.M.; Starsich, F.H.L.; Keitel, R.C.; Pratsinis, S.E. $Nd^{3+}$-Doped $BiVO_4$ luminescent nanothermometers of high sensitivity. *Chem. Commun.* **2019**, *55*, 7147–7150. [CrossRef] [PubMed]
3. Wang, X.; Wolfbeis, O.S.; Meier, R.J. Luminescent probes and sensors for temperature. *Chem. Soc. Rev.* **2013**, *42*, 7834. [CrossRef] [PubMed]
4. Dramićanin, M.D. Sensing temperature via downshifting emissions of lanthanide-doped metal oxides and salts. A review. *Methods Appl. Fluoresc.* **2016**, *4*, 42001. [CrossRef]
5. Dramićanin, M.D. *Luminescence Thermometry, Methods, Materials and Applications*; Woodhead Publishing: Sawston, UK, 2018; ISBN 9780081020296.
6. Ćirić, A.; Zeković, I.; Medić, M.; Antić, Ž.; Dramićanin, M.D. Judd-Ofelt modelling of the dual-excited single band ratiometric luminescence thermometry. *J. Lumin.* **2020**, *225*. [CrossRef]
7. Dramićanin, M.D. Trends in luminescence thermometry. *J. Appl. Phys.* **2020**, *128*, 40902. [CrossRef]
8. Ćirić, A.; Stojadinović, S.; Dramićanin, M.D. An extension of the Judd-Ofelt theory to the field of lanthanide thermometry. *J. Lumin.* **2019**, *216*. [CrossRef]
9. Wade, S.A.; Collins, S.F.; Baxter, G.W. Fluorescence intensity ratio technique for optical fiber point temperature sensing. *J. Appl. Phys.* **2003**, *94*, 4743. [CrossRef]
10. Ćirić, A.; Stojadinović, S.; Dramićanin, M.D. Custom-built thermometry apparatus and luminescence intensity ratio thermometry of $ZrO_2$:$Eu^{3+}$ and $Nb_2O_5$:$Eu^{3+}$. *Meas. Sci. Technol.* **2019**, *30*, 45001. [CrossRef]
11. Geitenbeek, R.G.; De Wijn, H.W.; Meijerink, A. Non-Boltzmann Luminescence in $NaYF_4$:$Eu^{3+}$: Implications for Luminescence Thermometry. *Phys. Rev. Appl.* **2018**, *10*, 1. [CrossRef]
12. Suta, M.; Antić, Ž.; Đorđević, V.; Kuzman, S.; Dramićanin, M.D.; Meijerink, A. Making $Nd^{3+}$ a Sensitive Luminescent Thermometer for Physiological Temperatures—An Account of Pitfalls in Boltzmann Thermometry Making $Nd^{3+}$ a Sensitive Luminescent Thermometer for Physiological Temperatures—An Account of Pitfalls in Boltzmann Thermomet. *Nanomaterials* **2020**, *10*, 543. [CrossRef] [PubMed]
13. Smentek, L. Judd-Ofelt Theory—The Golden (and the Only One) Theoretical Tool of f Electron Spectroscopy. *Comput. Methods Lanthan. Actin. Chem.* **2015**, 241–268. [CrossRef]
14. Judd, B.R. Optical Absorption Intensities of Rare-Earth Ions. *Phys. Rev.* **1962**, *127*, 750–761. [CrossRef]
15. Ofelt, G.S. Intensities of Crystal Spectra of Rare-Earth Ions. *J. Chem. Phys.* **1962**, *37*, 511–520. [CrossRef]
16. Görller-Walrand, C.; Binnemans, K. *Chapter 167 Spectral intensities of f-f transitions*; Elsevier: Amsterdam, The Netherlands, 1998; Volume 25, pp. 101–264.
17. Hehlen, M.P.; Brik, M.G.; Krämer, K.W. 50th anniversary of the Judd–Ofelt theory: An experimentalist's view of the formalism and its application. *J. Lumin.* **2013**, *136*, 221–239. [CrossRef]
18. Walsh, B.M. Judd-Ofelt theory: Principles and practices brian m. walsh. *Int. Sch. At. Mol. Spectrosc.* **2006**, 403–433. [CrossRef]
19. Reisfeld, R. Optical Properties of Lanthanides in Condensed Phase, Theory and Applications. *AIMS Mater. Sci.* **2015**, *2*, 37–60. [CrossRef]
20. Hamm, P.; Helbing, J.; Liu, G.; Jacquier, B.; Thermo, S.; Swart, H.C.; Terblans, J.J.; Ntwaeaborwa, O.M.; Kroon, R.E.; Coetsee, E.; et al. *Spectroscopic Properties of Rare Earths in Optical Materials*; Springer Series in Materials Science; Hull, R., Parisi, J., Osgood, R.M., Warlimont, H., Liu, G., Jacquier, B., Eds.; Springer: Berlin/Heidelberg, Germany, 2005; Volume 83, ISBN 3-540-23886-7.
21. Edelstein, N.M. Electronic Structure of f-Block Compounds. In *Organometallics of the f-Elements*; Springer Netherlands: Dordrecht, The Netherlands, 1979; pp. 37–79.
22. Carnall, W.T.; Crosswhite, H.; Crosswhite, H.M. *Energy Level Structure and Transition Probabilities in the Spectra of the Trivalent Lanthanides in $LaF_3$*; Argonne: Lemont, IL, USA, 1978.
23. Weber, M.J.; Varitimos, T.E.; Matsinger, B.H. Optical intensities of rare-earth ions in yttrium orthoaluminate. *Phys. Rev. B* **1973**, *8*, 47–53. [CrossRef]
24. Tanner, P.A. Some misconceptions concerning the electronic spectra of tri-positive europium and cerium. *Chem. Soc. Rev.* **2013**, *42*, 5090. [CrossRef] [PubMed]
25. Binnemans, K. Interpretation of europium(III) spectra. *Coord. Chem. Rev.* **2015**, *295*, 1–45. [CrossRef]

26. Ćirić, A.; Stojadinović, S.; Brik, M.G.; Dramićanin, M.D. Judd-Ofelt parametrization from emission spectra: The case study of the $Eu^{3+}$ $^5D_1$ emitting level. *Chem. Phys.* **2020**, *528*. [CrossRef]
27. Ćirić, A.; Stojadinović, S.; Sekulić, M.; Dramićanin, M.D. JOES: An application software for Judd-Ofelt analysis from $Eu^{3+}$ emission spectra. *J. Lumin.* **2019**, *205*, 351–356. [CrossRef]
28. Ćirić, A.; Stojadinović, S.; Dramićanin, M.D. Temperature and concentration dependent Judd-Ofelt analysis of $Y_2O_3$:$Eu^{3+}$ and $YVO_4$:$Eu^{3+}$. *Phys. B Condens. Matter* **2020**, *579*. [CrossRef]
29. Preda, E.; Stef, M.; Buse, G.; Pruna, A.; Nicoara, I. Concentration dependence of the Judd–Ofelt parameters of $Er^{3+}$ ions in $CaF_2$ crystals. *Phys. Scr.* **2009**, *79*, 035304. [CrossRef]
30. Carlos, L.D.; De Mello Donegá, C.; Albuquerque, R.Q.; Alves, S.; Menezes, J.F.S.; Malta, O.L. Highly luminescent europium(III) complexes with naphtoiltrifluoroacetone and dimethyl sulphoxide. *Mol. Phys.* **2003**, *101*, 1037–1045. [CrossRef]
31. dos Santos, B.F.; dos Santos Rezende, M.V.; Montes, P.J.R.; Araujo, R.M.; dos Santos, M.A.C.; Valerio, M.E.G. Spectroscopy study of $SrAl_2O_4$:$Eu^{3+}$. *J. Lumin.* **2012**, *132*, 1015–1020. [CrossRef]
32. Suta, M.; Meijerink, A. A Theoretical Framework for Ratiometric Single Ion Luminescent Thermometers—Thermodynamic and Kinetic Guidelines for Optimized Performance. *Adv. Theory Simul.* **2020**, *3*, 2000176. [CrossRef]
33. Villanueva-Delgado, P.; Biner, D.; Krämer, K.W. Judd–Ofelt analysis of β-$NaGdF_4$:$Yb^{3+}$,$Tm^{3+}$ and β-$NaGdF_4$:$Er^{3+}$ single crystals. *J. Lumin.* **2017**, *189*, 84–90. [CrossRef]
34. Krupke, W.F.; Gruber, J.B. Energy Levels of $Er^{3+}$ in $LaF_3$ and Coherent Emission at 1.61 μm. *J. Chem. Phys.* **1964**, *41*, 1225–1232. [CrossRef]
35. Weber, M.J. Probabilities for Radiative and Nonradiative Decay of $Er^{3+}$ in $LaF_3$. *Phys. Rev.* **1967**, *157*, 262–272. [CrossRef]
36. Sardar, D.K.; Bradley, W.M.; Perez, J.J.; Gruber, J.B.; Zandi, B.; Hutchinson, J.A.; Trussell, C.W.; Kokta, M.R. Judd–Ofelt analysis of the $Er^{3+}$ ($4f^{11}$) absorption intensities in $Er^{3+}$ –doped garnets. *J. Appl. Phys.* **2003**, *93*, 2602–2607. [CrossRef]
37. Buddhudu, S.; Bryant, F. Optical transitions of $Er^{3+}$:$La_2O_2S$ and $Er^{3+}$:$Y_2O_2S$. *J. Less Common Met.* **1989**, *147*, 213–225. [CrossRef]
38. Morrison, C.A.; Leavitt, R.P. Chapter 46 Spectroscopic properties of triply ionized. In *Handbook on the Physics and Chemistry of Rare Earths*; Elsevier: Amsterdam, The Netherlands, 1982; Volume 5, pp. 461–692.
39. Moustafa, S.Y.; Sahar, M.R.; Ghoshal, S.K. Spectroscopic attributes of $Er^{3+}$ ions in antimony phosphate glass incorporated with Ag nanoparticles: Judd-Ofelt analysis. *J. Alloys Compd.* **2017**, *712*, 781–794. [CrossRef]
40. Kaminskii, A.A.; Mironov, V.S.; Kornienko, A.; Bagaev, S.N.; Boulon, G.; Brenier, A.; Di Bartolo, B. New laser properties and spectroscopy of orthorhombic crystals $YAlO_3$:$Er^{3+}$. Intensity luminescence characteristics, stimulated emission, and full set of squared reduced-matrix elements |⟨α[SL]J| |U(t)| |α′[S′ L′]J′⟩|² for $Er^{3+}$ Ions. *Phys. Status Solidi* **1995**, *151*, 231–255. [CrossRef]
41. Chen, C.Y.; Sibley, W.A.; Yeh, D.C.; Hunt, C.A. The optical properties of $Er^{3+}$ and $Tm^{3+}$ in $KCaF_3$ crystal. *J. Lumin.* **1989**, *43*, 185–194. [CrossRef]
42. Weber, M.J. Radiative and Multiphonon Relaxation of Rare-Earth Ions in $Y_2O_3$. *Phys. Rev.* **1968**, *171*, 283–291. [CrossRef]
43. Kisliuk, P.; Krupke, W.F.; Gruber, J.B. Spectrum of $Er^{3+}$ in Single Crystals of $Y_2O_3$. *J. Chem. Phys.* **1964**, *40*, 3606–3610. [CrossRef]
44. Capobianco, J.A.; Kabro, P.; Ermeneux, F.S.; Moncorgé, R.; Bettinelli, M.; Cavalli, E. Optical spectroscopy, fluorescence dynamics and crystal-field analysis of $Er^{3+}$ in $YVO_4$. *Chem. Phys.* **1997**, *214*, 329–340. [CrossRef]
45. Reisfeld, R.; Katz, G.; Spector, N.; Jørgensen, C.K.; Jacoboni, C.; De Pape, R. Optical transition probabilities of $Er^{3+}$ in fluoride glasses. *J. Solid State Chem.* **1982**, *41*, 253–261. [CrossRef]
46. Souriau, J.C.; Borel, C.; Wyon, C.; Li, C.; Moncorgé, R. Spectroscopic properties and fluorescence dynamics of $Er^{3+}$ and $Yb^{3+}$ in $CaYAlO_4$. *J. Lumin.* **1994**, *59*, 349–359. [CrossRef]
47. Jamalaiah, B.C.; Suhasini, T.; Rama Moorthy, L.; Janardhan Reddy, K.; Kim, I.-G.; Yoo, D.-S.; Jang, K. Visible and near infrared luminescence properties of $Er^{3+}$-doped LBTAF glasses for optical amplifiers. *Opt. Mater.* **2012**, *34*, 861–867. [CrossRef]
48. Renuka Devi, A.; Jayasankar, C.K. Optical properties of $Er^{3+}$ ions in lithium borate glasses and comparative energy level analyses of $Er^{3+}$ ions in various glasses. *J. Non. Cryst. Solids* **1996**, *197*, 111–128. [CrossRef]
49. Merino, R.I.; Orera, V.M.; Cases, R.; Chamarro, M.A. Spectroscopic characterization of $Er^{3+}$ in stabilized zirconia single crystals. *J. Phys. Condens. Matter* **1991**, *3*, 8491–8502. [CrossRef]
50. Amin, J.; Dussardier, B.; Schweizer, T.; Hempstead, M. Spectroscopic analysis of $Er^{3+}$ transitions in lithium niobate. *J. Lumin.* **1996**, *69*, 17–26. [CrossRef]
51. Nachimuthu, P.; Jagannathan, R. Judd-Ofelt Parameters, Hypersensitivity, and Emission Characteristics of $Ln^{3+}$ ($Nd^{3+}$, $Ho^{3+}$, and $Er^{3+}$) Ions Doped in PbO-$PbF_2$ Glasses. *J. Am. Ceram. Soc.* **2004**, *82*, 387–392. [CrossRef]
52. Shinn, M.D.; Sibley, W.A.; Drexhage, M.G.; Brown, R.N. Optical transitions of $Er^{3+}$ ions in fluorozirconate glass. *Phys. Rev. B* **1983**, *27*, 6635–6648. [CrossRef]
53. Swapna, K.; Mahamuda, S.; Venkateswarlu, M.; Srinivasa Rao, A.; Jayasimhadri, M.; Shakya, S.; Prakash, G.V. Visible, Up-conversion and NIR (~1.5μm) luminescence studies of $Er^{3+}$ doped Zinc Alumino Bismuth Borate glasses. *J. Lumin.* **2015**, *163*, 55–63. [CrossRef]
54. Moorthy, L.R.; Jayasimhadri, M.; Saleem, S.A.; Murthy, D.V.R. Optical properties of $Er^{3+}$-doped alkali fluorophosphate glasses. *J. Non. Cryst. Solids* **2007**, *353*, 1392–1396. [CrossRef]
55. Sardar, D.K.; Gruber, J.B.; Zandi, B.; Hutchinson, J.A.; Trussell, C.W. Judd-Ofelt analysis of the $Er^{3+}$($4f^{11}$) absorption intensities in phosphate glass: $Er^{3+}$, $Yb^{3+}$. *J. Appl. Phys.* **2003**, *93*, 2041–2046. [CrossRef]

56. Lalla, E.A.; Konstantinidis, M.; De Souza, I.; Daly, M.G.; Martín, I.R.; Lavín, V.; Rodríguez-Mendoza, U.R. Judd-Ofelt parameters of $RE^{3+}$-doped fluorotellurite glass ($RE^{3+}$ = $Pr^{3+}$, $Nd^{3+}$, $Sm^{3+}$, $Tb^{3+}$, $Dy^{3+}$, $Ho^{3+}$, $Er^{3+}$, and $Tm^{3+}$). *J. Alloys Compd.* **2020**, *845*, 156028. [CrossRef]
57. Piao, R.; Wang, Y.; Zhang, Z.; Zhang, C.; Yang, X.; Zhang, D. Optical and Judd-Ofelt spectroscopic study of $Er^{3+}$-doped strontium gadolinium gallium garnet single-crystal. *J. Am. Ceram. Soc.* **2018**, jace.16114. [CrossRef]
58. Che, Y.; Zheng, F.; Dou, C.; Yin, Y.; Wang, Z.; Zhong, D.; Sun, S.; Teng, B. A promising laser crystal $Er^{3+}$:$YPO_4$ with intense multi-wavelength emission characteristics. *J. Alloys Compd.* **2020**, *157854*. [CrossRef]
59. Huy, B.T.; Sengthong, B.; Van Do, P.; Chung, J.W.; Ajith Kumar, G.; Quang, V.X.; Dao, V.-D.; Lee, Y.-I. A bright yellow light from a $Yb^{3+}$,$Er^{3+}$-co-doped $Y_2SiO_5$ upconversion luminescence material. *RSC Adv.* **2016**, *6*, 92454–92462. [CrossRef]
60. Pan, Y.; Gong, X.H.; Chen, Y.J.; Lin, Y.F.; Huang, J.H.; Luo, Z.D.; Huang, Y.D. Polarized spectroscopic properties of $Er^{3+}$:$BaGd_2(MoO_4)_4$ crystal. *Opt. Mater.* **2012**, *34*, 1143–1147. [CrossRef]
61. Lu, X.; You, Z.; Li, J.; Zhu, Z.; Jia, G.; Wu, B.; Tu, C. The optical properties of $Er^{3+}$ doped $NaY(MoO_4)_2$ crystal for laser applications around 1.5 μm. *J. Alloys Compd.* **2006**, *426*, 352–356. [CrossRef]
62. Lu, H.; Gao, Y.; Hao, H.; Shi, G.; Li, D.; Song, Y.; Wang, Y.; Zhang, X. Judd-Ofelt analysis and temperature dependent upconversion luminescence of $Er^{3+}$/$Yb^{3+}$ codoped $Gd_2(MoO_4)_3$ phosphor. *J. Lumin.* **2017**, *186*, 34–39. [CrossRef]
63. Huang, X.; Wang, G. Growth and optical characteristics of $Er^{3+}$:$LiLa(MoO_4)_2$ crystal. *J. Alloys Compd.* **2009**, *475*, 693–697. [CrossRef]
64. Huang, X.Y.; Lin, Z.B.; Zhang, L.Z.; Wang, G.F. Spectroscopic characteristics of $Er^{3+}$/$Yb^{3+}$:$LiLa(WO_4)_2$ crystal. *Mater. Res. Innov.* **2008**, *12*, 94–97. [CrossRef]
65. Kuleshov, N.V.; Lagatsky, A.A.; Podlipensky, A.V.; Mikhailov, V.P.; Kornienko, A.A.; Dunina, E.B.; Hartung, S.; Huber, G. Fluorescence dynamics, excited-state absorption, and stimulated emission of $Er^{3+}$ in $KY(WO_4)_2$. *J. Opt. Soc. Am. B* **1998**, *15*, 1205. [CrossRef]
66. Carnall, W.T.; Fields, P.R.; Rajnak, K. Electronic Energy Levels in the Trivalent Lanthanide Aquo Ions. I. $Pr^{3+}$, $Nd^{3+}$, $Pm^{3+}$, $Sm^{3+}$, $Dy^{3+}$, $Ho^{3+}$, $Er^{3+}$, and $Tm^{3+}$. *J. Chem. Phys.* **1968**, *49*, 4424–4442. [CrossRef]
67. Amotchkina, T.; Trubetskov, M.; Hahner, D.; Pervak, V. Characterization of e-beam evaporated Ge, $YbF_3$, ZnS, and $LaF_3$ thin films for laser-oriented coatings. *Appl. Opt.* **2020**, *59*, A40. [CrossRef]
68. Sell, J.A.; Fong, F.K. Oscillator strength determination in $LaCl_3$:$Pr^{3+}$ by photon upconversion. *J. Chem. Phys.* **1975**, *62*, 4161–4164. [CrossRef]
69. Imanaga, S.; Yokono, S.; Hoshina, T. Cooperative absorption in $Eu_2O_2S$. *J. Lumin.* **1978**, *16*, 77–87. [CrossRef]
70. Nigara, Y. Measurement of the Optical Constants of Yttrium Oxide. *Jpn. J. Appl. Phys.* **1968**, *7*, 404–408. [CrossRef]
71. Shi, H.-S.; Zhang, G.; Shen, H.-Y. Measurement of principal refractive indices and the thermal refractive index coefficients of yttrium vanadate. *J. Synth. Cryst.* **2001**, *30*, 85–88.
72. Pirzio, F.; Cafiso, S.D.D.D.; Kemnitzer, M.; Guandalini, A.; Kienle, F.; Veronesi, S.; Tonelli, M.; Aus der Au, J.; Agnesi, A. Sub-50-fs widely tunable Yb:$CaYAlO_4$ laser pumped by 400-mW single-mode fiber-coupled laser diode. *Opt. Express* **2015**, *23*, 9790. [CrossRef]
73. Wood, D.L.; Nassau, K. Refractive index of cubic zirconia stabilized with yttria. *Appl. Opt.* **1982**, *21*, 2978. [CrossRef] [PubMed]
74. Zelmon, D.E.; Small, D.L.; Jundt, D. Infrared corrected Sellmeier coefficients for congruently grown lithium niobate and 5 mol% magnesium oxide –doped lithium niobate. *J. Opt. Soc. Am. B* **1997**, *14*, 3319. [CrossRef]
75. Jayasimhadri, M.; Moorthy, L.R.; Saleem, S.A.; Ravikumar, R.V.S.S.N. Spectroscopic characteristics of $Sm^{3+}$-doped alkali fluorophosphate glasses. *Spectrochim. Acta Part A Mol. Biomol. Spectrosc.* **2006**, *64*, 939–944. [CrossRef]
76. Dorenbos, P.; Marsman, M.; Van Eijk, C.W.E.; Korzhik, M.V.; Mlnkov, B.I. Scintillation properties of $Y_2SiO_5$:Pr crystals 1. *Radiat. Eff. Defects Solids* **1995**, *135*, 325–328. [CrossRef]
77. Jaque, D.; Findensein, J.; Montoya, E.; Capmany, J.; Kaminskii, A.A.; Eichler, H.J.; Solé, J.G. Spectroscopic and laser gain properties of the $Nd^{3+}$:β'-$Gd_2(MoO_4)$ 3 non-linear crystal. *J. Phys. Condens. Matter* **2000**, *12*, 9699–9714. [CrossRef]
78. Huang, X.; Fang, Q.; Yu, Q.; Lü, X.; Zhang, L.; Lin, Z.; Wang, G. Thermal and polarized spectroscopic characteristics of $Nd^{3+}$:$LiLa(WO_4)_2$ crystal. *J. Alloys Compd.* **2009**, *468*, 321–326. [CrossRef]
79. Romanyuk, Y.E.; Borca, C.N.; Pollnau, M.; Rivier, S.; Petrov, V.; Griebner, U. Yb-doped $KY(WO_4)_2$ planar waveguide laser. *Opt. Lett.* **2006**, *31*, 53–55. [CrossRef] [PubMed]

# Heat-Induced Transformation of Luminescent, Size Tuneable, Anisotropic Eu:Lu(OH)$_2$Cl Microparticles to Micro-Structurally Controlled Eu:Lu$_2$O$_3$ Microplatelets

Madeleine Fellner, Alberto Soppelsa and Alessandro Lauria *

Laboratory for Multifunctional Materials, Department of Materials, ETH Zürich, Vladmir-Prelog-Weg 5, 8093 Zürich, Switzerland; madeleine.fellner@mat.ethz.ch (M.F.); soppelsa.alberto@gmail.com (A.S.)
* Correspondence: alessandro.lauria@mat.ethz.ch

**Abstract:** Synthetic procedures to obtain size and shape-controlled microparticles hold great promise to achieve structural control on the microscale of macroscopic ceramic- or composite-materials. Lutetium oxide is a material relevant for scintillation due to its high density and the possibility to dope with rare earth emitter ions. However, rare earth sesquioxides are challenging to synthesise using bottom-up methods. Therefore, calcination represents an interesting approach to transform lutetium-based particles to corresponding sesquioxides. Here, the controlled solvothermal synthesis of size-tuneable europium doped Lu(OH)$_2$Cl microplatelets and their heat-induced transformation to Eu:Lu$_2$O$_3$ above 800 °C are described. The particles obtained in microwave solvothermal conditions, and their thermal evolution were studied using powder X-ray diffraction, scanning electron microscopy (SEM), transmission electron microscopy (TEM), optical microscopy, thermogravimetric analysis (TGA), luminescence spectroscopy (PL/PLE) and infrared spectroscopy (ATR-IR). The successful transformation of Eu:Lu(OH)$_2$Cl particles into polycrystalline Eu:Lu$_2$O$_3$ microparticles is reported, together with the detailed analysis of their initial and final morphology.

**Keywords:** anisotropy; particle synthesis; luminescence; europium-doping; Lu(OH)$_2$Cl; Lu$_2$O$_3$

## 1. Introduction

Luminescent micro- or nanoparticles of wide band gap semiconductors are interesting building blocks for innovative functional materials with macroscopic dimensions, for example for optical ceramics [1–4]. This strategy may represent a valuable and cost-effective alternative to the growth of single crystals of the same material, especially when the growth of single crystals is limited with respect to accessible geometries, doping homogeneity or high temperatures [5,6]. Moreover, the intentional assembly of micro- or nanoparticles holds promises for generating particle-based macroscopic ceramics and composites, in which the particle structure and morphology can impart functionality to the assembly [7–10]. Indeed, polymer composites containing aligned alumina platelets, where the microstructure of the composite led to improved mechanical properties of the macroscopic sample were recently reported [7]. Moreover, by matching the refractive index of glass microplatelets and polymethyl methacrylate host, structurally similar materials proved to additionally gain optical transparency in the obtained composites [8,9]. A similar bottom-up approach can also be applied to other particle geometries. For example, aligned metal nanowires in a polymer matrix could be used to modify the optical properties of the composite, generating a dichroic material [10]. When functional micro- or nanomaterials are assembled without dispersing hosts, optical grade polycrystalline ceramics and composites for scintillation detection may be obtained, e.g., by using radioluminescent microparticles as building blocks [2,3,11,12]. In this scenario, doped rare earth sesquioxides (RE$_2$O$_3$) are appealing materials for phosphor or scintillation applications where light transmission is required, due to their wide band gap. Lutetia (Lu$_2$O$_3$) in particular is an interesting candidate for ionising radiation detection, lutetium being the heaviest rare earth, ensuring high stopping power against X-rays and gamma rays. $Lu_2O_3$ has a band gap of around 5.8 eV, well above the visible range, and it is an ideal host for optically active rare earth dopants as it allows for substitutional doping which leads to bright radioluminescence [1,6,13–15]. Indeed, the quantum efficiency of $Eu:Lu_2O_3$ can reach up to 90% as reported for materials synthesised by combustion reactions [14]. However, the synthesis of rare earth sesquioxide particles thorough low temperature methods like solvo- or hydrothermal syntheses cannot be easily achieved. While solvothermal conditions may lead to square rare earth oxide (e.g., $Gd_2O_3$) nanoplatelets of around 10 nm by using acetate precursors, hydrothermal syntheses typically yield, depending on the conditions chosen, hydroxyl chloride-, hydroxide-, and oxocarbonate-microparticles, like already observed in the case of Tb, Y, and La based materials [16–19]. Synthesising such particles, an additional thermal conversion step is needed in order to form the corresponding rare earth oxides, which may impact further processing of materials based on such particles. As a consequence, these materials may represent a valuable alternative to single crystals when used as intermediate building blocks toward rare earth oxide particle-based macroscopic composites and bulk materials, provided that profitable procedures of assembly, forming, processing, and conversion into rare earth oxides can be established in order to exploit the relatively easy and cost-effective synthesis of these intermediate materials.

Considering all the above it seems clear that the control and precise understanding of the reactivity and thermal transformation of rare earth-based particles is essential to determine their suitability as constituting elements of more complex multiparticle functional materials. In this work we describe the synthesis of size-tuneable, anisotropic $Eu:Lu(OH)_2Cl$ microplatelets, with special emphasis on the characterisation of their thermal evolution into highly luminescent $Eu:Lu_2O_3$, which could be an extremely versatile platform for multiparticle composites or ceramics for several optical, photonics, or scintillation applications [20,21].

## 2. Materials and Methods

Lutetium chloride (anhydrous, 99.99%, Sigma Aldrich, Buchs, Switzerland), benzyl alcohol (99.8%, Sigma Aldrich, Buchs, Switzerland) and Europium acetate (ABCR, Karlsruhe, Germany) were used as received without further purification.

### 2.1. Synthesis of $Lu(OH)_2Cl$ Microcrystals

Reactions were carried out in a microwave oven (CEM, Kamp-Lintfort, Germany) using 10 mL reaction tubes. In a typical synthesis, $LuCl_3$ (168.8 mg, 0.6 mmol) and $Eu(Ac)_3$ (6.6 mg, 0.02 mmol) were mixed with benzyl alcohol (5 mL) and sealed in an argon filled glovebox. The reaction mixtures were consequently heated in a microwave oven to either 200 °C for 1 min (sample A) or 150 °C for 5 min and 60 min (samples B and C, respectively). The resulting white precipitates were washed twice by dispersion in ethanol (2 × 6 mL) and diethyl ether (2 × 6 mL). Materials were calcined at either 500 or 1000 °C in air using a Carbolite furnace equipped with a quartz tubular chamber with a ramp rate of 10 °C/min.

### 2.2. Characterisation

Powder X-ray diffraction (XRD) was performed using a PANalytical Xpert Pro or Empyrean diffractometer using copper k radiation and an HTK 1200 high temperature chamber. Elemental analysis was carried out by the Laboratory of Organic Chemistry at ETH (Vladimir-Prelog-Weg 3, 8093 Zürich, Switzerland). For C, H, and N analysis a LECO TruSpec Micro (USA) system was used, while ion chromatography was employed to determine Cl. Scanning electron microscopy (SEM) was carried out using a Zeiss Leo Gemini 1530 microscope using a 3 keV electron beam. Thermogravimetric analysis (TGA) was performed using a Netzsch STA 449C instrument in the range from 25 to 1000 °C using a ramp of 10 °C min$^{-1}$. Transmission electron microscopy and electron diffraction was performed on a JEOL 2200fs, operating at 200 kV, and equipped with a Gatan heating holder for in situ high temperature analysis. Photoluminescence spectroscopy was measured on a Jasco FP-8500 fluorometer equipped with a solid sample holder using emission and excitation bandwidths of 2.5 nm. Attenuated Total Reflection Infrared (ATR-IR) spectra were recorded on a Bruker Alpha-P spectrometer on solid powder samples. Optical micrographs were collected after dispersing the particles in various solvents (water, ethanol, diethyl ether) in a petri-dish through a Leica DMIL LED inverted microscope. Images of fluorescing particles were captured through a Leica DM6000B microscope equipped with a colour camera and using a 254 nm Wood-lamp as light source. Particle sizes were obtained by measuring at least 50 particles as they appeared in SEM images.

## 3. Results and Discussion
### 3.1. Synthesis of Size-Tuneable Eu:Lu(OH)$_2$Cl Microparticles

Monoclinic lutetium dihydroxychloride (Lu(OH)$_2$Cl) anisotropic micro- and nanoparticles doped with 3 mol % were synthesised by microwave assisted non-aqueous solvothermal reactions using lutetium chloride as precursor in benzyl alcohol as a solvent. XRD analysis revealed the monoclininc crystal structure of the products, irrespective of the synthetic conditions chosen (Figure 1d) [17,22]. The product stoichiometry was further confirmed by microelemental analysis on the obtained powders, which showed a ratio of carbon (0.09), chlorine (1.0) and hydrogen (2.4) with respect to Lutetium (1.0). This result was consistent with the formation of the hydroxyl chloride, with the carbon and part of the hydrogen possibly related to traces of organic residuals at the particle surface, which often occur in solvothermal methods carried in organic media. The size of the particles could be easily controlled by tuning the synthetic parameters, namely temperature and reaction time. Mixtures heated to 200 °C for 1 min (sample A) yielded particles with average length of 8.4 ± 3.5 µm, width of 4.2 ± 1.3 µm, and a thickness of around 300 nm. At lower temperature (150 °C) 5 min of reaction were sufficient to observe the formation of smaller particles with an average length of 1.6 ± 0.5 µm, width of 0.5 ± 0.2 µm, and a thickness of around 150 nm (sample B). Longer reaction times (sample C) at lower temperature (150 °C, 60 min) further reduced the particle size to 260 ± 10 nm in length, 120 ± 5 nm in width, and thicknesses in the range 10–40 nm (Figure 1a–c). These findings revealed that higher reaction temperatures and shorter reaction times led to larger particles. Despite the growth mechanism not being fully understood, these results might indicate the initial formation of larger particles, followed by their fragmentation to form smaller particles with similar shape factors. A similar disassembly of larger particles into smaller constituent particles with identical composition has been reported elsewhere, e.g., for tungstite particles [23]. However, the exact mechanism of this non-classical crystal growth still has to be fully clarified and investigated in more depth.

All the Lu(OH)$_2$Cl particles in this work exhibited an anisotropic, hexagonal-shaped elongated platelet morphology which was retained in both polar and unpolar solvents (Figure 1, Figures S1 and S2). This is likely due to the anisotropic unit cell of monoclinic Lu(OH)$_2$Cl, which possesses three distinct crystallographic axes a, b and c along which different crystal growth rates may be expected, possibly resulting in the observed morphologies [22].

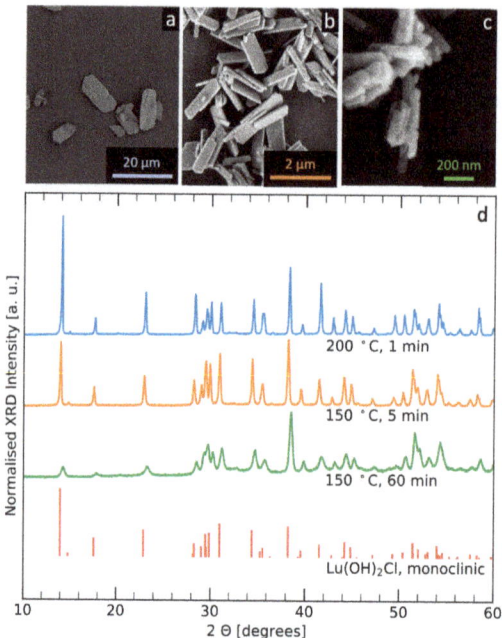

**Figure 1.** Role of the synthetic parameters in determining the particle morphology. (a–c) SEM micrographs of samples A, B, C respectively. (d) XRD of the samples displayed in the pictures. The diffractograms were normalised with respect to the peak at 37.9 2θ for clarity. The diffraction pattern for monoclinic Lu(OH)$_2$Cl (ICSD collection code 260838) is shown for reference.

### 3.2. Thermal Evolution and Formation of Eu:Lu$_2$O$_3$ Platelets

Eu:Lu(OH)$_2$Cl particles could be converted to Eu:Lu$_2$O$_3$ by calcination at 1000 °C [1,24]. The transformation of the particles' crystal structures upon exposure to heat was studied. TGA revealed three regions of weight loss upon heating up to 1000 °C, the different stages of the heat induced transformation were labelled with different colours, orange (<400 °C), green (400–650 °C) and red (>650 °C) (Figure 2a). XRD heated in situ (Figure 2b) showed that the initial monoclinic Lu(OH)$_2$Cl degraded to an intermediate product which can be at least partially associated with LuOCl before it fully turned into cubic Lu$_2$O$_3$ [22,25]. The last stage above 650 °C corresponded to the coalescence of the Lu$_2$O$_3$ crystallites. Based on the thermogravimetric analysis and the diffraction data, the stoichiometric transformations in the material upon annealing were tentatively proposed as follows:

$$2\,Lu(OH)_2Cl \rightarrow Lu(OH)_2Cl + LuOCl + H_2O \rightarrow Lu_2O_3 + 2\,HCl + H_2O, \tag{1}$$

It should be noted that the intermediate stage consisted of a mixture of species which is the reason why the diffractogram could not be fully assigned to a single specific crystal structure by XRD and ATR-IR (Figure 2b, Figure S3, Table S1). The effect of the annealing on the morphology of the particles could be further monitored during TEM experiments, where the samples were annealed in situ (Figure 2c–e). The evolution of the crystal structure of single microplatelets could be observed. Initially, the platelets exhibited lamellar structures along the edge, which disappeared when the Lu(OH)$_2$Cl was transformed to polycrystalline Lu$_2$O$_3$, above 800 °C (Figure S4). The lamellar periodicity in as synthesised microparticles was measured to be around 1 nm (Figure S5). This distance could not be correlated to any of the lattice parameters of Lu(OH)$_2$Cl. Therefore, its origin might be due to layers rich in oxygen, hydrogen and chlorine intercalated with layers rich in lutetium as it was observed in mixed crystal lamellar structures [26]. The overall morphology of the microplatelet was not affected by the heat treatment (Figure 2e). This is also shown by SEM micrographs of platelets after ex situ annealing at 1000 °C (Figure 2f, Figure S4). While the surface roughness of the platelets was enhanced due to the polycrystalline nature of the newly formed $Lu_2O_3$ and due to the change of density of the initial and final crystal structure, the overall initial platelet morphology was retained in all directions. Therefore, the transformation of Eu:Lu(OH)$_2$Cl platelets into Eu:Lu$_2$O$_3$ platelets represents a useful type of morphological control for cubic $Lu_2O_3$, appearing as a promising tool to design multiparticle assemblies which can be treated at high temperature without catastrophic shrinkage or structural rearrangement, which typically are the main source of difficulties in ceramic powder processing.

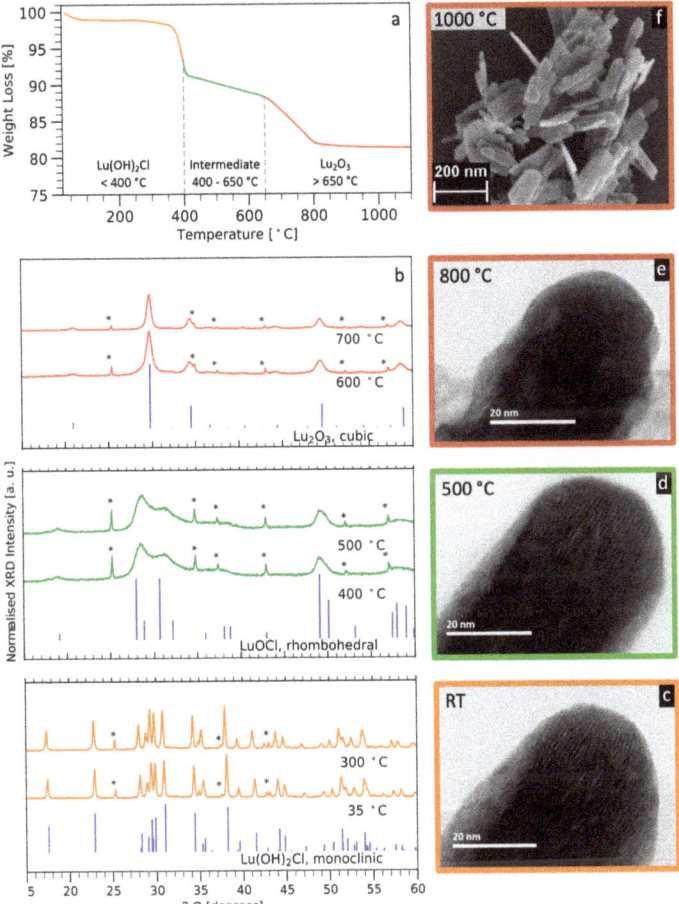

**Figure 2.** Thermal evolution of Eu:Lu(OH)$_2$Cl platelets (sample C). (**a**) TGA. (**b**) In situ high temperature XRD (peaks corresponding to the corundum substrate are marked with asterisks *). The colours orange, green and red label different temperature stages: <400, 400–650, >650 °C, respectively. Lu(OH)$_2$Cl ICSD collection code 260838, LuOCl PDF code 00-035-1344, Lu$_2$O$_3$ ICSD collection code 40471 are displayed for reference. (**c–e**) In situ heating TEM micrographs of the same platelet seen through its edge, recorded at room temperature, at 500 and at 800 °C, respectively. (**f**) SEM of powder calcined at 1000 °C.

## 3.3. Luminescence

Due to its full 4f-shell, $Lu^{3+}$ is an optically inactive rare earth ion. Consequently, $Lu_2O_3$ is a PL-silent material which can be activated by doping with optically active rare earth elements [27,28]. The PL spectra of untreated, semi-calcined, and calcined powders showed clear differences in terms of transitions ratios and intensities, as expected by considering the strict dependence on the lattice site geometry typically expressed by the emission profile of europium (Figure 3) [27,29]. The emission profile of Eu:Lu(OH)$_2$Cl particles was in good agreement with the one reported for Eu(OH)$_2$Cl [30]. A blue luminescence associated to organic side-products resulting from the polymerisation of benzyl alcohol could be observed in Eu:Lu(OH)$_2$Cl samples (Figure S6). Considering the lower crystal grade associated with broader XRD peaks of the intermediate compound (Figure 2b), one could expect a inhomogeneous broadening of the europium(III) emission [27]. However, this broadening is not very evident in the recorded PL spectra. The Eu:Lu$_2$O$_3$ particles obtained after calcination expressed bright red luminescence under UV excitation even after being redispersed in water (Figure S7).

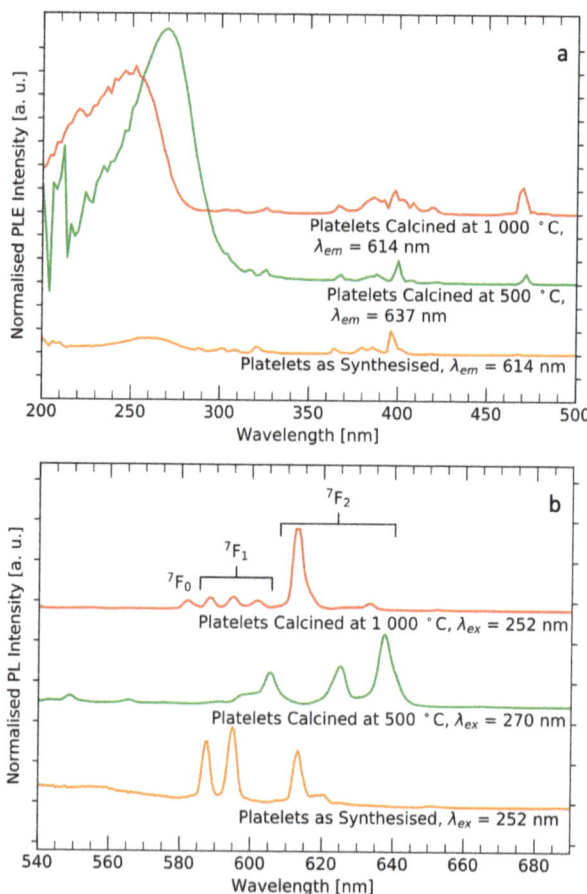

**Figure 3.** Effect of annealing on the $Eu^{3+}$ related luminescence. (a) PLE spectra of Eu:Lu(OH)$_2$Cl, the intermediate and the final product Eu:Lu$_2$O$_3$. Emission wavelengths were 614, 637 and 614 nm, respectively. Spectra were normalised at 394 nm. (b) PL of initial product, intermediate and final product, $\lambda_{ex}$ = 252, 270 and 252 nm, respectively. All spectra were normalised for the most intense peak.

## 4. Conclusions

We report the controlled, solvothermal synthesis of Lu(OH)$_2$Cl particles and the thermal evolution of the material rendering Lu$_2$O$_3$ above 800 °C. The particle size could be readily controlled by varying synthetic parameters such as temperature and time while the crystal structure remained the same. Composition and structure of the europium doped platelets however changed dramatically with annealing. The decomposition of Lu(OH)$_2$Cl to Lu$_2$O$_3$ was also reflected in the photoluminescence emission spectra of the initial and final microparticles. In summary, an up to now unknown level of morphology control of Eu:Lu(OH)$_2$Cl micromaterials which could be transformed to Eu:Lu$_2$O$_3$ was demonstrated. Since Lu$_2$O$_3$ is technologically important for applications such as X-ray and γ-ray detection, these results pave the way towards microstructurally controlled ceramic- and composite-materials [20,21].

**Supplementary Materials:** The following are available online at https://www.mdpi.com/article/10.3390/cryst11080992/s1, Supporting Information.pdf.

**Author Contributions:** Conceptualization, A.L.; Data curation, M.F. and A.L.; Investigation, M.F., A.S. and A.L.; Methodology, A.L.; Project administration, A.L.; Supervision, A.L.; Visualization, M.F., A.S. and A.L.; Writing—original draft, M.F.; Writing—review & editing, M.F. and A.L. All authors have read and agreed to the published version of the manuscript.

**Funding:** This research received no external funding.

**Acknowledgments:** The authors are grateful to ETH Zürich for the financial support and to Gabriele Ilari (Electron Microscopy Center, Swiss Federal Laboratories for Materials Science and Technology, Empa) for the TEM data. Furthermore, the Complex Materials group at the department of Materials at ETH Zürich is acknowledged for providing access to optical microscopes.

**Conflicts of Interest:** The authors declare no conflict of interest.

## References

1. Seeley, Z.M.; Kuntz, J.D.; Cherepy, N.J.; Payne, S.A. Transparent Lu$_2$O$_3$:Eu ceramics by sinter and HIP optimization. *Opt. Mater.* **2011**, *33*, 1721–1726. [CrossRef]
2. Satapathy, S.; Ahlawat, A.; Paliwal, A.; Singh, R.; Singh, M.K.; Gupta, P.K. Effect of calcination temperature on nanoparticle morphology and its consequence on optical properties of Nd:Y$_2$O$_3$ transparent ceramics. *CrystEngComm.* **2014**, *16*, 2723–2731. [CrossRef]
3. Serrano, A.; Caballero-Calero, O.; García, M.Á.; Lazić, S.; Carmona, N.; Castro, G.R.; Martín-González, M.; Fernández, J.F. Cold sintering process of ZnO ceramics: Effect of the nanoparticle/microparticle ratio. *J. Eur. Ceram. Soc.* **2020**, *40*, 5535–5542. [CrossRef]
4. Massera, J.; Gaussiran, M.; Gluchowski, P.; Lastusaari, M.; Hupa, L.; Petit, L. Processing and characterization of phosphate glasses containing CaAl$_2$O$_4$:Eu$^{2+}$,Nd$^{3+}$ and SrAl$_2$O$_4$:Eu$^{2+}$,Dy$^{3+}$ microparticles. *J. Eur. Ceram. Soc.* **2015**, *35*, 3863–3871. [CrossRef]
5. Vanecek, V.; Kral, R.; Paterek, J.; Babin, V.; Jary, V.; Hybler, J.; Kodama, S.; Kurosawa, S.; Yokota, Y.; Yoshikawa, A.; et al. Modified vertical Bridgman method: Time and cost effective tool for preparation of Cs$_2$HfCl$_6$ single crystals. *J. Cryst. Growth* **2020**, *533*, 125479. [CrossRef]
6. Guzik, M.; Pejchal, J.; Yoshikawa, A.; Ito, A.; Goto, T.; Siczek, M.; Lis, T.; Boulon, G. Structural investigations of Lu$_2$O$_3$ as single crystal and polycrystalline transparent ceramic. *Cryst. Growth Des.* **2014**, *14*, 3327–3334. [CrossRef]
7. Erb, R.M.; Libanori, R.; Rothfuchs, N.; Studart, A.R. Composites Reinforced in Three Dimensions by Using Low Magnetic Fields. *Science* **2012**, *335*, 199–204. [CrossRef]
8. Magrini, T.; Bouville, F.; Lauria, A.; Le Ferrand, H.; Niebel, T.P.; Studart, A.R. Transparent and tough bulk composites inspired by nacre. *Nat. Commun.* **2019**, *10*, 2794. [CrossRef]
9. Magrini, T.; Moser, S.; Fellner, M.; Lauria, A.; Bouville, F.; Studart, A.R. Transparent Nacre-like Composites Toughened through Mineral Bridges. *Adv. Funct. Mater.* **2020**, *30*, 2002149. [CrossRef]
10. Nydegger, M.; Deshmukh, R.; Tervoort, E.; Niederberger, M.; Caseri, W. Composites of Copper Nanowires in Polyethylene: Preparation and Processing to Materials with NIR Dichroism. *ACS Omega* **2019**, *4*, 11223–11228. [CrossRef] [PubMed]
11. Chernenko, K.A.; Gorokhova, E.I.; Eronko, S.B.; Sandulenko, A.V.; Venevtsev, I.D.; Wieczorek, H.; Rodnyi, P.A. Structural, Optical, and Luminescent Properties of ZnO:Ga and ZnO:In Ceramics. *IEEE Trans. Nucl. Sci.* **2018**, *65*, 2196–2202. [CrossRef]
12. Shmurak, S.Z.; Kedrov, V.V.; Klassen, N.V.; Shakhrai, O.A. Spectroscopy of composite scintillators. *Phys. Solid State* **2012**, *54*, 2266–2276. [CrossRef]

13. Kalyvas, N.; Liaparinos, P.; Michail, C.; David, S.; Fountos, G.; Wójtowicz, M.; Zych, E.; Kandarakis, I. Studying the luminescence efficiency of $Lu_2O_3$:Eu nanophosphor material for digital X-ray imaging applications. *Appl. Phys. A* **2012**, *106*, 131–136. [CrossRef]
14. Zych, E.; Meijerink, A.; De, C.; Donegá, M. Quantum efficiency of europium emission from nanocrystalline powders of $Lu_2O_3$: Eu. *J. Phys. Condens. Matter* **2003**, *15*, 5145–5155. [CrossRef]
15. Zych, E.; Trojan-Piegza, J. Low-Temperature Luminescence of $Lu_2O_3$: Eu Ceramics upon Excitation with Synchrotron Radiation in the Vicinity of Band Gap Energy. *Chem. Mater.* **2006**, *18*, 2194–2199. [CrossRef]
16. Cao, Y.C. Synthesis of Square Gadolinium-Oxide Nanoplates. *J. Am. Chem. Soc.* **2004**, *126*, 7456–7457. [CrossRef]
17. Mahajan, S.V.; Hart, J.; Hood, J.; Everheart, A.; Redigolo, M.L.; Koktysh, D.S.; Payzant, E.A.; Dickerson, J.H. Synthesis of $RE(OH)_2Cl$ and REOCl (RE=Eu, Tb) nanostructures. *J. Rare Earths* **2008**, *26*, 131–135. [CrossRef]
18. Fang, Y.-P.; Xu, A.-W.; You, L.-P.; Song, R.-Q.; Yu, J.C.; Zhang, H.-X.; Li, Q.; Liu, H.-Q. Hydrothermal Synthesis of Rare Earth (Tb, Y) Hydroxide and Oxide Nanotubes. *Adv. Funct. Mater.* **2003**, *13*, 955–960. [CrossRef]
19. Li, G.; Peng, C.; Zhang, C.; Xu, Z.; Shang, M.; Yang, D.; Kang, X.; Wang, W.; Li, C.; Cheng, Z.; et al. $Eu^{3+}/Tb^{3+}$-Doped $La_2O_2CO_3/La_2O_3$ Nano/Microcrystals with Multiform Morphologies: Facile Synthesis, Growth Mechanism, and Luminescence Properties. *Inorg. Chem.* **2010**, *49*, 10522–10535. [CrossRef]
20. Liu, J.; Liu, B.; Zhu, Z.; Chen, L.; Hu, J.; Xu, M.; Cheng, C.; Ouyang, X.; Zhang, Z.; Ruan, J.; et al. Modified timing characteristic of a scintillation detection system with photonic crystal structures. *Opt. Lett.* **2017**, *42*, 987–990. [CrossRef]
21. Knapitsch, A.; Auffray, E.; Fabjan, C.W.; Leclercq, J.L.; Lecoq, P.; Letartre, X.; Seassal, C. Photonic crystals: A novel approach to enhance the light output of scintillation based detectors. *Nucl. Instrum. Methods Phys. Res.* **2011**, *628*, 385–388. [CrossRef]
22. Zehnder, R.A.; Clark, D.L.; Scott, B.L.; Donohoe, R.J.; Palmer, P.D.; Runde, W.H.; Hobart, D.E. Investigation of the Structural Properties of an Extended Series of Lanthanide Bis-hydroxychlorides $Ln(OH)_2Cl$ (Ln = Nd − Lu, except Pm and Sm). *Inorg. Chem.* **2010**, *49*, 4781–4790. [CrossRef]
23. Olliges-Stadler, I.; Rossell, M.D.; Süess, M.J.; Ludi, B.; Bunk, O.; Pedersen, J.S.; Birkedal, H.; Niederberger, M. A comprehensive study of the crystallization mechanism involved in the nonaqueous formation of tungstite. *Nanoscale* **2013**, *5*, 8517. [CrossRef]
24. Thoř, T.; Rubešová, K.; Jakeš, V.; Cajzl, J.; Nádherný, L.; Mikolášová, D.; Kučerková, R.; Nikl, M. Lanthanide-doped $Lu_2O_3$ phosphors and scintillators with green-to-red emission. *J. Lumin.* **2019**, *215*, 116647. [CrossRef]
25. Brandt, G.; Diehl, R. Preparation, powder data and crystal structure of YbOCl. *Mater. Res. Bull.* **1974**, *9*, 411–419. [CrossRef]
26. Pinna, N.; Garnweitner, G.; Beato, P.; Niederberger, M.; Antonietti, M. Synthesis of Yttria-Based Crystalline and Lamellar Nanostructures and their Formation Mechanism. *Small* **2004**, *1*, 112–121. [CrossRef]
27. Binnemans, K. Interpretation of europium(III) spectra. *Coord. Chem. Rev.* **2015**, *295*, 1–45. [CrossRef]
28. Lauria, A.; Chiodini, N.; Fasoli, M.; Mihóková, E.; Moretti, F.; Nale, A.; Nikl, M.; Vedda, A. Acetate–citrate gel combustion: A strategy for the synthesis of nanosized lutetium hafnate phosphor powders. *J. Mater. Chem.* **2011**, *21*, 8975. [CrossRef]
29. Boyer, J.C.; Vetrone, F.; Capobianco, J.A.; Speghini, A.; Bettinelli, M. Variation of Fluorescence Lifetimes and Judd-Ofelt Parameters between Eu 3+ Doped Bulk and Nanocrystalline Cubic $Lu_2O_3$. *J. Phys. Chem. B* **2004**, *108*, 20137–20143. [CrossRef]
30. Mahajan, S.V.; Redígolo, M.L.; Koktysh, D.S.; Dickerson, J.H. Stepwise assembly of europium sesquioxide nanocrystals and nanoneedles. In Proceedings of the Nanophotonic Materials III, San Diego, CA, USA, 13–17 August 2006; Gaburro, Z., Cabrini, S., Eds.; Volume 6321, p. 632102.

# Article

# Charge Compensation in Europium-Doped Hafnia Nanoparticles: Solvothermal Synthesis and Colloidal Dispersion

Xavier H. Guichard, Francesco Bernasconi and Alessandro Lauria *

Laboratory for Multifunctional Materials, Department of Materials, ETH Zürich, Vladimir-Prelog-Weg 5, 8093 Zürich, Switzerland; xavier.guichard@mat.ethz.ch (X.H.G.); frbernas@student.ethz.ch (F.B.)
* Correspondence: alessandro.lauria@mat.ethz.ch

**Abstract:** Effective charge compensation of europium in hafnium oxide nanoparticles was achieved at low temperature, allowing high doping incorporation (up to 6 at.%) and enhanced luminescence. The efficiency of the incorporation and charge compensation was confirmed by scanning electron microscope energy dispersive X-ray spectroscopy and powder X-ray diffraction measurements. Despite the known polymorphism of hafnium oxide, when doped to a concentration above 3 at.%, only the pure monoclinic phase was observed up to 6 at.% of europium. Furthermore, the low-temperature solvothermal route allowed the direct formation of stable dispersions of the synthesized material over a wide range of concentrations in aqueous media. The dispersions were studied by diffuse light scattering (DLS) to evaluate their quality and by photoluminescence to investigate the incorporation of the dopants into the lattice.

**Keywords:** europium luminescence; charge compensation; niobium; microwave synthesis; hafnia; dispersion

## 1. Introduction

The cubic polymorph of hafnia can be stabilized at room temperature when sufficient doping with rare earth ions was achieved [1–6]. Indeed for doping levels exceeding 3 at.% the thermodynamically stable monoclinic phase of hafnium was not anymore the only favourable crystal structure, often leading to the simultaneous appearance of the cubic polymorph. A similar behaviour was also reported for zirconia with stabilization of cubic/tetragonal polymorph from 0.5 at.% on [7–11]. The origin of these structural modifications are attributed to the charge mismatch at the doping lattice site, and the consequent creation of defects by the substitutional doping. One proposed approach to avoid the effect of this defective dopant implies the simultaneous doping with elements with higher oxidation state than hafnium or zirconium, in order to compensate for the charge mismatch induced by the substitution of $Hf^{4+}$ or $Zr^{4+}$ by only trivalent $RE^{3+}$ ions [2,7,9,11,12]. This strategy was successfully applied to hafnia and zirconia by using $Nb^{5+}$ or $Ta^{5+}$ as co-dopants and a significant improvement of the $Eu^{3+}$ related emission was reported with increase ranging from 5 to 50 times [2,7,12]. However, if the dopants incorporation is not well controlled, a only partial charge compensation may occur, and indeed often the remaining cubic/tetragonal phase was reported at $Eu^{3+}$ concentrations of few at.% [2,7,9,11–13]. It would therefore be desirable to obtain higher control over the doping incorporation, and on its effects on the crystal lattice symmetry, in order to obtain phase pure functional nanocrystals. Charge compensation was demonstrated to be effective also to improve the incorporation of functional dopants in titania synthesized by solvothermal methods, increasing the tolerance of the host toward trivalent ions [14]. Moreover, the synthetic pathways explored to produce charge compensated RE doped hafnia nanocrystals, usually imply high temperatures (above 300 °C), and further treatments at around 1000 °C, or higher, are often needed to optimise the dopant incorporation [9,11,13]. These treatments typically induce a certain degree of particle agglomeration, strongly limiting the possibility to exploit these systems to obtain stable dispersions. In fact, in order to enable easier and more versatile approaches for advanced additive manufacturing and further processing of particle-based materials, colloidal syntheses requiring lower temperatures, seem to represent a more suitable strategy.

In this work, a solvothermal synthetic approach, for the incorporation of $Eu^{3+}$ ions in hafnia nanocrystals at relatively low temperatures (namely 260 °C) was studied, aiming at maximizing the light output of $Eu^{3+}$ fluorescence. Moreover, the realization of stable dispersions was investigated, in the attempt to formulate high quality functional inks suitable for a wide variety of processes and applications.

The incorporation of the dopants (charge compensation and luminescence activator ions) was studied by analysing the particles photoluminescence (PL) properties. At the same time, the effects of the charge compensation on the lattice symmetry of the obtained nanocrystals was monitored by means of powder X-ray diffraction (PXRD) and scanning transmission electron microscopy (STEM) coupled with energy dispersive X-ray spectroscopy (EDX). The attribution of characteristic excitation pathways of europium in charge compensated colloidal nano-systems, compared to literature data on $ZrO_2$, was also discussed.

## 2. Materials and Methods

### 2.1. Preparation of $HfO_2$:Eu Dispersions

The doped $HfO_2$ nanoparticles were prepared through a slightly modified microwave assisted solvothermal synthesis reported elsewhere [15]. Benzyl alcohol solutions of each precursor: $HfCl_4$ 1 M and $EuCl_3$, $NbCl_5$, $TaCl_5$ 0.1 M were prepared to be used as stock solutions. Each precursor was initially dissolved in five molar equivalents of methanol and stirred until the solution turned clear, before the final addition of solvent required to reach the desired concentration. Subsequently in a 10 mL microwave glass tube the different stock solution were mixed in the appropriate quantity to reach the desired composition with a total amount of precursor of 5 mmol. For the synthesis of hafnia with 3 at.% $Eu^{3+}$, the stock solutions of $HfCl_4$ was mixed with that of $EuCl_3$ (485 and 150 µL, respectively) then the mixture was completed to 1.5 mL with benzyl alcohol and stirred. Finally, 2.5 mL of a 0.5M HCl solution in benzyl alcohol were used to yield to a total volume of 4 mL of clear reacting mixture. The vial was sealed with a Teflon lid and heated in a Discover SP CEM microwave (Kamp-Lintfort, Germany), with the subsequent steps 100 °C, 1 min; 200 °C, 1 min; 260 °C, 14 min. During these steps the pressure rose up to a final maximum value of 17 bars. After the synthesis, the formation of a white product was observed. The reaction mixture was then washed twice with diethyl ether and the product was collected by centrifugation (4000 rmp, 10 min). The centrifuged particles were dispersed in 5 mL of Millipore water yielding to a transparent dispersion. Part of the dispersion was dried for characterization. The particles were readily dispersible irrespective of the doping concentration and no precipitation was observed upon dilution of the particles in water.

### 2.2. Characterisation Methods

All X-ray diffraction (XRD) measurements were performed on powders in reflection mode (Cu Kα radiation at 45 kV and 40 mA) on Empyrean diffractometer from PANalytical (Almelo, Netherlands). STEM and TEM was performed on a FEI Talos F200X (ThermoFisher Scientific, Zurich, Switzerland) operated at 200 kV. STEM analyses were carried out with a bright field detector (BF STEM) and were accompanied by the high resolution energy dispersive spectroscopy (EDS) using the SuperX integrated EDS-system with four silicon drift detectors (SDDs). The EDS-STEM analyses were performed with a probe size of 0.5 nm. The samples were prepared by dropping 10 µL of water dispersion of nanocrystals (100 µg/mL) on lacey carbon Au grids. Photoluminescence (PL) and photoluminescence excitation (PLE) measurements were performed using a 90°-optical geometry Jasco FP-8500 spectrofluorometer (Tokyo, Japan). All photoluminescence measurement were collected at room temperature on water dispersions of particles in 10 × 10 mm$^2$ quartz cuvettes (Thorlabs, Bergkirchen, Germany). Dynamic light scattering (DLS) of diluted particle water dispersions (100 µg*mL$^{-1}$) was collected on a Zetasizer NS instrument (Malvern Panalytical, Malvern, UK) in backscattering mode (scattering angle 173°) at a temperature of 25 °C.

## 3. Results

Readily dispersible doped and co-doped HfO$_2$ nanocrystals were obtained right after the synthesis. The initial treatments of the precursors with methanol indeed allowed a homogeneous dissolution of the precursor in benzyl alcohol, and it also prevented the formation of side products of the solvent polymerization during the forthcoming microwave reaction [16,17]. In addition, the presence of HCl in the reactive mixture proved to be essential for obtaining high quality dispersions.

### 3.1. Structural Analysis of the Nanocrystals

Figure 1 shows X-ray diffraction, recorded on 3 at.% Eu$^{3+}$ HfO$_2$; 6 at.% Eu$^{3+}$ and 6 at.% Nd$^{5+}$ HfO$_2$; and 6 at.% Eu$^{3+}$ and 6 at.% Ta$^{5+}$ HfO$_2$ nanopowder samples. The expected monoclinic structure is observed for the 3 at.% doped HfO$_2$, as well as for the samples doped with 6 at.% of Eu$^{3+}$ and charge compensated with either Ta$^{5+}$ or Nb$^{5+}$ showing nearly identical diffractograms, where we could not find traces of cubic phase. Besides identical crystal geometry, the samples show similar broadening of the diffraction peaks, indicating that the crystal size is little affected by the different kind and amount of dopants. In fact, the Scherrer analysis held on the (111) peak at 31.5 degrees led to quite similar values of the average crystal size for Eu doped and Eu/Nb co-doped particles, namely 8.1 and 8.5 ± 1.5 nm, respectively, while slightly smaller sizes of 7.3 ± 1.5 nm were obtained for for the Eu/Ta co-doped sample. These results are further supported by TEM investigations (see Figure S1 in the Supplementary Materials for representative micrographs of Eu/Nb co-doped HfO$_2$) where nanocrystals with diameters ranging between 5 and 10 nm could be observed.

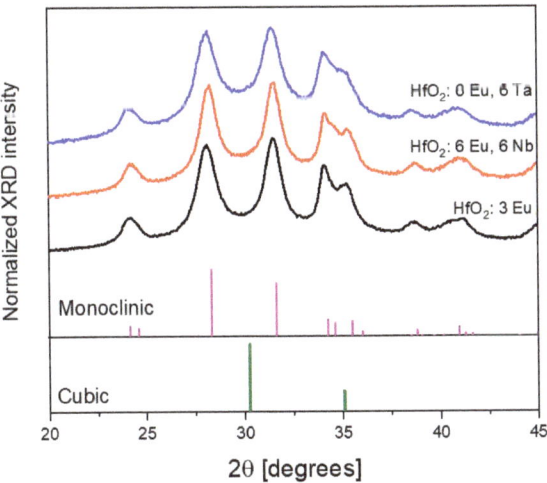

**Figure 1.** PXRD of HfO$_2$: 3 Eu (black curve), HfO$_2$: 6 Eu, 6 Nb (red curve), HfO$_2$: 6 Eu, 6 Ta (blue curve). The pattern of monoclinic (pink bars, PDF 04-007-8630) and cubic (green bars, PDF 04-002-0037) HfO$_2$ are shown as reference.

## 3.2. STEM-EDX Elemental Analysis

STEM analysis on co-doped HfO$_2$ nanocrystals evidences the occurrence of particle agglomerates of 10 to 40 nm. Moreover, the elemental analysis presented as EDX mappings in Figure 2 shows dopants homogeneously distributed in the particles, irrespective of the nature of the co-dopant. The results of the quantitative EDX analysis, reported in Table 1, account for the high doping efficiency of the solvothermal method presented in this work. This can be inferred based on the good agreement between measured and nominal compositions. Indeed, the concentration of Nb and Eu are quite similar to the nominal ones, while slightly higher values are measured for Ta. This disagreement might derive from some overdoping, which can not be excluded, although its most probable origin could lie in the partial overlap of the EDX signals of Ta and Hf (Figure S2), leading to lower accuracy of the quantitative dermination of this element in hafnia by this method.

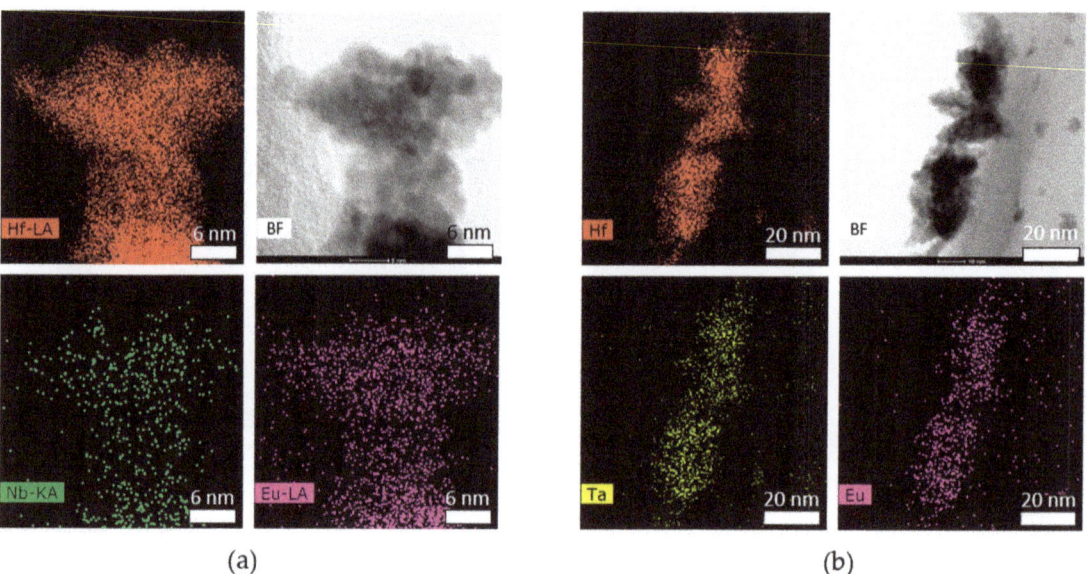

**Figure 2.** STEM of co-doped nanoparticles. Elemental mapping and bright field (BF) stem image of the HfO$_2$: 6 Eu, 6 Nb (**a**) and HfO$_2$: 6 Eu, 6 Ta samples (**b**).

**Table 1.** EDX quantitative elemental analysis on charge compensated co-doped HfO$_2$ nanopowders.

| Element | HfO$_2$: 6 Eu, 6 Nb | HfO$_2$: 6 Eu, 6 Ta |
|---|---|---|
| Hf | Nominal 88.0 measured 88.2 ± 1 at.% | Nominal 88.0 measured 85.7 ± 1 at.% |
| Eu | Nominal 6.0 measured 5.8 ± 1 at.% | Nominal 6.0 measured 6.1 ± 1 at.% |
| Nd | Nominal 6.0 measured 6.0 ± 1 at.% | - |
| Ta | - | Nominal 6.0 measured 8.2 ± 1 at.% |

### 3.3. Evaluation of the Particle Dispersion

Figure 3a shows digital images of water dispersions of the obtained nanocrystals. The dispersions are stable without further particle functionalization right after the washing stage. No signs of sedimentation could be observed after weeks assessing the stability of the particles in water. This is in agreement with previous observations on TiO$_2$ nanoparticles obtained through a similar method where positively charged particles could be obtained, also showing high stability in water due to their strong electrostatic repulsion [18]. Figure 3b shows DLS hydrodynamic size distributions, which lie in the 10–50 nm range for all samples, without formation of larger agglomerates. These values are in good agreement with TEM observations (Figure 2 and Figure S1), where also 10–40 nm clusters of smaller primary particles could be recognised.

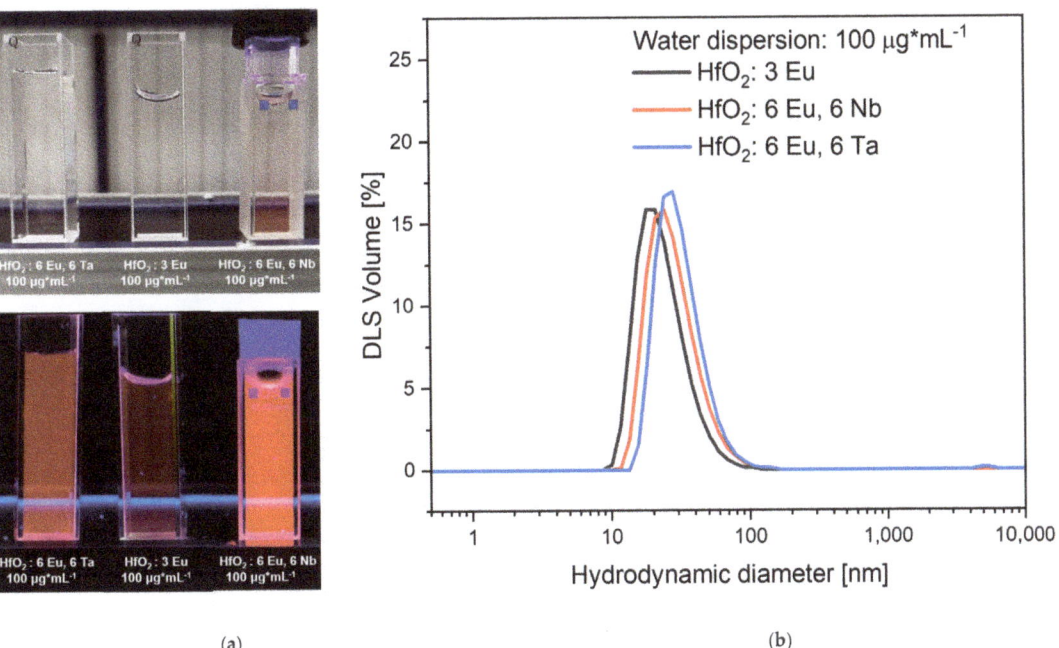

**Figure 3.** High quality dispersions of doped and co-doped HfO$_2$ nanocrystals. (a) Digital images of dispersions with particle concentration of 100 µg*mL$^{-1}$ illuminated by the 254 nm line of a Wood-lamp with (top panel) and without (bottom panel) ambient light. (b) Hydrodynamic diameter distribution of HfO$_2$ nanoparticles dispersed in water as measured by dynamic light scattering.

### 3.4. Optical Charcterization

Photoluminescence excitation (PLE) spectra obtained monitoring the Eu$^{3+}$ luminescence on particle dispersions are presented in Figure 4a, revealing the dominance of broad excitation bands located at wavelengths lower than 300 nm. The sharp intra-center excitation lines of Eu$^{3+}$ occur between 300 and 465 nm (Figure 4a, inset). While these intra-center excitation are negligibly affected by the presence of co-dopants, and compatible with the expected intrinsic transitions of Eu$^{3+}$ ions in hafnia, the high energy bands in the UVC spectral region are strongly affected by the charge compensation. Indeed, the main band at 205 nm of the Eu doped sample, is coupled to an additional band centered at around 225 nm in the Ta$^{5+}$ doped sample. In the case of the nanocrystals co-doped with Nb$^{5+}$, the PLE is dominated by a broad band at around 260 nm. However, these different excitation channels lead to the typical PL (Figure 4b) of Eu$^{3+}$ ions in monoclinic HfO$_2$, mostly unaltered by the co-dopants. The only observable difference is the minor change in the $^5D_0 \rightarrow {}^7F_1$ transition (see Figure S3) where the charge compensated samples shows a stronger splitting of the energy sublevels which can be attributed to an increase in the strength of the crystal field [19,20].

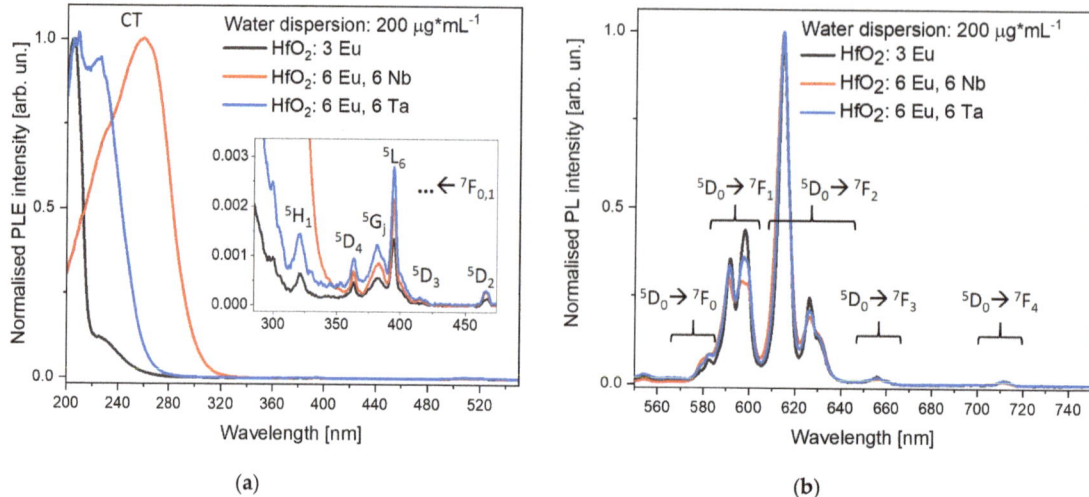

**Figure 4.** Optical characterization of water dispersions of $HfO_2$: Eu charge compensated and non-charge compensated samples. (**a**) normalized PL spectra of $HfO_2$: 3 Eu (black line), $HfO_2$: 6 Eu, 6 Nb (red line), and $HfO_2$: 6 Eu, 6 Ta (blue line). All spectra were collected by monitoring the emission at 610 nm. Inset: zoom in the 300–500 nm spectral range. (**b**) PL spectra of $HfO_2$: 3 Eu, $HfO_2$: 6 Eu, 6 Ta, and $HfO_2$: 6 Eu, 6 Nb, excited at 205, 225, and 255 nm, respectively.

When rising the particle concentration in dispersion, a linear response of the PL intensity is observed up to a loading of 200 µg*mL$^{-1}$ (Figure S4), as expected for low concentration regime of unperturbed systems. At higher particle concentration, detrimental scattering and absorption phenomena should be expected. Indeed, the recorded emission increases but deviates from the linear trend, likely due to the light diffusion and absorption of the incident beam, before it reaches the center of the sampling cuvette concequently reducing the detected intensity. This feature is of quite high importance, especially when considering that the PL output of the dispersion could be also increased by rising the doping level at constant particle loading. Indeed, when the nominal dopant concentration is increased, a higher PL output is observed (Figure 5a), as expected for a higher activator concentration. Additionally, the shape of the emission was not altered by variations in doping levels, as shown in the inset. However, it should be noted that the intensity of the $HfO_2$: Eu, Nb emission increases linearly with the dopant content through the whole concentration range according to the PL integrals reported in Figure 5b (calculated in the spectral range between 515–745 nm).

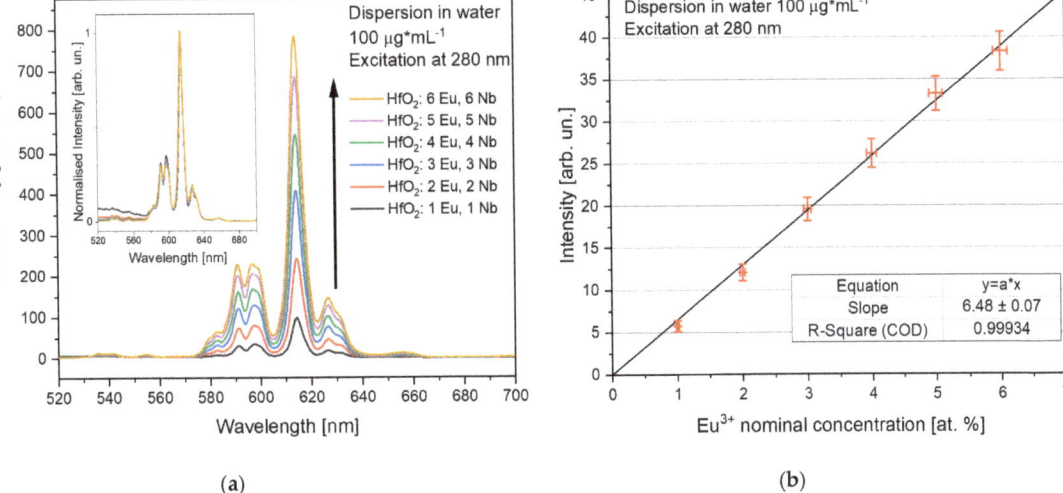

**Figure 5.** PL of $Eu^{3+}$ doped $HfO_2$ nanocrystals co doped by $Nb^{5+}$ as a function of the doping level. (**a**) PL spectra recorded by exciting at 280 nm. Inset: PL normalised at its maximum. (**b**) Integrated intensities as a function of $Eu^{3+}$ nominal concentration and linear fit.

## 4. Discussion

$Eu^{3+}$ related emission is highly dependent on its local environment [19,21]. In fact, the PL spectra of europium can be used as a direct indication of its site symmetry. In Figure 4b the emission of all samples corresponds to that expected for $Eu^{3+}$ in monoclinic $HfO_2$ [1,19]. It is therefore possible to argue that the observed emissions originates from $Eu^{3+}$ laying inside the nanocrystals, i.e., that the europium locates inside the crystals irrespective of the presence of charge compensating ions.

In absence of charge compensation, Eu-doped $HfO_2$ shows an intense excitation band near 5.85 eV, which could be related to the host free exciton [12]. When charge compensation takes place, additional excitation bands appear, strongly depending on the nature of the co-dopant, demonstrating the equally effective incorporation of the pentavalent Ta and Nb ions into the lattice during the synthesis. Such excitation bands at around 260 nm and 230 nm measured for $HfO_2$: 6 Eu, 6 Nb and $HfO_2$: 6 Eu, 6 Ta, respectively, well agree with similar trends reported by Yin et al. [7] in $ZrO_2$:Eu,Nb and $ZrO_2$:Eu,Ta. Other works on similarly doped $HfO_2$ by Wiatrowska et al. [12] observed similar features, which were interpreted as the $O^{2-} \rightarrow M^{5+}$ (M = Ta and Nb) charge transfer transitions. Since the $Eu^{3+}$ f-f emissions in the monoclinic site are effectively excited through such charge transfer states, we may infer that $Eu^{3+}$ ions as well as $Nb^{5+}$ and $Ta^{5+}$ lie within the hafnia lattice in near proximity. EDX elemental analysis (Figure 2 and Figure S1) also supports this conclusion, showing the presence of both dopants in the same locations, i.e., within the hafnia particles and in the proportion expected based on the nominal composition set during the synthesis. All these results, point at the effective incorporation of the dopants through this synthetic route and to the strong association of activator ions and co-dopants at very close distance.

At last, the direct proportionality between the europium content and the intensity of the emission combined with the consistency of the spectral shape shown in Figure 5 proves both the quantitative incorporation in the monoclinic host and that the doping sites are not quenched at higher europium concentration. This effect is often not observed in the literature where, for similar concentration ranges, higher $Eu^{3+}$ concentration induces a minimal PL increase [1,3,7,8,22]. Altogether, the good incorporation of co-dopants seems to effectively reduce non radiative decay pathways which might be the reason behind such early onset of luminescence quenching in other systems.

## 5. Conclusions

In this work a modified solvothermal synthesis of $Eu^{3+}$ doped $HfO_2$ was reported. The structural effects of europium trivalent ions, usually responsible for the occurrence of even small amounts of cubic phase already at concentrations of few at.%, could be avoided through charge compensation by co-doping the materials with equal amounts of pentavalent tantalum or niobium ions. The obtained materials show phase pure monoclinic structure of $HfO_2$, even at unprecedented $Eu^{3+}$ concentration values up to 6 at.%. The compositional analysis and luminescence studies suggest a quite effective incorporation of both the luminescence activator ions and the co-dopant, lying in close proximity. Through a wide range of $Eu^{3+}$ doping, not only the stabilization of the monoclinic polymorph was achieved, but also a linear dependence of the bright RE-related red emission, suggesting the role of the co-dopants in the reduction of non radiative decay pathways.

The colloidal synthesis proposed here, not requiring high temperatures and leading to very small agglomerates, led to particle dispersions showing high stability over several weeks. This feature enables the design of functional inks with tunable excitation based on the nature of the co-dopant, and with controlled luminescence intensity based either on the RE concentration or on the particle loading. These features represent a further improvement toward employing luminescent hafnia nanocrystals for several advanced processes and for many practical applications.

**Supplementary Materials:** The following are available online at https://www.mdpi.com/article/10.3390/cryst11091042/s1, Figure S1: TEM micrographs of $HfO_2$: 6 Eu, 6 Nb, Figure S2: EDX spectra of $HfO_2$:6 Eu, 6 Nb and $HfO_2$: 6 Eu, 6 Ta, Figure S3: Enlarged PL spectra showing the splitting of the $^7F_1$ level of $Eu^{3+}$, Figure S4: Optical characterization of the $HfO_2$:6 Nb, 6 Eu water dispersion at different particle loadings.

**Author Contributions:** Conceptualization X.H.G. and A.L.; investigation X.H.G. and F.B.; writing—original draft preparation X.H.G. and A.L.; writing—review and editing X.H.G., F.B. and A.L.; supervision A.L. All authors have read and agreed to the published version of the manuscript.

**Funding:** This research received no external funding.

**Data Availability Statement:** The data that support the findings of this study are available from the authors upon request.

**Acknowledgments:** The authors are grateful to ETH Zurich for financial support. They also acknowledge the Scientific Center for Optical and Electron Microscopy (ScopeM) of ETH Zurich for providing the electron microscopy facilities and Alla Sologubenko for help on TEM analyses.

**Conflicts of Interest:** The authors declare no conflict of interest.

## References

1. Lauria, A.; Villa, I.; Fasoli, M.; Niederberger, M.; Vedda, A. Multifunctional Role of Rare Earth Doping in Optical Materials: Nonaqueous Sol–Gel Synthesis of Stabilized Cubic $HfO_2$ Luminescent Nanoparticles. *ACS Nano* **2013**, *7*, 7041–7052. [CrossRef]
2. Laganovska, K.; Bite, I.; Zolotarjovs, A.; Smits, K. Niobium enhanced europium ion luminescence in hafnia nanocrystals. *J. Lumin.* **2018**, *203*, 358–363. [CrossRef]
3. LeLuyer, C.; Villanueva-Ibañez, M.; Pillonnet, A.; Dujardin, C. $HfO_2$:X (X = $Eu^{3+}$, $Ce^{3+}$, $Y^{3+}$) Sol Gel Powders for Ultradense Scintillating Materials. *J. Phys. Chem. A* **2008**, *112*, 10152–10155. [CrossRef]
4. Gerken, L.R.H.; Keevend, K.; Zhang, Y.; Starsich, F.H.L.; Eberhardt, C.; Panzarasa, G.; Matter, M.T.; Wichser, A.; Boss, A.; Neels, A.; et al. Lanthanide-Doped Hafnia Nanoparticles for Multimodal Theranostics: Tailoring the Physicochemical Properties and Interactions with Biological Entities. *ACS Appl. Mater. Interfaces* **2019**, *11*, 437–448. [CrossRef]
5. Villanueva-Ibañez, M.; Le Luyer, C.; Marty, O.; Mugnier, J. Annealing and doping effects on the structure of europium-doped $HfO_2$ sol-gel material. *Opt. Mater. (Amst.)* **2003**, *24*, 51–57. [CrossRef]
6. Fujimori, H.; Yashima, M.; Sasaki, S.; Kakihana, M.; Mori, T.; Tanaka, M.; Yoshimura, M. Cubic-tetragonal phase change of yttria-doped hafnia solid solution: High-resolution X-ray diffraction and Raman scattering. *Chem. Phys. Lett.* **2001**, *346*, 217–223. [CrossRef]

7. Yin, X.; Wang, Y.; Wan, D.; Huang, F.; Yao, J. Red-luminescence enhancement of $ZrO_2$-based phosphor by codoping $Eu^{3+}$ and $M^{5+}$ (M = Nb, Ta). *Opt. Mater. (Amst.)* **2012**, *34*, 1353–1356. [CrossRef]
8. Hui, Y.; Zou, B.; Liu, S.; Zhao, S.; Xu, J.; Zhao, Y.; Fan, X.; Zhu, L.; Wang, Y.; Cao, X. Effects of $Eu^{3+}$-doping and annealing on structure and fluorescence of zirconia phosphors. *Ceram. Int.* **2015**, *41*, 2760–2769. [CrossRef]
9. Smits, K.; Olsteins, D.; Zolotarjovs, A.; Laganovska, K.; Millers, D.; Ignatans, R.; Grabis, J. Doped zirconia phase and luminescence dependence on the nature of charge compensation. *Sci. Rep.* **2017**, *7*, 44453. [CrossRef] [PubMed]
10. Bugrov, A.N.; Smyslov, R.Y.; Zavialova, A.Y.; Kopitsa, G.P.; Khamova, T.V.; Kirilenko, D.A.; Kolesnikov, I.E.; Pankin, D.V.; Baigildin, V.A.; Licitra, C. Influence of Stabilizing Ion Content on the Structure, Photoluminescence and Biological Properties of $Zr_{1-x}Eu_xO_{2-0.5x}$ Nanoparticles. *Crystals* **2020**, *10*, 1038. [CrossRef]
11. Smits, K.; Sarakovskis, A.; Grigorjeva, L.; Millers, D.; Grabis, J. The role of Nb in intensity increase of Er ion upconversion luminescence in zirconia. *J. Appl. Phys.* **2014**, *115*, 213520. [CrossRef]
12. Wiatrowska, A.; Zych, E. Modeling Luminescent Properties of $HfO_2$:Eu Powders with Li, Ta, Nb, and V Codopants. *J. Phys. Chem. C* **2012**, *116*, 6409–6419. [CrossRef]
13. Kiisk, V.; Puust, L.; Mändar, H.; Ritslaid, P.; Rähn, M.; Bite, I.; Jankovica, D.; Sildos, I.; Jaaniso, R. Phase stability and oxygen-sensitive photoluminescence of $ZrO_2$:Eu,Nb nanopowders. *Mater. Chem. Phys.* **2018**, *214*, 135–142. [CrossRef]
14. Primc, D.; Zeng, G.; Leute, R.; Walter, M.; Mayrhofer, L.; Niederberger, M. Chemical Substitution—Alignment of the Surface Potentials for Efficient Charge Transport in Nanocrystalline $TiO_2$ Photocatalysts. *Chem. Mater.* **2016**, *28*, 4223–4230. [CrossRef]
15. De Keukeleere, K.; De Roo, J.; Lommens, P.; Martins, J.C.; Van Der Voort, P.; Van Driessche, I. Fast and Tunable Synthesis of $ZrO_2$ Nanocrystals: Mechanistic Insights into Precursor Dependence. *Inorg. Chem.* **2015**, *54*, 3469–3476. [CrossRef] [PubMed]
16. Olliges-Stadler, I.; Rossell, M.D.; Niederberger, M. Co-operative Formation of Monolithic Tungsten Oxide-Polybenzylene Hybrids via Polymerization of Benzyl Alcohol and Study of the Catalytic Activity of the Tungsten Oxide Nanoparticles. *Small* **2010**, *6*, 960–966. [CrossRef] [PubMed]
17. Braendle, A.; Perevedentsev, A.; Cheetham, N.J.; Stavrinou, P.N.; Schachner, J.A.; Mösch-Zanetti, N.C.; Niederberger, M.; Caseri, W.R. Homoconjugation in poly(phenylene methylene)s: A case study of non-π-conjugated polymers with unexpected fluorescent properties. *J. Polym. Sci. Part B Polym. Phys.* **2017**, *55*, 707–720. [CrossRef]
18. Frantz, C.; Lauria, A.; Manzano, C.V.; Guerra-Nuñez, C.; Niederberger, M.; Storrer, C.; Michler, J.; Philippe, L. Nonaqueous Sol-Gel Synthesis of Anatase Nanoparticles and Their Electrophoretic Deposition in Porous Alumina. *Langmuir* **2017**, *33*, 12404–12418. [CrossRef] [PubMed]
19. Binnemans, K.; Görller-Walrand, C. A simple model for crystal field splittings of the $^7F_1$ and $^5D_1$ energy levels of $Eu^{3+}$. *Chem. Phys. Lett.* **1995**, *245*, 75–78. [CrossRef]
20. Bünzli, J.-C.G.; Chauvin, A.-S.; Vandevyver, C.D.B.; Bo, S.; Comby, S. Lanthanide Bimetallic Helicates for in Vitro Imaging and Sensing. *Ann. N. Y. Acad. Sci.* **2008**, *1130*, 97–105. [CrossRef] [PubMed]
21. Binnemans, K. Interpretation of europium(III) spectra. *Coord. Chem. Rev.* **2015**, *295*, 1–45. [CrossRef]
22. Gupta, S.K.; Reghukumar, C.; Kadam, R.M. $Eu^{3+}$ local site analysis and emission characteristics of novel $Nd_2Zr_2O_7$:Eu phosphor: Insight into the effect of europium concentration on its photoluminescence properties. *RSC Adv.* **2016**, *6*, 53614–53624. [CrossRef]

Article

# Calcium Tungstate Doped with Rare Earth Ions Synthesized at Low Temperatures for Photoactive Composite and Anti-Counterfeiting Applications

Soung-Soo Yi and Jae-Yong Jung *

Division of Materials Science and Engineering, Silla University, Busan 46958, Korea; ssyi@silla.ac.kr
* Correspondence: eayoung21@naver.com; Tel.: +82-51-999-5465

**Abstract:** A precursor was prepared using a co-precipitation method to synthesize crystalline calcium tungstate. The prepared precursor was dried in an oven at 80 °C for 18 h. The dried powders, prepared without a heat treatment process, were observed in XRD analysis to be a crystalline $CaWO_4$ phase, confirming that the synthesis of crystalline $CaWO_4$ is possible even at low temperature. To use this crystalline $CaWO_4$ as a light emitting material, rare earth ions were added when preparing the precursor. The $CaWO_4$ powders doped with terbium ($Tb^{3+}$) and europium ($Eu^{3+}$) ions, respectively, were also observed to be crystalline in XRD analysis. The luminescence of the undoped $CaWO_4$ sample exhibited a wide range of 300~600 nm and blue emission with a central peak of 420 nm. The $Tb^{3+}$-doped sample showed green light emission at 488, 545, 585, and 620 nm, and the $Eu^{3+}$-doped sample showed red light emission at 592, 614, 651, and 699 nm. Blue, green, and red $CaWO_4$ powders with various luminescence properties were mixed with glass powder and heat-treated at 600 °C to fabricate a blue luminescent PiG disk. In addition, a flexible green and red light-emitting composite was prepared by mixing it with a silicone-based polymer. An anti-counterfeiting application was prepared by using the phosphor in an ink, which could not be identified with the naked eye but can be identified under UV light.

**Keywords:** photoluminescence; co-precipitation; anti-counterfeiting; $CaWO_4$

## 1. Introduction

Rare earth ion doped luminescent materials are attracting a lot of attention because of their various applications as materials in lighting, leisure luminescent materials, and luminescent diodes [1–3]. In particular, the crystalline material tungstic acid is a suitable host material for doping rare earth ions because of its excellent thermal stability, and high energy transfer efficiency from tungsten ions to activator ions [4,5]. In general, tungstic acid hosts are classified into two groups according to their crystal structure: scheelite [6] and wolframites [7]. Representative materials are $BaWO_4$, $SrWO_4$, $CaWO_4$, $PbWO_4$, $MgWO_4$, $CdWO_4$, and $ZnWO_4$ [8–10]. Among them, calcium tungstate ($CaWO_4$) is $Ca^{2+}$ and WO4 with the coordination numbers 8 and 4 [11]. There is a scheelite structure composed of $Ca^{2+}$ and $WO_4^{2-}$ [12]. Because $CaWO_4$ with these characteristics also exhibits excellent optical properties and high chemical stability, it is widely applicable to phosphors for X-Ray augmentation screens, fluorescent lamps, light emitting diodes, scintillators, field emission displays, and white LEDs. In addition, phosphors made by doping rare earth ions with $CaWO_4$ as the host have the advantage of strong luminescence intensity with a narrow bandgap, caused by energy transfer between the 4f-4f shells of the doped rare earth ions, emitting light at various wavelengths [13,14]. Oh et al. reported a crystalline $CaWO_4$ synthesis method in which calcium chloride ($CaCl_2$) and sodium tungstate ($Na_2WO_4 \cdot 2H_2O$) in a molar ratio of 1:1 was dried at 100 °C for 12 h and exposed to microwaves (2.45 GHz, 1250 W, 15 min) after reheating at 600 °C [15]. To synthesize $CaWO_4$, Phurangr et al. prepared 0.005 mole of calcium nitrate ($Ca(NO_3)_2$) and sodium tungstate ($Na_2WO_4 \cdot 2H_2O$)

and dissolved them in 15 mL of ethylene glycol. This solution was put in an autoclave and heated for 20 min using a microwave (600 W), and studies on $CaWO_4$ crystallinity, chemical bond formation, and surface shape have been reported [16]. Du et al. prepared calcium carbonate, tungsten oxide, and dysprosium oxide in a chemically quantitative ratio then pulverized and kneaded the compound in a mortar. The mixture was placed in an alumina crucible and sintered at 1100 °C for 6 h in air to synthesize crystalline $CaWO_4$. In addition, a phosphor having light emission characteristics at 572 nm by adding the dysprosium ion was presented [17].

Previous studies have mainly synthesized $CaWO_4$ by supplying additional energy using high temperature or microwaves. Alternatively, it would be practically valuable to use relatively little energy during synthesis and to expand the utility of the phosphor powder.

In this study, a precursor was prepared by co-precipitation with calcium nitrate and sodium tungstate and drying at 80 °C to synthesize crystalline $CaWO_4$ white powder. It can be potentially used as a light emitting material by doping with rare earth ions such as terbium ($Tb^{3+}$) and europium ($Eu^{3+}$) in order to impart various luminescent properties.

The synthesized $CaWO_4$ phosphor was mixed with glass powder and a silicone-based polymer to prepare a disk as a flexible composite light emitter under UV light. It has possible application in the field of anti-counterfeiting when used in a solution, since it cannot be observed with the naked eye and can only be confirmed using UV light.

## 2. Materials and Methods

### 2.1. Crystalline $CaWO_4$ Synthesized at Low Temperature

The starting materials were Calcium nitrate ($Ca(NO_3)_2$), Sodium tungstate ($Na_2WO_4$), Turbium(III) nitrate hydrate ($Tb(NO_3)_3 \cdot xH_2O$, $Tb^{3+}$) and Europium(III) nitrate hydrate ($Eu(NO_3)_3 \cdot xH_2O$, $Eu^{3+}$).

A total of 1 mmol of $Ca(NO_3)_2$ was dissolved in beaker 'A' containing 50 mL (80 °C) of distilled water. $Na_2WO_4$ was put in beaker 'B', under the same conditions as in beaker 'A' and dissolved (Figure 1). The solution in beaker 'B' after being completely dissolved was slowly poured into beaker 'A' while stirring and maintained for about 30 min. After that, a white powder was recovered using a centrifuge. The white powder was prepared by rinsing with distilled water three times to remove the remaining sodium. The white powder was dried in an oven at 80 °C for 16 h to investigate its crystallinity and luminescent properties. In addition, $Tb(NO_3)_3 \cdot xH_2O$ or $Eu(NO_3)_3 \cdot xH_2O$ (0.05 mol%) ions were added to beaker 'A' during the co-precipitation reaction to impart luminescent properties [18]. The experiment was carried out at 25 °C and 55% humidity.

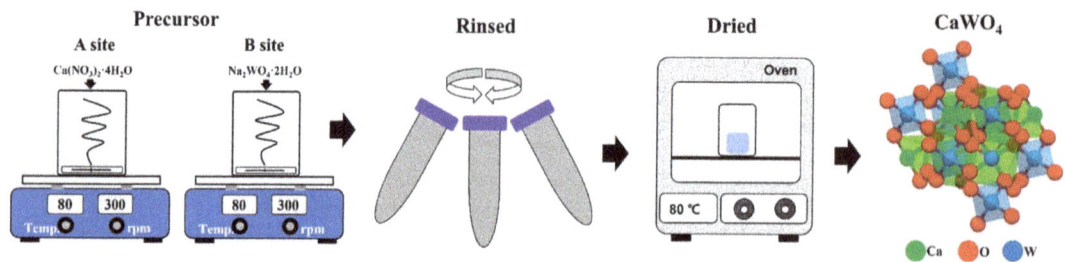

**Figure 1.** Schematic of the $CaWO_4$ synthesis procedure.

### 2.2. Characterization

Structural characterizing was performed by X-ray diffraction (XRD, Rigaku Ultima IV) with Cu Kα radiation ($\lambda$ = 1.5406 Å). The chemical composition of samples was studied by X-ray photoelectron spectroscopy (XPS; Thermo Fisher Scientific, Gloucester, UK) using Al-Kα lines. The C1s at 284.6 eV was used to calibrate the peak position of the insulating samples. The surface morphology was investigated using a field emission scanning electron microscope (FE-SEM, SU-8220, Hitach, Tokyo, Japan). The photoluminescence spectra were obtained using a fluorescence spectrophotometer (Scinco, FS-2, Seoul, Korea) with a 150 W Xenon lamp as the excitation source and a photomultiplier tube operating at 350 V.

### 2.3. Fabrication of Photoactive Composite and Anti-Counterfeiting Application

$CaWO_4$ powder was mixed with glass frits ($BaO$-$ZnO$-$B_2O_3$-$SiO_2$) at a weight ratio of 1:3 wt%. The mixed powder was placed in a metal mold and pressed with a press to prepare a round disk. A disk which absorbs UV light and emits light was finally produced by heat treatment in an air atmosphere at 600 °C for 5 h. In addition, a composite that was flexible and emits light by absorbing UV light was prepared by mixing a silicon-based polymer (Polydimethylsiloxane (PDMS)) and $CaWO_4$:$RE^{3+}$ (RE = Tb, Eu) powder. For anti-counterfeiting application, a solution was prepared by adding 10 wt% polyvinylpyrolidone (PVP, M.W. = 14,000) and 1 wt% of the synthesized powder to 10 mL of ethanol. The prepared solution, which cannot be confirmed with the naked eye and can only be confirmed with UV light, was stamped and painted on a banknote [19].

## 3. Results and Discussion
### 3.1. Structural and Morphology

Figure 2a shows the XRD analysis of the crystallinity and structure of the $CaWO_4$ white powder prepared by the co-precipitation method after dried in an oven at 80 °C. The synthesized $CaWO_4$ was consistent with the ICDD card (NO. 01-085-0433) and tetragonal structure of scheelite [20]. In addition, phases (101), (112), (204), and (312), which are the main peaks, were identified. It has been shown that the synthesis of crystalline $CaWO_4$ is possible at a low temperature and a simple process without a heat treatment process. The crystallinity and structure of $CaWO_4$:$Tb^{3+}$ and $CaWO_4$:$Eu^{3+}$ powder doped with rare earth ions were also confirmed in the same manner as pure $CaWO_4$. There was no secondary phase of $CaWO_4$ due to rare earth doping. However, as shown in Figure 2b, when the lattice spacing was compared with the main peak (112) phase, it showed a change due to the addition of rare earth. For pure $CaWO_4$, the lattice spacing was 0.277 nm. The lattice spacing of the rare earth doped $CaWO_4$:$Tb^{3+}$ (0.298 nm) and $CaWO_4$:$Eu^{3+}$ (0.279 nm) were increased. This is thought to be the result of doping with rare earth ions with relatively large ionic radii ($r(Tb^{3+})$ = 0.92 Å, $r(Eu^{3+})$ = 0.95 Å) in the $CaWO_4$ lattice [21].

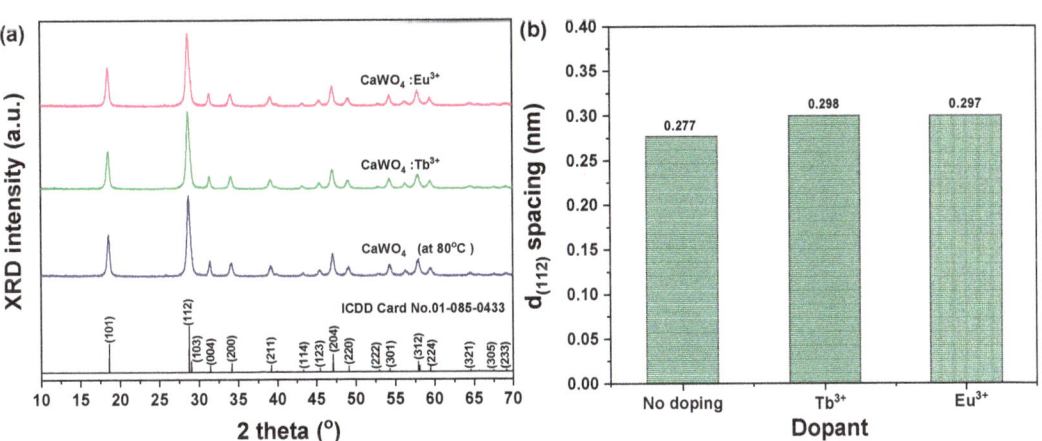

Figure 2. (a) XRD patterns and (b) change in $d_{(112)}$ spacing; $CaWO_4$, $CaWO_4$:$Tb^{3+}$, and $CaWO_4$:$Eu^{3+}$.

In addition, the size of each sample particle was investigated by substituting the half width and peak position of the main peak (112) in Scherrer's equation [22]. $CaWO_4$ samples were calculated at 76 nm, $CaWO_4$:$Tb^{3+}$ at 85 nm, and $CaWO_4$:$Eu^{3+}$ at 90 nm.

Titipun et al. synthesized $CaMoO_4$, $SrMoO_4$, $CaWO_4$, and $SrWO_4$ using the co-precipitation method at room temperature. The $MXO_4$ (M = Ca and Sr, X = Mo and W) nanoparticles precipitated—$M^{2+}$ cations as electron pair acceptors (Lewis acid) and reacted with $XO_4{}^{2-}$ anions as electron pair donors (Lewis base). The reaction between these two species ($M^{2+}{\leftarrow}{:}XO_4{}^{2-}$) proceeded to produce bonding. The lowest molecular orbital energy of the Lewis acid interacted with the highest molecular orbital energy of the Lewis base, and $MXO_4$ nanoparticles were finally synthesized [23]. It is thought that the $CaWO_4$ powder synthesized at low temperature in this study can also be synthesized without additional energy supply, as in the previous case.

In addition, Puneet et al. identified the oxide phase of rare earth ions doped in a synthesized $CaWO_4$ lattice through synchrotron X-ray diffraction analysis [24]. In this study, when the doped rare earth ions were calculated using a single unit cell of $CaWO_4$, it was calculated that the doped amount was about $1.59 \times 10^{19}$ RE atoms/$cm^3$ (RE = $Tb^{3+}$, $Eu^{3+}$).

The size and surface morphology of the synthesized crystalline $CaWO_4$, $CaWO_4{:}Tb^{3+}$, and $CaWO_4{:}Eu^{3+}$ particles were observed by FE-SEM. In addition, EDS mapping was performed to confirm the components of the synthesized samples, as shown in Figure 3. The size of the synthesized particles was observed to be about 5 μm and spherical at low magnification regardless of doping with rare earth ions, but when observed at high magnification, smaller particles of about 75 nm ($CaWO_4$), 83 nm ($CaWO_4{:}Tb^{3+}$), and 86 nm ($CaWO_4{:}Eu^{3+}$) were observed to be agglomerated.

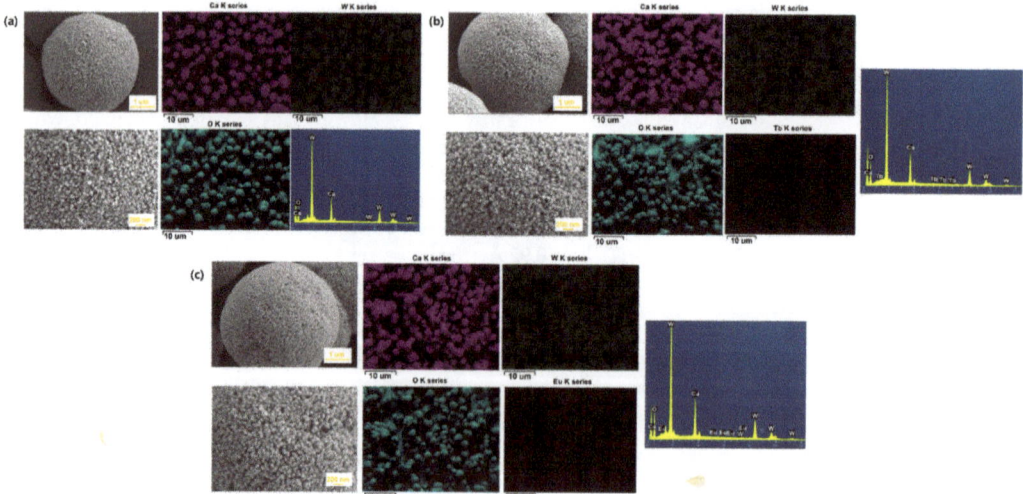

**Figure 3.** SEM-EDS analysis; (**a**) $CaWO_4$, (**b**) $CaWO_4{:}Tb^{3+}$, and (**c**) $CaWO_4{:}Eu^{3+}$.

In addition, in the rare-earth-doped $CaWO_4{:}Tb^{3+}$ (Figure 3b) and $CaWO_4{:}Eu^{3+}$ (Figure 3c) samples, each rare-earth component was confirmed, and it was confirmed that the rare-earth ions were evenly distributed without agglomeration.

*3.2. Chemical States and Phtoluminescence Proeprties*

Figure 4 shows the XPS measurements used to determine the chemical state of the synthesized $CaWO_4$, $CaWO_4{:}Tb^{3+}$, and $CaWO_4{:}Eu^{3+}$. Ca 2p, W 4f, and O 1s were confirmed as shown in Figure 4a. A trace amount of Na 1s was detected. This is thought to be due to sodium tungsten in the starting material, and it is thought to be a leftover that was not removed during the washing process when preparing the precursor.

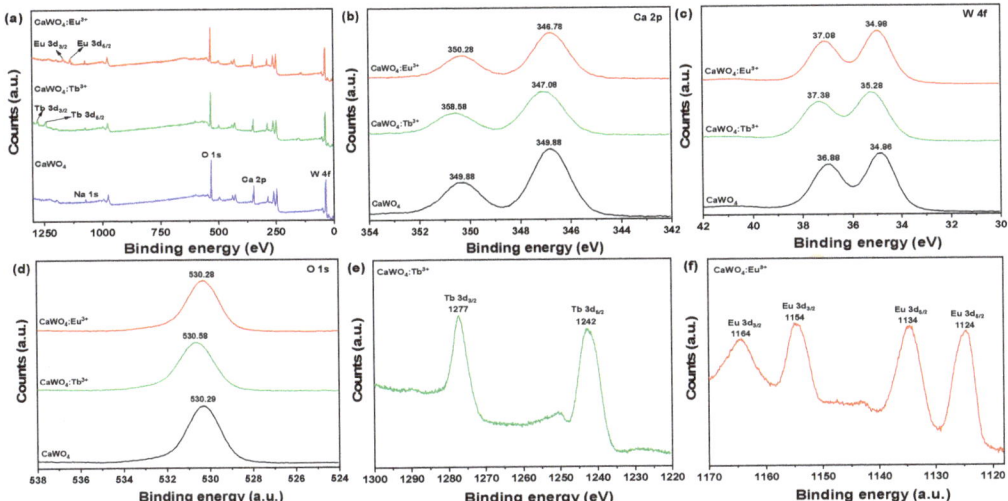

**Figure 4.** XPS spectra of (**a**) survey, (**b**) Ca 2p, (**c**) W 4f, (**d**) O 1s, (**e**) Tb 3d, and (**f**) Eu 3d.

The peaks of Ca 2p had core binding energies of 346.78 eV ($2P_{3/2}$) and 349.88 eV ($2P_{1/2}$), indicating that Ca is in the +2 oxidation state [25]. At the peak of W 4f, the detected core binding energies were 34.86 eV ($4f_{7/2}$) and 36.88 eV ($4f_{5/2}$), which is considered to be the +6 state of W [26]. The peak of the binding energy of O 1s was detected as 530.29 eV, which is considered to indicate the crystal lattice oxygen with increasing binding energy [27].

In the rare earth-doped sample ($CaWO_4:Tb^{3+}$, $CaWO_4:Eu^{3+}$), the binding energy of the Ca 2p (Figure 4b), W 4f (Figure 4c), and O 1s (Figure 4d) components was slightly changed. This change in binding energy is considered due to rare earth doping, and is related to the change in the lattice spacing observed in the XRD result and SEM-EDS component analysis, which means that the rare earth is doped in $CaWO_4$. In addition, the respective binding energy spectra were observed in the samples doped with $Tb^{3+}$ and $Eu^{3+}$. In Figure 4c, binding energies of 1277 eV (Tb $3d_{3/2}$) and 1242 eV (Tb $3d_{5/2}$) were observed, and in Figure 4f, binding energy peaks of 1164 (Eu $3d_{3/2}$), 1154 (Eu $3d_{3/2}$), 1134 (Eu $3d_{5/2}$), and 1124 (Eu $3d_{5/2}$) eV were obtained. This indicates the presence of rare earth ions in the +3 oxidation state following the synthesis of the sample [27].

The excitation and emission spectra of $CaWO_4$ are shown in Figure 5a. A signal having an excitation wavelength of 254 nm peak was detected, and a blue signal having a broad bandwidth of 420 nm was observed in the emission spectrum. This is thought to be due to the transfer of ions from the 2p orbital of oxygen to the 5d orbital of the vacant $W^{6+}$ [28–31]. The absorption and emission spectra of $CaWO_4:Tb^{3+}$ powder are shown in Figure 5b.

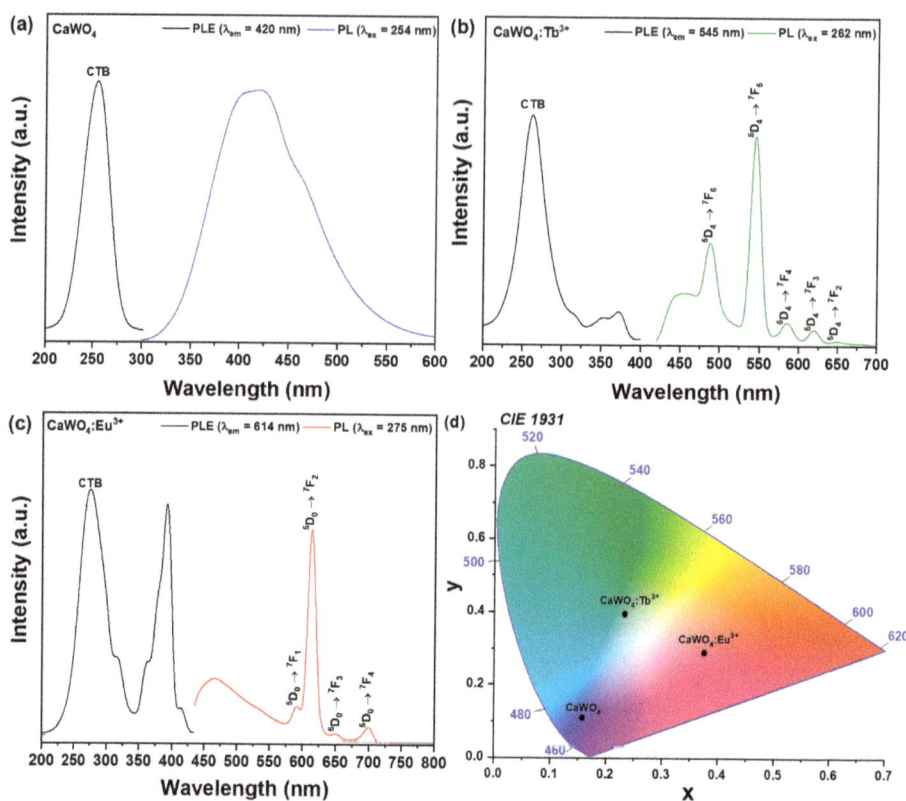

**Figure 5.** Luminescence spectra of (**a**) $CaWO_4$, (**b**) $CaWO_4:Tb^{3+}$, and (**c**) $CaWO_4:Eu^{3+}$, and (**d**) CIE coordination.

The photoluminescence excitation (PLE) spectrum controlled with a photoluminescence (PL) wavelength of 545 nm has a peak at 262 nm and is widely distributed over a 200 ~ 300 nm region, which is an absorption spectrum by charge transfer band (CTB) generated between $O^{-2}$-$W^{6+}$ and $O^{-2}$-$Tb^{3+}$ ions [32]. The relatively weak absorption signals between 330 and 400 nm are the 4f-4f transition signals of $Tb^{3+}$ ions [33]. The peak signals at 350 and 372 nm were generated by the $^7F_6 \rightarrow {^5G_5}$ and $^7F_6 \rightarrow {^5G_6}$ transition signals, respectively. When excited with the strongest absorption wavelength of 262 nm, the PL spectra of $CaWO_4:Tb^{3+}$ were observed at 488 ($^5D_4 \rightarrow {^7F_6}$), 545 ($^5D_4 \rightarrow {^7F_5}$), 585 ($^5D_4 \rightarrow {^7F_4}$), 620 ($^5D_4 \rightarrow {^7F_3}$), and 648 ($^5D_4 \rightarrow {^7F_2}$) nm. Among these emission peaks, the green emission spectrum by the magnetic dipole transition at 545 nm was about 2.4 times stronger than the blue emission intensity by the electric dipole transition at 488 nm. This means that the $Tb^{3+}$ ions located in the parent crystal are located at sites of inversion symmetry [34].

In $CaWO_4:Eu^{3+}$ the PLE spectrum controlled with a PL wavelength of 614 nm has a peak at 275 nm and is widely distributed over a 200–340 nm region (Figure 5b), which is the absorption spectrum by CTB generated between $O^{-2}$-$W^{6+}$ and $O^{-2}$-$Eu^{3+}$ ions [35]. The relatively weak absorption signals between 350 and 440 nm are the 4f-4f transition signals of $Eu^{3+}$ ions [34]. The peak signals at 362, 391, and 414 nm were generated by the $^7F_0 \rightarrow {^5D_4}$, $^7F_0 \rightarrow {^5G_2}$, and $^7F_0 \rightarrow {^5L_6}$ transition signals, respectively. When excited with the strongest absorption wavelength of 275 nm, the PL spectra of $CaWO_4:Eu^{3+}$ were observed at 592 ($^5D_0 \rightarrow {^7F_1}$), 614 ($^5D_0 \rightarrow {^7F_2}$), 650 ($^5D_0 \rightarrow {^7F_3}$), and 700 ($^5D_0 \rightarrow {^7F_4}$) nm. Here, 614 nm is a spectrum by electric dipole transition, 592 nm is a magnetic dipole transition, and 650 and 700 nm are electric dipole signals [36].

At this time, the PL intensity of 614 nm was observed to be about 6 times stronger than that of 592 nm, which means that $Eu^{3+}$ ions in the parent lattice are located at non-inverting symmetric sites. The ratio between the intensity of the red emission and the intensity of the orange emission is also called an asymmetry ratio [37]. The color coordinates (CIE 1931) according to the emission spectra of $CaWO_4$, $CaWO_4:Tb^{3+}$, and $CaWO_4:Eu^{3+}$ are shown in Figure 5c. They were found to be located at the blue, green, and red coordinates, respectively.

### 3.3. Photoactive Composite and Anti-Counterfeiting Application

Figure 6 shows the application of the synthesized phosphors as a photoreactive composite for anti-counterfeiting. The disk composite made by mixing glass powder and $CaWO_4$ showed no reaction in daylight but showed a blue light emission in response to UV light [38]. In addition, the composite made by mixing with a silicone-based polymer (PDMS) could be flexibly bent, and each unique color was realized in UV light. These materials are thought to be applicable to the display and laser industries.

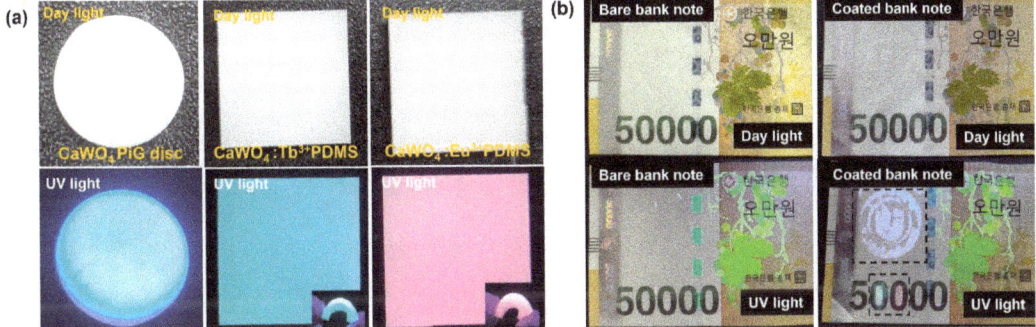

**Figure 6.** Photograph of (a) phosphors in glass disc and flexible composite, (b) phosphors painted on bank notes for anti-counterfeiting.

In addition, a solution made by mixing PVP polymer and $CaWO_4$ phosphor with ethanol was stamped and applied to banknotes. The location could not be recognized with the naked eye, and the shape and unique color could be confirmed only by UV light, suggesting that the synthesized phosphor can be applied to anti-counterfeiting.

### 4. Conclusions

Calcium nitrate and sodium tungstate were dissolved in distilled water, and crystalline $CaWO_4$ white powder was synthesized without a high-temperature heat treatment process by co-precipitation. $CaWO_4:Tb^{3+}$ and $CaWO_4:Eu^{3+}$ phosphors were synthesized by doping trace amounts of $Tb^{3+}$ or $Eu^{3+}$ ions, respectively, for use as various light-emitting materials. The synthesized powders were confirmed to have a crystalline scheelite structure from XRD results, and a secondary phase resulting from rare earth doping was not observed. However, it was found that the lattice spacing was slightly changed by doping with rare earth with a relatively large ionic radius. The particles synthesized by SEM-EDS analysis were about 75–85 nm in size and were aggregated in a spherical shape. Ca, W, and O were confirmed in the component analysis, and it was found that the doped Tb and Eu were evenly distributed by mapping. The binding energies of Ca 2p, W 4f, and O 1s were identified by XPS component analysis, and the signals of Tb 3d and Eu 3d energy binding by rare earth doping were detected. In the PLE and PL spectra, when each specimen was excited with CTB, blue emission was observed over a wide area of 420 nm for $CaWO_4$, green emission at 545 nm for $CaWO_4:Tb^{3+}$, and red emission from $CaWO_4:Eu^{3+}$ at 614 nm. The synthesized phosphor was mixed with glass powder and PDMS polymer to prepare a disk-shaped and flexible composite that can be applied to a display. In addition, it was suggested that the solution made by mixing with PVP polymer can be applied to anti-counterfeiting because it is impossible to visually confirm when applied to banknotes and can only be confirmed by UV light.

**Author Contributions:** Conceptualization, J.-Y.J.; data curation, J.-Y.J. and S.-S.Y.; formal analysis, J.-Y.J.; funding acquisition, S.-S.Y.; investigation, J.-Y.J. and S.-S.Y.; methodology, J.-Y.J.; project administration, J.-Y.J.; software, J.-Y.J.; supervision, J.-Y.J.; visualization, J.-Y.J.; writing—original draft, J.-Y.J. and S.-S.Y.; writing—review and editing, J.-Y.J. All authors have read and agreed to the published version of the manuscript.

**Funding:** This research received no external funding.

**Institutional Review Board Statement:** Not applicable.

**Data Availability Statement:** The data presented in this study are available in the database of the authors at the Faculty of Materials Science and Engineering.

**Acknowledgments:** This research was supported by Basic Science Research Program through the National Research Foundation of Korea (NRF) funded by the Ministry of Education (NRF-2020R1F1A1072616).

**Conflicts of Interest:** The authors declare no conflict of interest.

## References

1. Liu, C.; Pokhrel, S.; Tessarek, C.; Li, H.; Schowalter, M.; Rosenauer, A.; Eickhoff, M.; Li, S.; Mädler, L. Rare-Earth-Doped Y4Al2O9 Nanoparticles for Stable Light-Converting Phosphors. *ACS Appl. Nano Mater.* **2019**, *3*, 699. [CrossRef]
2. Withnall, R.; Silver, J. Physics of Light Emission from Rare Earth-Doped Phosphors. *Handb. Vis. Disp. Technol.* **2015**, *1*, 1019–1028. [CrossRef]
3. Kim, D. Recent Developments in Lanthanide-Doped Alkaline Earth Aluminate Phosphors with Enhanced and Long-Persistent Luminescence. *Nanomaterials* **2021**, *11*, 723. [CrossRef] [PubMed]
4. Parhi, P.; Karthik, T.N.; Manivannan, V. Synthesis and characterization of metal tungstates by novel solid-state metathetic approach. *J. Alloy Compd.* **2008**, *465*, 380–386. [CrossRef]
5. Zhang, J.; Zhu, X.; Huang, H.; Zhang, Y. Synthesis of Crystallized BaWO$_4$ Nanorods in a Microemulsion System. *Adv. Chem. Eng. Sci.* **2017**, *7*, 228–234. [CrossRef]
6. Kuzmin, A.; Purans, J. Local atomic and electronic structure of tungsten ions in AWO$_4$ crystals of scheelite and wolframite types. *Radiat. Meas.* **2001**, *33*, 583–586. [CrossRef]
7. Smith, J.; Desgreniers, S.; Tse, J.; Klug, D. High-pressure phase transition observed in barium hydride. *J. Appl. Phys.* **2007**, *102*, 043520. [CrossRef]
8. Liao, J.; Qiu, B.; Wen, H.; Chen, J.; You, W.; Liu, L. Synthesis process and luminescence properties of Tm$^{3+}$ in AWO$_4$ (A = Ca, Sr, Ba) blue phosphors. *J. Alloy. Compd.* **2009**, *487*, 758–762. [CrossRef]
9. Sota, E.N.; Che Ros, F.; Hassan, J. Synthesis and Characterisation of AWO4 (A=Mg, Zn) Tungstate Ceramics. *J. Phys. Conf. Ser.* **2018**, *1083*, 12002. [CrossRef]
10. Cooper, T.; De Leeuw, N. A combined ab initio and atomistic simulation study of the surface and interfacial structures and energies of hydrated scheelite: Introducing a CaWO4 potential model. *Surf. Sci.* **2003**, *531*, 159–176. [CrossRef]
11. Oliveira, M.C.; Gracia, L.; Nogueira, I.C.; do Carmo Gurgel, M.F.; Mercury, J.M.R.; Longo, E.; Andrés, J. Synthesis and morphological transformation of BaWO$_4$ crystals: Experimental and theoretical insights. *Ceram. Int.* **2016**, *42*, 10913–10921. [CrossRef]
12. Zhang, H.; Liu, T.; Zhang, Q.; Wang, X.; Yin, J.; Song, M.; Guo, X. First-principles study on electronic structures of BaWO$_4$ crystals containing F-type color centers. *J. Phys. Chem. Solids* **2008**, *69*, 1815–1819. [CrossRef]
13. Yekta, S.; Sadeghi, M.; Babanezhad, E. Synthesis of CaWO$_4$ nanoparticles and its application for the adsorption-degradation of organophosphorus cyanophos. *J. Water Process. Eng.* **2016**, *14*, 19–27. [CrossRef]
14. Balestrieri, M.; Colis, S.; Gallart, M.; Schmerber, G.; Ziegler, M.; Gilliot, P.; Dinia, A. Photoluminescence properties of rare earth (Nd, Yb, Sm, Pr)-doped CeO$_2$ pellets prepared by solid-state reaction. *J. Mater. Chem. C Mater. Opt. Electron. Devices* **2015**, *3*, 7014–7021. [CrossRef]
15. Oh, W.; Chong, Y.; Park, C.S. Lim Solid-State Metathetic Synthesis and Photoluminescence of Calcium Tungstate Particles Assisted by Cyclic Microwave Irradiation. *Asian J. Chem.* **2012**, *24*, 3319.
16. Phuruangrat, A.; Thongtem, T.; Thongtem, S. Synthesis, characterisation and photoluminescence of nanocrystalline calcium tungstate. *J. Exp. Nanosci.* **2010**, *5*, 263–270. [CrossRef]
17. Li, J.; Zhang, J.; Hao, Z.; Zhang, X.; Zhao, J.; Luo, Y. Upconversion properties and dynamics study in Tm$^{3+}$ and Yb$^{3+}$ codoped CaSc$_2$O$_4$ oxide material. *J. Appl. Phys.* **2013**, *113*, 223507. [CrossRef]

18. Jung, J.; Kim, J.; Shim, Y.; Hwang, D.; Son, C.S. Structure and Photoluminescence Properties of Rare-Earth ($Dy^{3+}$, $Tb^{3+}$, $Sm^{3+}$)-Doped $BaWO_4$ Phosphors Synthesized via Co-Precipitation for Anti-Counterfeiting. *Materials* **2020**, *13*, 4165. [CrossRef]
19. Jung, J.; Yi, S.; Hwang, D.; Son, C. Structure, Luminescence, and Magnetic Properties of Crystalline Manganese Tungstate Doped with Rare Earth Ion. *Materials* **2021**, *14*, 3717. [CrossRef]
20. Wang, S.; Gao, H.; Sun, G.; Li, Y.; Wang, Y.; Liu, H.; Chen, C.; Yang, L. Structure characterization, optical and photoluminescence properties of scheelite-type CaWO4 nanophosphors: Effects of calcination temperature and carbon skeleton. *Opt. Mater.* **2020**, *99*, 109562. [CrossRef]
21. Jia, Y.Q. Crystal radii and effective ionic radii of the rare earth ions. *J. Solid State Chem.* **1991**, *95*, 184–187. [CrossRef]
22. Bokuniaeva, A.O.; Vorokh, A.S. Estimation of particle size using the Debye equation and the Scherrer formula for polyphasic $TiO_2$ powder. *J. Phys. Conf. Ser.* **2019**, *1410*, 012057. [CrossRef]
23. Thongtem, T.; Kungwankunakorn, S.; Kuntalue, B.; Phuruangrat, A.; Thongtem, S. Luminescence and absorbance of highly crystalline $CaMoO_4$, $SrMoO_4$, $CaWO_4$ and $SrWO_4$ nanoparticles synthesized by co-precipitation method at room temperature. *J. Alloy Compd.* **2010**, *506*, 475–481. [CrossRef]
24. Kaur, P.; Khanna, A.; Singh, M.N.; Sinha, A.K. Structural and optical characterization of Eu and Dy doped $CaWO_4$ nanoparticles for white light emission. *J. Alloy Compd.* **2020**, *834*, 154804. [CrossRef]
25. Maheshwary, M.; Singh, B.P.; Singh, R.A. Color tuning in thermally stable $Sm^{3+}$-activated $CaWO_4$ nanophosphors. *New J. Chem.* **2015**, *39*, 4494–4507. [CrossRef]
26. Terohid, S.; Heidari, S.; Jafari, A.; Asgary, S. Effect of growth time on structural, morphological and electrical properties of tungsten oxide nanowire. *Appl. Phys. A* **2018**, *124*, 1–9. [CrossRef]
27. Pawlak, D.A.; Ito, M.; Oku, M.; Shimamura, K.; Fukuda, T. Interpretation of XPS O (1s) in Mixed Oxides Proved on Mixed Perovskite Crystals. *J. Phys. Chem. B* **2002**, *106*, 504–507. [CrossRef]
28. Fang, Z.B.; Wang, J.J.; Yang, X.F.; Chen, T.H.; Ni, H.N.; Li, Z.B.; Tan, Y.S. Reliability Study of the Band Gap of Rare Earth Oxides Measured by XPS Spectra. *Key Eng. Mater.* **2013**, *562–565*, 891–895.
29. Park, K.; Ahn, H.; Nguyen, H.; Jang, H.; Mho, S. Optical Properties of $Eu_2(WO_4)_3$ and $Tb_2(WO_4)_3$ and of $CaWO_4$ Doped with $Eu^{3+}$ or $Tb^{3+}$—Revisited. *J. Korean Phys. Soc.* **2008**, *53*, 2220–2223. [CrossRef]
30. Wang, W.; Yang, P.; Gai, S.; Niu, N.; He, F.; Lin, J. Fabrication and luminescent properties of $CaWO_4$: $Ln^{3+}$ nanocrystals. *J. Nanoparticle Res.* **2010**, *12*, 2295. [CrossRef]
31. Cavalli, E.; Boutinaud, P.; Mahiou, R.; Bettinelli, M.; Dorenbos, P. Luminescence Dynamics in $Tb^{3+}$-Doped $CaWO_4$ and $CaMoO_4$ Crystals. *Inorg. Chem.* **2010**, *49*, 4916–4921. [CrossRef]
32. Sun, S.; Wu, L.; Yi, H.; Wu, L.; Ji, J.; Zhang, C.; Zhang, Y.; Kong, Y.; Xu, J. Energy transfer between $Ce^{3+}$ and $Tb^{3+}$ and the enhanced luminescence of a green phosphor $SrB_2O_4$:$Ce^{3+}$, $Tb^{3+}$, Na. *Opt. Mater. Express* **2016**, *6*, 1172. [CrossRef]
33. Alaria, J.; Borisov, P.; Dyer, M.S.; Manning, T.D.; Lepadatu, S.; Cain, M.G.; Mishina, E.D.; Sherstyuk, N.E.; Ilyin, N.A.; Hadermann, J.; et al. Engineered spatial inversion symmetry breaking in an oxide heterostructure built from isosymmetric room-temperature magnetically ordered components. *Chem. Sci.* **2014**, *5*, 1599–1610. [CrossRef]
34. Nayak, P.; Nanda, S.S.; Sharma, R.K.; Dash, S. Tunable luminescence of $Eu^{3+}$-activated $CaWO_4$ nanophosphors via $Bi^{3+}$ incorporation. *Luminescence* **2020**, *35*, 1068–1076. [CrossRef]
35. Malta, O.L.; Ribeiro, S.J.L.; Faucher, M.; Porcher, P. Theoretical intensities of 4f-4f transitions between Stark levels of the Eu3+ ion in crystals. *J. Phys. Chem. Solids* **1991**, *52*, 587–593. [CrossRef]
36. Savinov, M.; Nuzhnyy, D.; Ležaić, M.; Eckel, S.; Kamba, S.; Laufek, F.; Spaldin, N.A.; Knížek, K.; Vaněk, P.; Lamoreaux, S.K.; et al. A multiferroic material to search for the permanent electric dipole moment of the electron. *Nat. Mater.* **2010**, *9*, 649–654. [CrossRef]
37. Kolesnikov, I.E.; Povolotskiy, A.V.; Mamonova, D.V.; Kolesnikov, E.Y.; Kurochkin, A.V.; Lähderanta, E.; Mikhailov, M.D. Asymmetry ratio as a parameter of $Eu^{3+}$ local environment in phosphors. *J. Rare Earths* **2018**, *36*, 474–481. [CrossRef]
38. Jung, J.Y.; Shim, Y.; Son, C.S.; Kim, Y.; Hwang, D. Boron Nitride Nanoparticle Phosphors for Use in Transparent Films for Deep-UV Detection and White Light-Emitting Diodes. *ACS Appl. Nano Mater.* **2021**, *4*, 3529–3536. [CrossRef]

Article

# Thermal Stability and Radiation Tolerance of Lanthanide-Doped Cerium Oxide Nanocubes

Kory Burns [1,2], Paris C. Reuel [3], Fernando Guerrero [3], Eric Lang [1], Ping Lu [1], Assel Aitkaliyeva [2], Khalid Hattar [1,*] and Timothy J. Boyle [3,*]

1. Center for Integrated Nanotechnologies, Sandia National Laboratories, P.O. Box 8500, Albuquerque, NM 87185, USA; kdburns@sandia.gov (K.B.); ejlang@sandia.gov (E.L.); plu@sandia.gov (P.L.)
2. Department of Materials Science and Engineering, University of Florida, Gainesville, FL 32611, USA; aitkaliyeva@mse.ufl.edu
3. Advanced Materials Laboratory, Sandia National Laboratories, Albuquerque, NM 87106, USA; preuel@sandia.gov (P.C.R.); 2guerrerof2@gmail.com (F.G.)
* Correspondence: khattar@sandia.gov (K.H.); tjboyle@sandia.gov (T.J.B.); Tel.: +1-(505)-272-7625 (T.J.B.)

**Citation:** Burns, K.; Reuel, P.C.; Guerrero, F.; Lang, E.; Lu, P.; Aitkaliyeva, A.; Hattar, K.; Boyle, T.J. Thermal Stability and Radiation Tolerance of Lanthanide-Doped Cerium Oxide Nanocubes. *Crystals* **2021**, *11*, 1369. https://doi.org/10.3390/cryst11111369

Academic Editors: Alessandra Toncelli and Željka Antić

Received: 1 October 2021
Accepted: 5 November 2021
Published: 11 November 2021

**Publisher's Note:** MDPI stays neutral with regard to jurisdictional claims in published maps and institutional affiliations.

**Copyright:** © 2021 by the authors. Licensee MDPI, Basel, Switzerland. This article is an open access article distributed under the terms and conditions of the Creative Commons Attribution (CC BY) license (https://creativecommons.org/licenses/by/4.0/).

**Abstract:** The thermal and radiation stability of free-standing ceramic nanoparticles that are under consideration as potential fillers for the improved thermal and radiation stability of polymeric matrices were investigated by a set of transmission electron microscopy (TEM) studies. A series of lanthanide-doped ceria (Ln:CeO$_x$; Ln = Nd, Er, Eu, Lu) nanocubes/nanoparticles was characterized as synthesized prior to inclusion into the polymers. The Ln:CeO$_x$ were synthesized from different solution precipitation (oleylamine (ON), hexamethylenetetramine (HMTA) and solvothermal (t-butylamine (TBA)) routes. The dopants were selected to explore the impact that the cation has on the final properties of the resultant nanoparticles. The baseline CeO$_x$ and the subsequent Ln:CeO$_x$ particles were isolated as: (i) ON-Ce (not applicable), Nd (34.2 nm), Er (27.8 nm), Eu (42.4 nm), and Lu (287.4 nm); (ii) HMTA-Ce (5.8 nm), Nd (6.6 nm), Er (370.0 nm), Eu (340.6 nm), and Lu (287.4 nm); and (iii) TBA-Ce (4.1 nm), Nd (5.0 nm), Er (3.8 nm), Eu (7.3 nm), and Lu (3.8 nm). The resulting Ln:CeO$_x$ nanomaterials were characterized using a variety of analytical tools, including: X-ray fluorescence (XRF), powder X-ray diffraction (pXRD), TEM with selected area electron diffraction (SAED), and energy dispersive X-ray spectroscopy (EDS) for nanoscale elemental mapping. From these samples, the Eu:CeO$_x$ (ON, HMTA, and TBA) series were selected for stability studies due to the uniformity of the nanocubes. Through the focus on the nanoparticle properties, the thermal and radiation stability of these nanocubes were determined through in situ TEM heating and ex situ TEM irradiation. These results were coupled with data analysis to calculate the changes in size and aerial density. The particles were generally found to exhibit strong thermal stability but underwent amorphization as a result of heavy ion irradiation at high fluences.

**Keywords:** nanoparticles; in situ TEM; ion irradiation; thermal annealing; ceria

## 1. Introduction

Lanthanum oxide (LnO$_x$) nanomaterials are of interest as potential fillers in polymeric coatings to protect internal electronic components and circuits from high temperature and ionization exposure (i.e., rad-hard). This is mainly due to their high Z number, which helps to prevent the ionization of energy and variable oxidation state, which can be accessed when exposed to ionizing radiation-reducing Compton and Auger effects [1]. Computational studies using WinXCom software indicate that g-photons (1 keV–100 GeV) interacting with LnO$_x$ (Ln–La, Ce, Pr, Nd, Eu, Dy, Sm, Er, Yb, Lu) are effective shielding agents [2,3]. A number of systems have demonstrated their utility, such as LnO$_x$ (La [4], Ce, Nd, Sm)-doped borate glasses, which were found to effect γ-ray shielding [5]. To overcome the brittle nature of the ceramic oxide materials, the development of composite coatings (polymers with LnO$_x$ nanofillers) allows for a flexible, stable layer that is also rad-hard.

While proving to be an effective shielding agent against electromagnetic radiation, ceria ($CeO_2$) has been shown to be very sensitive to ion irradiation [6]. In ion–matter interaction theory, an impinging ion penetrates a target material and transfers its energy by both elastic and inelastic collisions with the target nuclei and target electrons, respectively. In previous reports, high doses of 2 MeV Au ion irradiation produced significant disorder on the Ce sub-lattice caused by inelastic collisions [6]. This is in direct contrast to 2.5 MeV electron beam irradiations, where elastic collisions produced isolated defects that were easily repaired by thermal annealing [7]. This suggests that dopant atoms within $CeO_2$ may be warranted to enhance radiation stability and prevent significant lattice disorder upon interaction with external stimuli. There are several candidate Ln dopants that have shown promise when utilized in electronic components from high radiation environments, in part due to their high threshold energies [8]. The Ln cations discussed in this report were selected due to their availability, representation across the Ln series, diverse oxidation states available, and diverse applications: $EuO_x$ is used as an activation ion for color television phosphor, $ErO_x$ is used as a gate dielectric in CMOS logic circuits in space environments, $NdO_x$ is used as a ceramic capacitor, and $LuO_x$ is used as a laser crystal for solid-state lasers [3].

For this work, we were interested in exploiting ceria ($CeO_x$) beyond the wide range of applications for which it is already used [9,10], including: catalysts [11,12], biomedical applications [1,12–14], actinide surrogates [15], pigments [16], sunscreen [12,13,17], and many other applications, as well as less frequently investigated rad-hard coatings. As mentioned above, studies have shown that doping Ln cations into the lattice of $CeO_x$ nanoparticles imparts changes to their final properties, including improved radiative recombination, reduced oxidation catalytic activity, enhanced UV-absorption capacity, and the promotion of intermediate electronic levels in the bandgap [12,14,17,18]. Herein, a study of lanthanide doped-$CeO_x$ (Ln:$CeO_x$) nanomaterials was undertaken to evaluate the thermal and radiation stability of the nanoparticles through in situ Transmission Electron Microscopy (TEM) techniques [19].

In an effort to explore the properties of Ln:$CeO_x$ nanoparticles, a series of nanoparticles was generated using oleylamine (ON) [20], hexamethylenetetramine (HMTA) [21], and t-butylamine (TBA) [22] as surfactants. The lanthanide cations used for dopants included Nd, Eu, Er, and Lu, which were selected to represent the series and are well known to occupy alternative oxidation states. The doping was verified by Scanning Transmission Electron Microscopy (STEM)-based EDS elemental mapping. The structural evolution from annealing and irradiation was verified using the unique in situ ion irradiation TEM ($I^3$TEM) available at Sandia National Laboratories. The results of the synthesis route, ligand, dopant, and resulting particulates were evaluated.

## 2. Experimental

Nanoparticle preparations were conducted on the bench-top under ambient atmospheric conditions using chemicals obtained and used as received from Aldrich Chemical Company, Inc. (Milwaukee, WI, USA). The lanthanide chloride hydrates ([$LnCl_3 \bullet 6H_2O$]) and nitrates ([$Ln(NO_3)_3 \bullet 6H_2O$]) were synthesized in-house using the appropriate metal and concentrated acid ((aq) HCl or (aq) $HNO_3$. All of the Ln precursors structures were verified by single-crystal X-ray experiments. Three routes to nanocubes or nanorods were evaluated, with modifications noted below: (i) oleylamine (ON) [17], (ii) hexamethylenetetramine (HMTA) [18], or (iii) t-butylamine (TBA) [19]. All the generated nanoparticle samples (ON, HMTA, TBA) were characterized by a variety of analytical methods: X-ray Fluorescence (XRF), Powder X-ray Diffraction (pXRD), Fourier Transformed InfraRed Spectroscopy (FTIR), and Dynamic Light Scattering (DLS). Full details of the instrumentation and product characterization are available in the Supplementary Materials. General descriptions of the three different routes to $CeO_2$ and Ln:$CeO_x$, where Ln = Eu, Er, and Lu are presented below. The yields were not determined due to the presence of excess surfactants. A summary of the analytical characterization can be found in Table 1.

Table 1. Summarized results of XRF, pXRD, ATR-FTIR, and DLS for $CeO_2$, $Nd:CeO_2$, $Eu:CeO_2$, $Er:CeO_2$, and $Lu:CeO_2$. The ATR-FTIR spectrum bends and stretches are further noted with medium (m), strong (s), weak (w), broad (br). Graphical representations of the data shown can be found in the Supplementary Materials.

|  | $CeO_2$ | $Nd:CeO_x$ | $Eu:CeO_x$ | $Er:CeO_x$ | $Lu:CeO_x$ |
|---|---|---|---|---|---|
| XRF | Ce (major) Eu (minor) Cl (minor) | Ce (major) Nd (minor) Cl (minor) Zr (minor) | Ce (major) Eu (minor) Cl (minor) | Ce (major) Er (minor) | Ce (major) Lu (minor) |
| pXRD | (25 °C) PDF 00-067-0123 $CeO_2$ ceria ǀ Cerium Oxide | (600 °C) PDF 00-067-0123 $CeO_2$ ceria ǀ Cerium Oxide | (600 °C) PDF 00-067-0122 $CeO_2$ ceria ǀ Cerium Oxide | (600 °C) PDF 00-067-0123 $CeO_2$ ceria ǀ Cerium Oxide | (600 °C) PDF 00-067-0123 $CeO_2$ ceria ǀ Cerium Oxide |
| ATR-FTIR ($cm^{-1}$) | 3317.93 (m) 2920.58 (s) 2851.09 (m) 2163.86 (w) 2037.01 (w) 1613.62 (m) 1460.77 (w) 1411.48 (m) 1064.97 (w) 965.24 (w) | 3355.01 (m, br) 2923.12 (m) 2853.10 (m) 2583.18 (w) 2164.12 (w) 2079.88 (w) 1980.97 (w) 1467.17 (s) 1393.84 (s) 1260.87 (w) 1083.29 (w) 843.62 (m) 802.36 (m) 708.90 (m) | 3340.37 (m) 3003.64 (w) 2920.74 (s) 2851.20 (m) 2162.81 (w) 2035.73 (w) 1494.59 (m) 1460.04 (s) 1404.08 (m) 1079.23 (w) 965.75 (w) 841.10 (w) 720.57 (m) | 3323.03 (s) 3213.14 (s) 2920.59 (s) 2851.22 (m) 2214.27 (w) 1624.61 (s) 1462.78 (w) 1410.58 (m) 1049.92 (w) 966.31 (w) 719.16 (s) | 3315.26 (s) 3210.01 (s) 2922.62 (m) 2852.34 (w) 2216.86 (w) 1618.16 (s) 1411.97 (m) 1117.11 (w) 966.47 (w) 699.03 (m) |
| DLS (nm) | 257 (100%) | 295 (100%) | 257 (54%) 901 (46%) | 343 (100%) | 518 (100%) |

X-ray Fluorescence (XRF). A ThermoFisher ARL (West Palm Beach, FL, USA) Quant'X EDXRF, DLS Spectrometer utilizing UniQuant software was used for all the analyses. The system uses a Fundamental Parameters approach based on the Sherman equation for the direct measurement of elemental concentrations based on integrated fluorescent peak intensities. In air, using a medium count rate, a single–repetition, multi-scan excitation (C Thin (5 kV, 60 s); Al (12 kV, 100 s), Pd Thick (28 kV, 100 s); Cu Thick (50 kV, 100 s)) was used to evaluate each sample.

Powder X-ray Diffraction (pXRD). The powders were collected on a Bruker D8 Avance diffractometer employing Cu $K_\alpha$ radiation (1.5406 Å) and a RTMS X'Celerator detector. The data were collected over a 2θ range of 5–70° at a scan rate of 0.083°/s and a zero-background holder was employed. The XRD patterns were analyzed using Bruker EVA software and indexed using the Powder Diffraction File PDF-4 + 2013.

Fourier Transformed InfraRed Spectroscopy (FTIR). The FTIR spectral data were collected on a Bruker Vector 22 MIR Spectrometer in an atmosphere of flowing nitrogen using an ATR powder attachment.

Dynamic Light Scattering (DLS). The DLS data were collected on a Malvern Instruments Zetasizer Nanoseries (NanoZS). All the samples were dispersed by sonication for 10 min at 50–60 Hz in their respective solvents (ON–methanol, Parr–tol, HMTA–water). Following the sonication, the samples were diluted and loaded into 1 cm glass cuvettes. A total of 10 sets of scans was performed for each sample to generate an average size distribution by intensity.

- Oleylamine (ON) [17]. In a round-bottomed flask, $(CeCl_3 \bullet 6H_2O)$ and any dopant $(LnCl_3 \bullet 6H_2O)$ where Ln = Nd, Eu, Er, and Lu were added to ON. After heating the reaction to 90 °C and stirring for 1 h, the reaction was then warmed to 265 °C. The reaction turned from pale brown to black. After cooling to room temperature, the solution was collected by centrifugation, washed with ethanol and hexanes and used without further manipulations.

- Hexamethylenetetramine (HMTA) [18]. In a round-bottomed flask, $(Ce(NO_3)_3 \bullet 6H_2O)$ and any dopant $(Ln(NO_3)_3 \bullet 6H_2O)$ where Ln = Nd, Eu, Er, and Lu were dissolved in water. An equal volume of concentrated HMTA was added at room temperature and the reaction was allowed to stir (24 h). The precipitate was collected by centrifugation, washed with hexanes and used without further manipulations.
- t-butylamine (TBA) [19]. In an Ehrlyenmeyer flask, $(Ce(NO_3)_3 \bullet 6H_2O)$ and any dopant $(Ln(NO_3)_3 \bullet 6H_2O)$ where Ln = Nd, Eu, Er, and Lu were dissolved in water and poured into a Teflon™-lined Parr-bomb™. To this mixture, toluene, oleic acid, and t-butylamine were added; the sample was sealed and heated at 180 °C for 12 h. The precipitate was collected by centrifugation, washed with hexanes, and used without further manipulations.

Transmission Electron Microscopy (TEM). The TEM samples were prepared by making a slurry of the powders isolated (vide supra) in methanol, sonicating them for 10 min at 50–60 Hz, and then dropping one drop of solution onto a lacey carbon TEM grid and allow for the solution to volatilize.

The TEM micrographs and selected area electron diffraction (SAED) patterns were collected using a highly modified JEOL 2100 TEM operating at an accelerating voltage of 200 keV. A FEI Titan™ G2 80–200 STEM with a Cs probe corrector and ChemiSTEM™ technology (X-FEG™ and SuperX™ EDS with four windowless silicon drift detectors) operated at 200 kV was used for compositional and structural analysis using EDS spectral imaging and high-angle annular dark-field (HAADF) imaging. The EDS spectral imaging was acquired as a series of frames, where the same region was scanned multiple times with a total acquisition time of over 30 min. An electron probe of size of about 0.13 nm, with a convergence angle of 18.1 mrad, and a current of ~75 pA was used for the EDS acquisition. The Ce L lines and Eu L lines were used for constructing the EDS maps of Ce and Eu, respectively. Since the Ce L lines overlap with the Eu L lines significantly, the EDS spectra were deconvoluted pixel-by-pixel using pure spectra of Ce and Eu as references before map construction. The HAADF images were recorded under similar optical conditions using an annular detector with a collection range of 60–160 mrad. The thermal anneals were performed utilizing the Gatan Heating stage in a JEOL 2100 up to 500 °C.

Heavy Ion Irradiation. The ex situ ion irradiations were performed on previously prepared TEM samples for the powders listed from the preparation routes noted above. The specimens were subjected to 15 MeV $Au^{4+}$ ion irradiation at room temperature using a 6 MV HVEE EN tandem accelerator. The ion fluence range included $1 \times 10^{13}$ and $1 \times 10^{15}$ $cm^{-2}$.

## 3. Results and Discussion

The interest in the addition of $Ln:CeO_x$ nanomaterial into polymers as a means through which to improve their properties (thermal and radiation tolerance) led us to prepare and characterize the particles prior to more complex studies. The dopants (none, Nd, Eu, Er, Lu) were selected to explore the size and electronic impact on the final behavior of the resulting $Ln:CeO_x$. Three selected paths focused on synthesis routes that could produce morphologically varied particles, involving different processing parameters, amenability processing for scale up, and different amine surfactants. Details concerning the: Section 3.1. Synthesis and Characterization of the nanoparticles, Section 3.2. Dopant Mapping for Eu, Section 3.3. Thermal Stability, and Section 3.4. Radiation Stability are presented sequentially in the sections below.

## 3.1. Synthesis and Characterization

The synthesis of nanoparticles following three routes (ON [17], HMTA [18], and TBA [19]) were undertaken as described in the Section 2, with full details available in the Supplementary Materials. The results of the syntheses are presented, followed by in situ and ex situ TEM thermal and ion irradiation measurements, respectively.

*Oleylamine (ON).* For the ON preparation, the resulting precipitates were independently collected by centrifugation. The green (unprocessed), undoped Ce particles were found to be $CeO_2$ by pXRD patterns. For the dopants, initial efforts focused on whether the dopant had precipitated with, or within, the $CeO_x$ matrix. The XRF analysis confirmed the presence of each of the different cations (Nd, Eu, Er, and Lu). However, there was also a trace amount of Cl that was associated with the starting precursor ligands. Noted in the $Nd:CeO_x(ON)$ sample was the presence of Zr, which was attributed to a mis-assignment or external contamination. The pXRD patterns of the doped species were in agreement with several different phases of $CeO_x$ along with numerous, minor unidentified peaks which were attributed to the organic surfactants. This assignment proved valid when the samples were thermally treated and the lower 'organic' peaks were lost and phase pure $CeO_2$ was identified. The FTIR analyses of all these samples verified that ON was present on each sample. Further analysis of the particulates available in solution were determined by the DLS measurements. For these samples, the particles were found to range from 257–518 nm in size. The larger size noted for the Eu dopant was attributed to the clustering of the smaller particles in solution. In order to assess the size of the crystallites formed, a TEM analysis was undertaken. For the undoped species, a polymeric matrix was found to surround the particles. The source of this is not known at this time, but its presence made high-resolution identification (bright field imaging and SAED) of the final particles difficult. Nonetheless, fine particles in the order of ~1–2 nm in diameter were observed. By contrast, the doped-$CeO_x$ specimens were found to have formed nanocubes that were ~30 nm in dimension. Further, these particles were found to be highly crystalline, as determined by the SAED analyses.

*Hexamethylenetetramine (HMTA).* Similarly, the XRF of HMTA showed each powder had the proper Ln cations present. The pXRD patterns of the green undoped Ce as well as the doped Nd matched the phase pure $CeO_2$. For the rest of the $Ln:CeO_x(HMTA)$ samples, again, a number of different phases of cerium oxide with minor HMTA species were present. Similarly, heat treatment produced phase pure $CeO_2$ and the loss of the 'organic' peaks. The FTIR clearly showed HMTA present on each of the samples. The DLS analyses showed a wide range of particles available from 33–463 nm. Again, Eu displayed two different sizes of particulate. The TEM analysis confirmed the small (~2–6 nm) octahedral-shaped particles for the undoped species, while the doped particles featured similarly shaped species, with sizes ranging from 80–200 nm.

*t-butylamine (TBA).* The substantially smaller amount of material made available by the Parr bomb approach using TBA made characterization much more difficult. However, for each sample, XRF showed the proper cations were present and that most phases were $CeO_2$ with Lu having an additional carbonate phase. For each, similar FTIR data were compiled and were consistent with the TBA and the other species that were used in the reduction of the nitrate precursor. The DLS data were very consistent in comparison, ranging only from 23–68 nm in size, with both Nd and Er demonstrating some cluster formation (222 and 424 nm, respectively). TEM analysis showed crystallite samples that were in the order of ~2–7 nm in size for the doped and undoped species.

For each of these routes, the nanoparticles were isolated, with only a few featuring residual starting ligands. The pXRD patterns revealed that organic moieties that were consistent with the various surfactants (ON, HMTA, TBA) for the different processes. The heat treatment of the samples led to the formation of phase pure $CeO_2$. The DLS showed much larger particles were present, but this was attributed to clustering. The crystal size was confirmed to be nanoparticles by the TEM analysis for each sample, as shown in Figure 1: (i) ON-Ce (not applicable), Nd (34.2 nm), Er (27.8 nm), Eu (42.4 nm), Lu

(287.4 nm); (ii) HMTA-Ce (5.8 nm), Nd (6.6 nm), Er (370.0 nm), Eu (340.6 nm), Lu (287.4 nm); and (iii) TBA-Ce (4.1 nm), Nd (5.0 nm), Er (3.8 nm), Eu (7.3 nm), Lu (3.8 nm). The ON all appeared to be below 50 nm in size and formed cubes, except for the Lu species, which were an order of magnitude larger in size. In comparison, the HMTA particles were similar for Ce and Nd (~5–6 nm) but the other doped species were again an order of magnitude larger. Further, these particles were irregular round species. Finally, the TBA nanocubes all proved to be <8 nm.

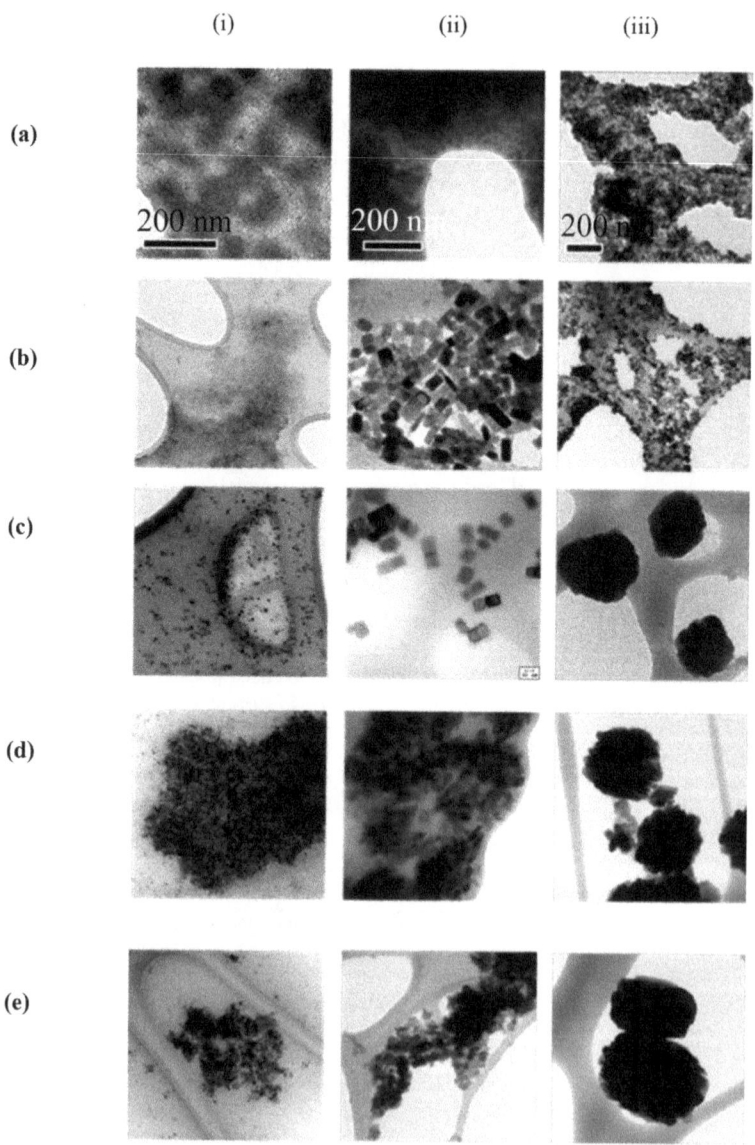

**Figure 1.** TEM images of column (i) TBA, (ii) ON, and (iii) HMTA-generated nanoparticles of column (a) $CeO_x$, (b) $Nd:CeO_x$, (c) $Eu:CeO_x$, (d) $Er:CeO_x$, and (e) $Lu:CeO_x$.

The samples were analyzed by XRF, which confirmed that the various dopants and Ce were present. In some instances, minor residual cations were observed, but in general, the cations introduced were isolated in the precipitate. The pXRD patterns confirmed for all the samples that $CeO_2$ was formed, with no indication of minor dopant oxide phases. This implies that the Ln dopants were successfully incorporated into the $CeO_2$ lattice. TEM elemental mapping was used to verify this concept (vide infra). The FTIR data revealed the presence of the various surfactants (HDA, ON, and TBA) on the isolated particles. It is difficult to distinguish between bound and free ligands, but it is believed that after extensive washing, the remaining bends and shifts are associated with the bound surfactants. DLS data revealed that sub-micron species were present in solution, which is consistent with the TEM data collected (vide infra). See Table 1 for a summary of the various results obtained on each sample and Supplementary Materials for the raw data.

## 3.2. Dopant Mapping for Eu

Based on the TEM images of the dopants across the different preparation routes, the variability in their oxidation states and their more centric sizes, the Eu:$CeO_x$ samples were selected for further study by HAADF imaging and EDS mapping (ON, HMTA, and TBA in Figures 2–4, respectively). While the above data (see Table 1) confirm the presence of the desired Eu dopant, its size, and its initial phase assignments, it was critical to verify the that Eu dopant was within the $CeO_x$ matrix; thus, elemental mapping was undertaken.

For the ON sample, the $CeO_x$(ON) exhibited a much finer particle size (<2 nm; Figure 2a) than the doped Eu:$CeO_x$(ON) samples (20–50 nm; Figure 2b). The EDS mapping for the Eu:$CeO_x$(ON) species verified that the Eu-dopant was present within the $CeO_x$ matrix (Figure 2c). Notably, several minor contaminates were identified that were not involved in the synthesis, such as a Si-based polymeric matrix. Additionally, Cl particles consistent with the XRF studies were found within the particles that most likely originated from the starting precursors. As shown in Figure 3, the $CeO_x$(HMTA) particles were found to feature an average size of ~4 nm, whereas the Eu:$CeO_x$(HMTA) were found to be considerably larger (~300 nm). The EDS mapping showed the uniform distribution of Eu within the large Eu:$CeO_x$(HMTA) particles (Figure 3c). As in the Eu:$CeO_x$(ON), Cl contamination was observed, along with other smaller impurities (Fe and Si), the presence of which is unaccounted for as they were not used in the part of the synthesis; however, as these elements were not identified by the XRF studies, it is believed an external contamination source may have accounted for their presence. The $CeO_x$(TBA) and Eu:$CeO_x$(TBA) samples were determined to be cubes 2–5 nm in size (Figure 4a,b) and the distribution of Eu within the Eu:$CeO_x$ was verified by the EDS maps (Figure 4c).

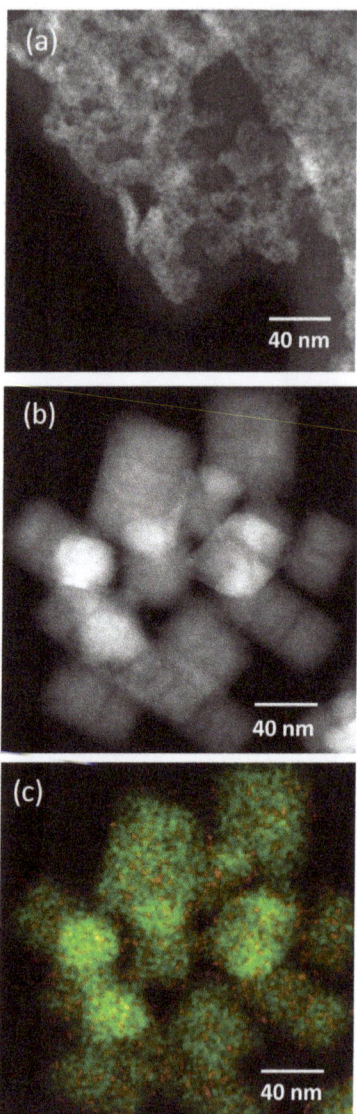

**Figure 2.** STEM HAADF imaging and EDS mapping of nanoparticles generated by ON: (**a**) HAADF of $CeO_x$ and (**b**) HAADF of $Eu:CeO_x$; (**c**) Ce and Eu L map of the $Eu:CeO_x$ NPs (Eu L—**red**, Ce–**green**).

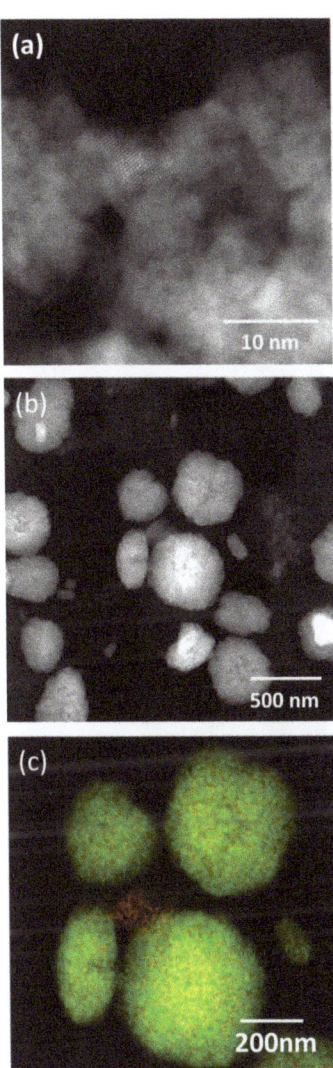

**Figure 3.** STEM HAADF imaging and EDS mapping of nanoparticles generated by HMTA: (**a**) HAADF of CeO$_x$ and (**b**) HAADF of Eu:CeO$_x$; (**c**) Ce and Eu L map of the Eu:CeO$_x$ NPs (Eu L—**red**, Ce L—**green**).

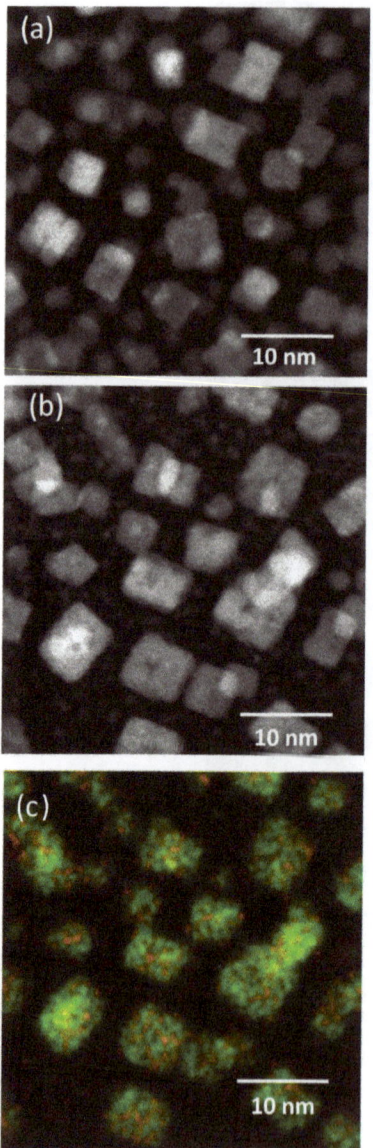

**Figure 4.** STEM HAADF imaging and EDS mapping of nanoparticles generated by TBA: (**a**) HAADF of CeO$_x$ and (**b**) HAADF of Eu:CeO$_x$; (**c**) Ce L and Eu L map of the Eu:CeO$_x$ NPs (Eu L—**red**, CeL—**green**).

*3.3. Thermal Stability*

In situ thermal annealing experiments were carried out on all the samples (ON, HMTA, TBA), as shown in Figures 5–7, respectively. Each specimen was heated from room temperature to 500 °C at an average heating rate of 0.5 °C/s. For the ON specimens (CeO$_x$(ON); Figure 5), the samples proved to be unstable under the electron beam, which was attributed to a high impurity concentration and the formation of complex oxides during the preparation route. This was emphasized further during annealing, and it was not possible to perform a statistical analysis. Further, the Eu:CeO$_x$(ON) sample displayed an initial increase in areal density, followed by a decrease from 100–200 °C. In this temperature regime, sputtering dominated the response of the material, which was likely due to an increase in momentum exchange between the atoms in the solid, which was accelerated by an increase in temperature, and thus an increase in mobility. The CeO$_x$(HMTA) samples were found to be mainly glommed together over the grid; nonetheless, well separated nanoparticles were observed in several areas, and these sections were used for thermal investigation. Upon heating, the nanoparticles that were in the highly concentrated regions were found to 'de-cluster', which was attributed to the thermal-induced mobility. It was also noted that the overall size of the nanoparticle material increased through the heating experiment. This was not due to sintering but, instead, to the emergence of larger nanoparticles from the room temperature cluster areas.

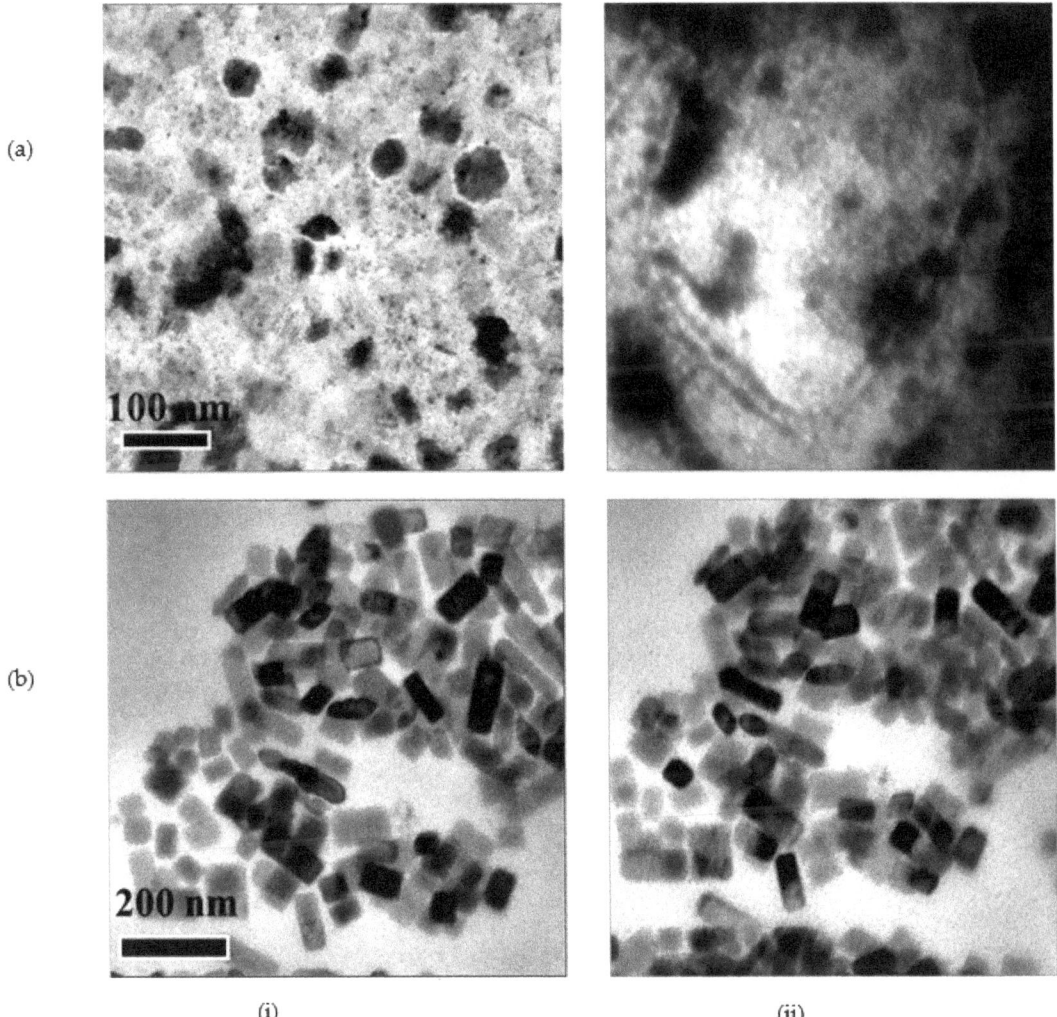

**Figure 5.** TEM images of nanoparticles generated from ON preparation route during thermal annealing: (i) 25 °C and (ii) 500 °C where (a) CeO$_x$ (scale bar = 100 nm) and (b) Eu: CeO$_x$ (scale bar = 200 nm).

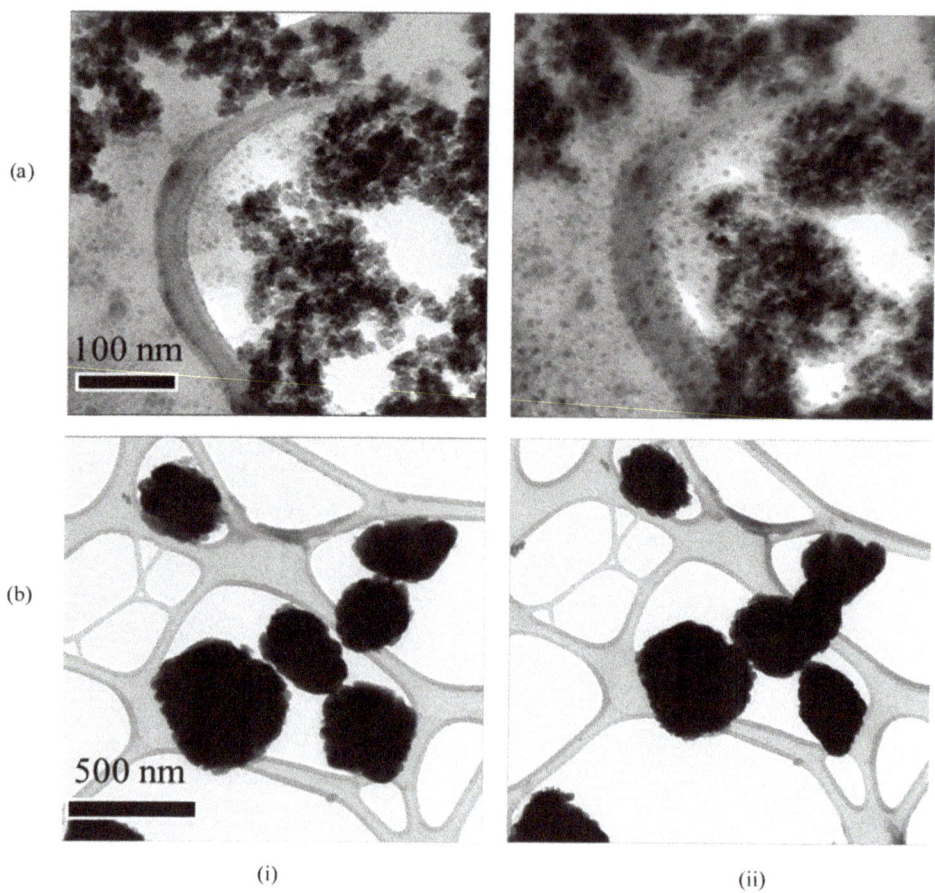

**Figure 6.** TEM images of nanoparticles generated from HMTA preparation route during thermal annealing: (i) 25 °C and (ii) 500 °C where (a) $CeO_x$ (scale bar = 100 nm) and (b) $Eu:CeO_x$ (scale bar = 500 nm).

The sizes of the $Eu:CeO_x$(HMTA) were unchanged over the temperature range investigated; however, the distance between the particles decreased as the organic ligands were thermally decomposed, allowing in situ clustering. Finally, the $CeO_x$(TBA) particles, analyzed in the same fashion, displayed the highest areal density of any particle studied coupled with the smallest average size. These nanoparticles showed a relatively low mobility under annealing conditions but still agglomerated at higher temperatures. For the $Eu:CeO_x$(TBA) samples, the particles appeared to destabilize at 200–300 °C, as suggested by the reduced areal density in comparison to the room temperature value. Table 2 provides a summary of the results from the thermal annealing experiments. In order to compare the various changes induced by the thermal study, the sizes of the samples were normalized and a plot of these versus the temperature was generated. As can be seen in Table 2, the $Eu:CeO_x$ nanocubes appeared to be more resistant to deformation compared to the $CeO_x$ samples. This implies that Ln dopants may be modulating the thermal resistance of nanocubes.

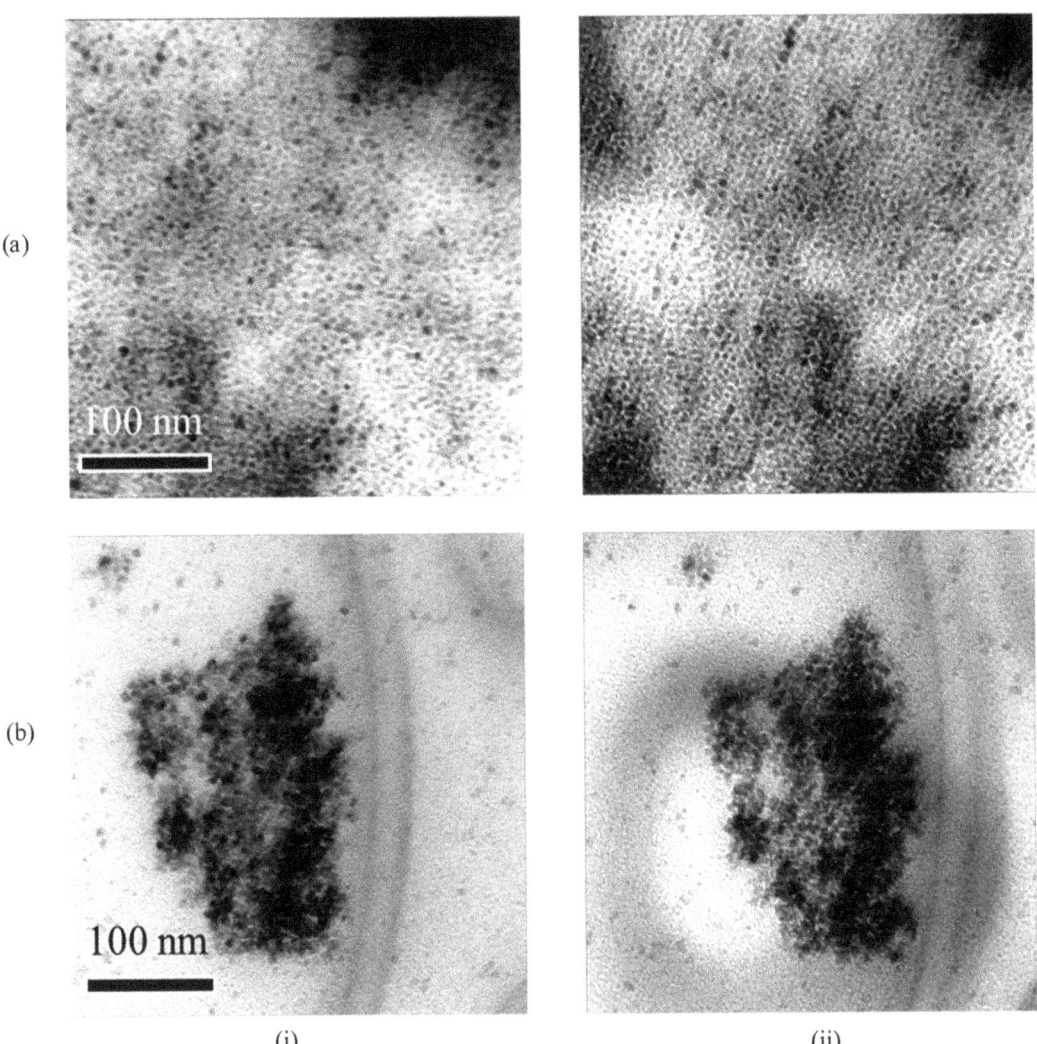

**Figure 7.** TEM images of nanoparticles generated from TBA preparation route during thermal annealing: (i) 25 °C and (ii) 500 °C where (**a**) CeO$_x$ (scale bar = 100 nm) and (**b**) Eu:CeO$_x$ (scale bar = 100 nm).

**Table 2.** Aerial density ($\mu m^{-2}$) and size (nm) versus temperature (°C) for (**a**) ON, (**b**) HMTA, and (**c**) TBA preparation of $CeO_2$ and $Eu:CeO_2$.

| Prep Route | Temp (°C) | Sample | Aerial Density ($\mu m^{-2}$) | Size (nm) | Sample | Aerial Density ($\mu m^{-2}$) | Size (nm) |
|---|---|---|---|---|---|---|---|
| ON | 25 | $CeO_2$ | N/A | N/A | $Eu:CeO_2$ | 190.63 | 42.4 |
|  | 100 |  |  |  |  | 201.06 | 42.34 |
|  | 200 |  |  |  |  | 189.14 | 42.42 |
|  | 300 |  |  |  |  | 177.23 | 42.42 |
|  | 400 |  |  |  |  | 171.27 | 42.43 |
|  | 500 |  |  |  |  | 160.85 | 42.45 |
| HMTA | 25 | $CeO_2$ | 2919 | 5.77 | $Eu:CeO_2$ | 2.204 | 340.57 |
|  | 100 |  | 3216 | 5.75 |  | 2.204 | 340.61 |
|  | 200 |  | 3797 | 5.79 |  | 2.204 | 340.6 |
|  | 300 |  | 4932 | 5.84 |  | 2.204 | 340.59 |
|  | 400 |  | 4676 | 5.81 |  | 2.204 | 340.6 |
|  | 500 |  | 4649 | 5.81 |  | 2.204 | 340.62 |
| TBA | 25 | $CeO_2$ | 15,882 | 4.07 | $Eu:CeO_2$ | 4593 | 7.3 |
|  | 100 |  | 16,445 | 4.04 |  | 4931 | 7.27 |
|  | 200 |  | 15,519 | 4.09 |  | 5026 | 7.23 |
|  | 300 |  | 15,501 | 4.09 |  | 4342 | 7.31 |
|  | 400 |  | 14,818 | 4.1 |  | 3910 | 7.34 |
|  | 500 |  | 14,083 | 4.12 |  | 3850 | 7.34 |

### 3.4. Irradiation Stability

The resistance of these nanoparticles to irradiation was evaluated by exposing the samples to different ion fluences. The TEM images of the ON, HDA, and TBA nanoparticles are shown in Figures 8–10, respectively.

For the $CeO_x$(ON) nanoparticles (see Figure 8i(a–c)), a decrease in crystallinity with increasing fluence was observed. This is consistent with the high concentration of defects that altered and/or destroyed the translational periodicity of the material. It was noted that at higher fluences, the ion irradiation also appeared to deplete the $SiO_x$ layer beneath the nanoparticles, while simultaneously forcing the aggregation of the particle. By contrast, in the Eu-doped analog of Figure 8ii(a–c), remarkable radiation resistance was observed in the nanoparticles, with diffraction patterns (DP) revealing high crystallinity through the highest fluence studied. What appear to be ablation pits formed from sputtering events on the nanoparticle surface are easily visible. These surface defects are highlighted with arrows in Figure 8ii(c).

Figure 9i(a–c) shows the $CeO_x$ (HMTA) preparation route. As can be observed, the polycrystalline green materials migrated towards a single crystalline/amorphous phase during exposure to various levels of ion irradiation. This process was likely akin to a recrystallization in the nanocubes with the simultaneous reduction in crystallinity at higher fluences. In Figure 9ii(a–c), the Eu-doped analog, it appears that the particles began to be sputtered at higher fluences, causing an overall reduction in the size of the nanoparticles. Additionally, the DP was amorphous where this process dominated the response of the material.

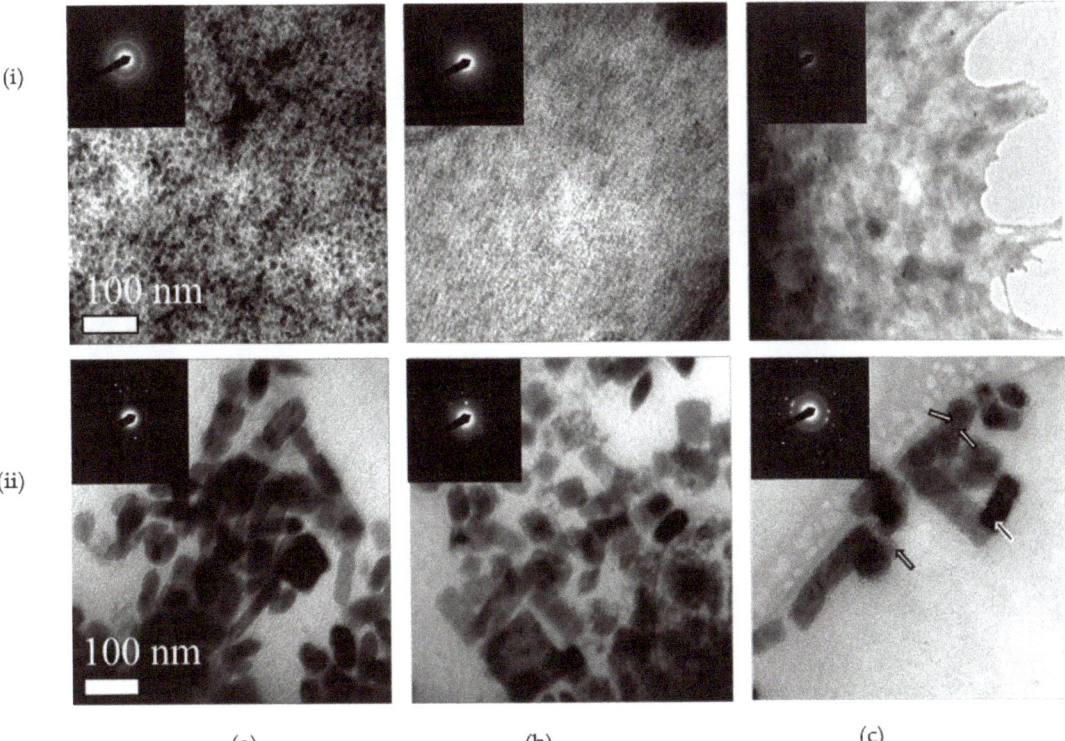

**Figure 8.** TEM images and SAED patterns of (i) $CeO_x$ and (ii) $Eu:CeO_x$ nanoparticles generated from ON preparation route during ion irradiation: (**a**) green, (**b**) $1 \times 10^{13}$ cm$^{-2}$, and (**c**) $1 \times 10^{15}$ cm$^{-2}$. Respective SAED patterns are displayed in the inset of each micrograph. Scale bars are 100 nm in all micrographs.

The $CeO_x$(TBA) nanoparticles, under the same ion exposure as that noted above, are shown in Figure 10i(a–c). As can be observed, the samples agglomerated as a result of destabilization from the ion interaction. This was further verified by the DP'S transition from a polycrystalline to an amorphous state. Lastly, in Figure 10ii(a–c), the Eu-doped particles remained separated at higher fluences and displayed a small amount of crystalline behavior at the highest fluence studied. These doped nanoparticle retained their structural integrity better than $CeO_x$(TBA) through ion irradiation.

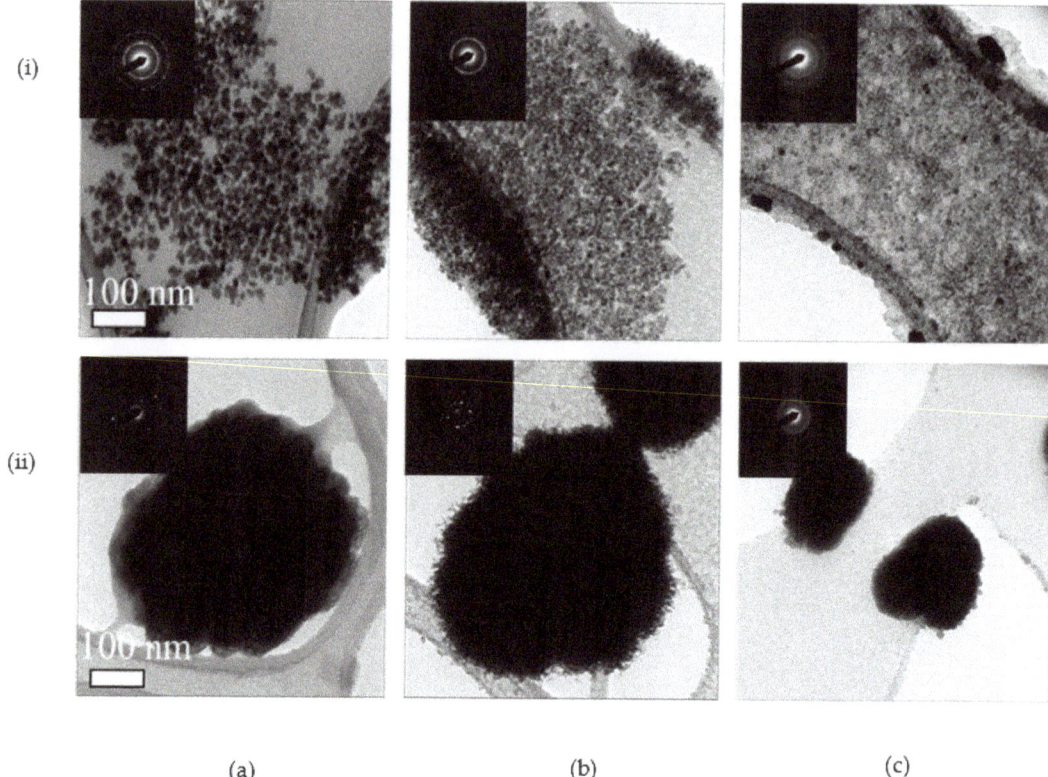

**Figure 9.** TEM images and SAED patterns of (i) CeO$_x$ and (ii) Eu:CeO$_x$ nanoparticles generated from HMTA preparation route during ion irradiation: (a) green, (b) 1 × 10$^{13}$ cm$^{-2}$, and (c) 1 × 10$^{15}$ cm$^{-2}$. Respective SAED patterns are displayed in the inset of each micrograph. Scale bars are 100 nm in all micrographs.

The high thermal stability of the particles was expected of these systems, given the high melting temperature (2400 °C) of the CeO$_x$, which resulted in homologous temperatures of 0.21. Although well above the operating temperature of many applications, this temperature regime shows that no significant nanoscale-enhanced sintering is observed in these systems. The poor irradiation stability observed in this work is similar to that seen in crystalline tungstate nanoparticles irradiated with heavy ions [23,24]. This in stark contrast to other amorphous ceramic nanoparticles and core@shell nanoparticles that have been explored, which suggests that the amorphous phase may be extremely important for radiation stability [25,26]. As such, future work will explore the role of both phase and doping on the stability of ceramic nanoparticles for both free-standing nanoparticles, as shown in this study, as well as those embedded in composite matrixes.

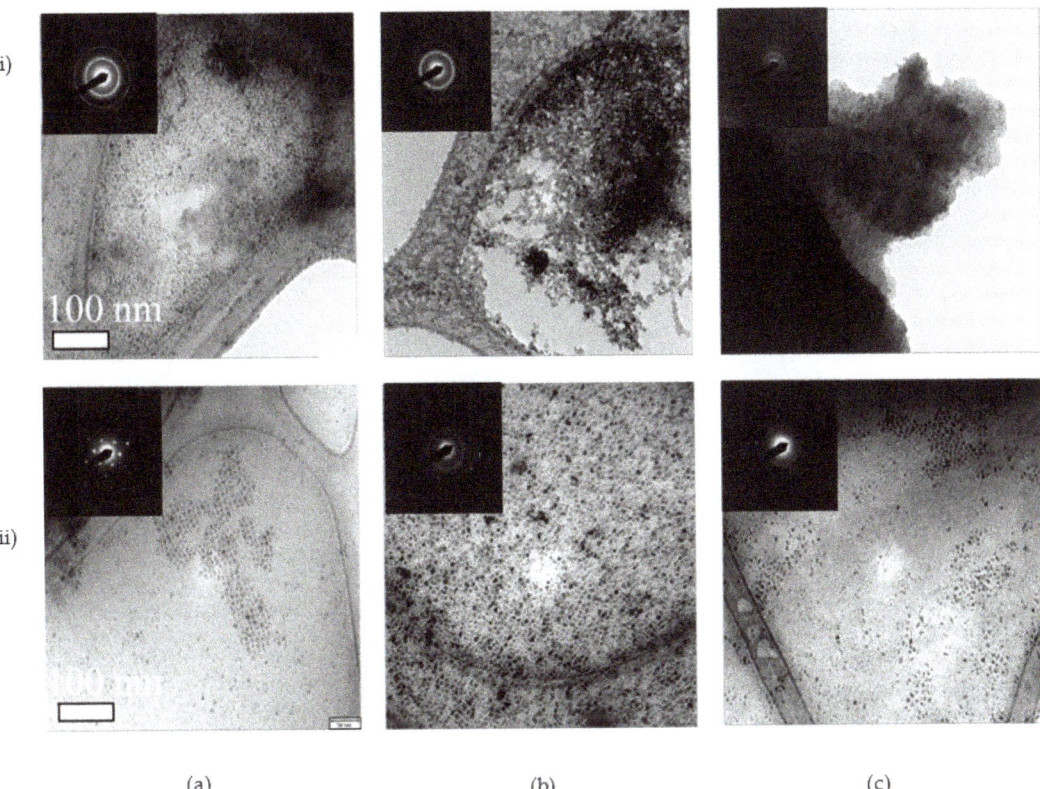

**Figure 10.** TEM images of (**i**) CeO$_x$ and (**ii**) Eu:CeO$_x$ nanoparticles generated from TBA preparation route during ion irradiation: (**a**) green, (**b**) $1 \times 10^{13}$ cm$^{-2}$, and (**c**) $1 \times 10^{15}$ cm$^{-2}$. Respective SAED patterns are displayed in the inset of each micrograph. Scale bars are 100 nm in all micrographs.

## 4. Summary and Conclusions

The synthesis of nanoparticles of CeO$_x$ and Ln:CeO$_x$ (Ln = Nd, Eu, Er, and Lu) was performed by using ON, HMTA, or TBA. The ON and TBA proved to be nanocubes whereas the HMTA produced larger rounded particulates. Elemental mapping and XRF studies confirmed the presence of the lanthanides, which were randomly distributed within a CeO$_x$ matrix. Thermal analyses of the undoped and Eu-doped CeO$_x$ samples for each preparative route revealed that the Ln-doped species appeared to be more resistant to thermal assaults in comparison to the undoped samples. In the ion stability studies, it was found that all of the samples displayed poor phase stability when exposed to 15 MeV Au ion irradiation. However, the Ln-doped species appeared to be slightly more stable than the homometallic nanoparticles. This variation suggests a potential difference in the active microstructural mechanisms during thermal annealing and ion irradiation. These results suggest that doping with alternative lanthanide cations in CeO$_x$ could serve as a good shielding agent against external stimuli.

**Supplementary Materials:** The following are available online at https://www.mdpi.com/article/10.3390/cryst11111369/s1. Figure S1: Analytical data for generated CeOx (a) XRF [(i) ON, (ii) HMTA, (iii) TBA], (b) pXRD RT [(i) ON, (ii) HMTA, (iii) TBA], (b*) pXRD 600 °C [(i) ON, (ii) HMTA, (iii) TBA], (c) FTIR [(i) ON, (ii) HMTA, (iii) TBA], (d) DLS [(i) ON, (ii) HMTA, (iii) TBA], (e) TGA/DSC [(i) ON, (ii) HMTA, (iii) TBA]; Figure S2: Analytical data for generated Nd:CeOx (a) XRF [(i) ON, (ii) HMTA,

(iii) TBA], (b) pXRD RT [(i) ON, (ii) HMTA, (iii) TBA], (b*) pXRD 600 °C [(i) ON, (ii) HMTA, (iii) TBA], (c) FTIR [(i) ON, (ii) HMTA, (iii) TBA], (d) DLS [(i) ON, (ii) HMTA, (iii) TBA]; Figure S3: Analytical data for generated Eu:CeOx (a) XRF [(i) ON, (ii) HMTA, (iii) TBA], (b) pXRD RT [(i) ON, (ii) HMTA, (iii) TBA], (b*) pXRD 600 °C [(i) ON, (ii) HMTA, (iii) TBA], (c) FTIR [(i) ON, (ii) HMTA, (iii) TBA], (d) DLS [(i) ON, (ii) HMTA, (iii) TBA], (e) TGA/DSC [(i) ON, (ii) HMTA, (iii) TBA]; Figure S4: Analytical data for generated Er:CeOx (a) XRF [(i) ON, (ii) HMTA, (iii) TBA], (b) pXRD RT [(i) ON, (ii) HMTA, (iii) TBA], (b*) pXRD 600 °C [(i) ON, (ii) HMTA, (iii) TBA], (c) FTIR [(i) ON, (ii) HMTA, (iii) TBA], (d) DLS [(i) ON, (ii) HMTA, (iii) TBA]; Figure S5: Analytical data for generated Lu:CeOx (a) XRF [(i) ON, (ii) HMTA, (iii) TBA], (b) pXRD RT [(i) ON, (ii) HMTA, (iii) TBA], (b*) pXRD 600 °C [(i) ON, (ii) HMTA, (iii) TBA], (c) FTIR [(i) ON, (ii) HMTA, (iii) TBA], (d) DLS [(i) ON, (ii) HMTA, (iii) TBA].

**Author Contributions:** Conceptualization, T.J.B. and K.H.; methodology, T.J.B., K.H. and K.B.; software, K.B.; validation, T.J.B. and K.H.; formal analysis, T.J.B., K.H., K.B., P.C.R. and F.G.; investigation, T.J.B.; resources, T.J.B., K.H. and P.L.; data curation, K.B., P.C.R., F.G. and P.L.; writing—original draft preparation, T.J.B. and K.B.; writing—review and editing, T.J.B., K.H., K.B., P.C.R., F.G., E.L., P.L. and A.A.; visualization, K.B. and P.C.R.; supervision, T.J.B. and K.H.; project administration, T.J.B. and K.H.; funding acquisition, T.J.B. and K.H. All authors have read and agreed to the published version of the manuscript.

**Funding:** This work was supported by Sandia National Laboratories.

**Data Availability Statement:** Not applicable.

**Acknowledgments:** This work was performed, in part, at the Center for Integrated Nanotechnologies, an Office of Science User Facility operated for the U.S. Department of Energy (DOE) Office of Science. Sandia National Laboratories is a multimission laboratory managed and operated by National Technology & Engineering Solutions of Sandia, LLC, a wholly owned subsidiary of Honeywell International, Inc., for the U.S. DOE's National Nuclear Security Administration under contract DE-NA-0003525. The views expressed in the article do not necessarily represent the views of the U.S. DOE or the United States Government.

**Conflicts of Interest:** The authors declare no conflict of interest.

# References

1. Baker, C.H. Harnessing cerium oxide nanoparticles to protect normal tissue from radiation damage. *Transl. Cancer Res.* **2013**, *2*, 343–358.
2. Issa, S.A.M.; Sayyed, N.I.; Zaid, N.H.N.; Matori, K.A. Effect of $Bi_2O_3$ in borate-tellurite-silicate glass system for development of gamma-rays shielding materials. *Results Phys.* **2018**, *9*, 206–210. [CrossRef]
3. Niranjan, R.S.; Rudraswamy, B.; Dhananjaya, N. Effective atomic number, electron density and kerma of gamma radiation for oxides of lanthanides. *Pramana* **2012**, *78*, 451–458. [CrossRef]
4. Al-Hadeethia, Y.; Sayyedc, M.I.; Kaewkhaod, J.; Raffah, B.M.; Almalkia, R.; Rajaramakrishnad, R.; Hussein, M.A. Chalcogenide glass-ceramics for radiation shielding applications. *Ceram. Int.* **2020**, *46*, 5380–5386.
5. Abd-Allah, W.M.; Fayad, A.M.; Saudi, H.A. Structure and Physical Properties of $Al_2O_3$ Nanofillers Embedded in Poly(Vinyl Alcohol). *Opt. Quantum Electron.* **2019**, *51*, 1–14.
6. Graham, J.T.; Zhang, Y.; Weber, W.J. Irradiation-induced defect formation and damage accumulation in single crystal $CeO_2$. *J. Nucl. Mater.* **2018**, *498*, 400–408. [CrossRef]
7. Costantini, J.M.; Lelong, G.; Guillaumet, M.; Gourier, D. Recovery of damage in electron-irradiated ceria. *J. Appl. Phys.* **2021**, *129*, 155901. [CrossRef]
8. Sorenson, J.J.; Tieu, E.; Morse, M.D. Bond dissociation energies of lanthanide sulfides and selenides. *J. Chem. Phys.* **2021**, *154*, 124307. [CrossRef]
9. Dhall, A.; Self, W. Cerium Oxide Nanoparticles: A Brief Review of Their Synthesis Methods and Biomedical Applications. *Antioxidants* **2018**, *7*, 97. [CrossRef]
10. Xu, C.; Qu, X. Cerium oxide nanoparticle: A remarkably versatile rare earth nanomaterial for biological applications. *NPG Asia Mater.* **2014**, *6*, e90. [CrossRef]
11. Wang, G.; Peng, Q.; Li, Y. Lanthanide-Doped Nanocrystals: Synthesis, Optical-Magnetic Properties, and Applications. *Accounts Chem. Res.* **2010**, *44*, 322–332. [CrossRef]
12. Wang, Z.; Quan, Z.; Lin, J. Remarkable Changes in the Optical Properties of $CeO_2$ Nanocrystals Induced by Lanthanide Ions Doping. *Inorg. Chem.* **2007**, *46*, 5237–5242. [CrossRef] [PubMed]
13. Lin, Y.-H.; Shen, L.-J.; Chou, T.-H.; Shih, Y.-H. Synthesis, Stability, and Cytotoxicity of Novel Cerium Oxide Nanoparticles for Biomedical Applications. *J. Clust. Sci.* **2021**, *32*, 405–413. [CrossRef]

14. Kumar, A.; Babu, S.; Karakoti, A.S.; Schulte, A.; Seal, S. Luminescence Properties of Europium-Doped Cerium Oxide Nanoparticles: Role of Vacancy and Oxidation States. *Langmuir* **2009**, *25*, 10998–11007. [CrossRef] [PubMed]
15. Hubbard, J.A.; Boyle, T.J.; Sepper, E.T.; Brown, A.; Settecerri, T.; Santarpia, J.L.; Kotula, P.; McKenzie, B.; Lucero, G.A.; Lemieux, L.J.; et al. Airborne Release Fractions from Surrogate Nuclear Waste Fires Containing Lanthanide Nitrates and Depleted Uranium Nitrate in 30% Tributyl Phosphate in Kerosene. *Nucl. Technol.* **2021**, *207*, 103–118. [CrossRef]
16. Olegárioa, R.C.; de Souzaa, E.C.F.; Borges, J.F.M.; da Cunhac, J.B.M.; de Andrade, A.V.C.; Antunesa, S.R.M.; Antunes, C.A. Synthesis and characterization of $Fe^{3+}$ doped cerium–praseodymium oxide pigments. *Dye. Pigment.* **2013**, *97*, 113–117. [CrossRef]
17. Yabe, S.; Yamashita, M.; Momose, S.; Tahira, K.; Yoshida, S.; Li, R.; Yin, S.; Sato, T. Synthesis and UV-shielding properties of metal oxide doped ceria via soft solution chemical processes. *Int. J. Inorg. Mater.* **2001**, *3*, 1003–1008. [CrossRef]
18. Cabral, A.C.; Cavalcante, L.S.; Deus, R.C.; Longo, E.; Simoes, A.Z.; Moura, F. Photoluminescence properties of praseodymium doped cerium oxide nanocrystals. *Ceram. Int.* **2014**, *40*, 4445–4453. [CrossRef]
19. Hattar, K.; Bufford, D.C.; Buller, D.L. Concurrent in situ ion irradiation transmission electron microscope. *Nucl. Instrum. Methods Phys. Res. Sect. B Beam Interact. Mater. At.* **2014**, *338*, 56–65. [CrossRef]
20. Yu, T.; Park, Y.I.; Kang, M.-C.; Joo, J.; Par, J.K.; Won, H.Y.; Hyeon, K.J.J.T. Large-Scale Synthesis of Water Dispersible Ceria Nanocrystals by a Simple Sol–Gel Process and Their Use as a Chemical Mechanical Planarization Slurry. *Eur. J. Inorg. Chem.* **2008**, *2008*, 855–858. [CrossRef]
21. Zhang, F.; Chan, S.-W.; Spanier, J.E.; Apak, E.; Jin, Q.; Robinson, R.D.; Herman, I.P. Cerium oxide nanoparticles: Size-selective formation and structure analysis. *Appl. Phys. Lett.* **2002**, *80*, 127–129. [CrossRef]
22. Yang, S.; Gao, L. Controlled Synthesis and Self-Assembly of $CeO_2$ Nanocubes. *J. Am. Chem. Soc.* **2006**, *128*, 9330–9331. [CrossRef] [PubMed]
23. Branson, J.V.; Hattar, K.; Rossi, P.; Vizkelethy, G.; Powell, C.J.; Hernandez-Sanchez, B.; Doyle, B. Ion beam characterization of advanced luminescent materials for application in radiation effects microscopy. *Nucl. Instrum. Methods Phys. Res. Sect. B Beam Interact. Mater. At.* **2011**, *269*, 2326–2329. [CrossRef]
24. Hoppe, S.M.; Hattar, K.; Boyle, T.J.; Villone, J.; Yang, P.; Patrick Doty, F.; Hernandez-Sanchez, B.A. Application of in-situ ion irradiation TEM and 4D tomography to advanced scintillator materials. In *Penetrating Radiation Systems and Applications XIII*; International Society for Optics and Photonics: Bellingham, WA, USA, 2012; Volume 8509.
25. Blair, S.J.; Muntifering, B.R.; Chan, R.O.; Barr, C.M.; Boyle, T.J.; Hattar, K. Unexpected radiation resistance of core/shell ceramic oxide nanoparticles. *Mater. Today Commun.* **2018**, *17*, 109–113. [CrossRef]
26. Kiani, M.T.; Hattar, K.; Wendy Gu, X. In situ TEM Study of Radiation Resistance of Metallic Glass–Metal Core–Shell Nanocubes. *ACS Appl. Mater. Interfaces* **2020**, *12*, 40910–40916. [CrossRef] [PubMed]

Article

# Low-Temperature Magnetic and Magnetocaloric Properties of Manganese-Substituted Gd$_{0.5}$Er$_{0.5}$CrO$_3$ Orthochromites

Neeraj Panwar [1,*], Kuldeep Singh [1], Komal Kanwar [1], Yugandhar Bitla [1], Surendra Kumar [2] and Venkata Sreenivas Puli [3,4]

[1] Department of Physics, Central University of Rajasthan Bandarsindri, Ajmer 305817, Rajasthan, India; 2019phdph006@curaj.ac.in (K.S.); 2017phdph02@curaj.ac.in (K.K.); bitla@curaj.ac.in (Y.B.)
[2] Department of Physics, Sri Venkateswara College, University of Delhi, Dhaula Kuan, New Delhi 110021, Delhi, India; surendra.kumar@svc.ac.in
[3] National Research Council, Washington, DC 20001, USA; venkata.puli.ctr@us.af.mil
[4] Materials and Manufacturing Directorate, Air Force Research Laboratory, 3005 Hobson Way, Wright Patterson Air Force Base, OH 45433, USA
* Correspondence: neerajpanwar@curaj.ac.in

**Abstract:** Rare-earth chromites have been envisioned to replace gas-based refrigeration technology because of their promising magnetocaloric properties at low temperatures, especially in the liquid helium temperature range. Here, we report the low-temperature magnetic and magnetocaloric properties of Gd$_{0.5}$Er$_{0.5}$Cr$_{1-x}$Mn$_x$O$_3$ ($x$ = 0, 0.1, 0.2, 0.3, 0.4 and 0.5) rare-earth orthochromites. The Néel transition temperature ($T_N$) was suppressed from 144 K for Gd$_{0.5}$Er$_{0.5}$CrO$_3$ to 66 K for the Gd$_{0.5}$Er$_{0.5}$Cr$_{0.5}$Mn$_{0.5}$O$_3$ compound. Furthermore, magnetization reversal was observed in the magnetization versus temperature behavior of the Gd$_{0.5}$Er$_{0.5}$Cr$_{0.6}$Mn$_{0.4}$O$_3$ and Gd$_{0.5}$Er$_{0.5}$Cr$_{0.5}$Mn$_{0.5}$O$_3$ compounds at 100 Oe applied magnetic field. The magnetic entropy change ($-\Delta S$) value varied from 16.74 J/kg-K to 7.46 J/kg-K, whereas the relative cooling power (RCP) ranged from 375.94 J/kg to 220.22 J/kg with a Mn ion concentration at 5 T field and around 7.5 K temperature. The experimental results were substantiated by a theoretical model. The present values of the magnetocaloric effect are higher than those of many undoped chromites, manganites and molecular magnets in the liquid helium temperature range.

**Keywords:** rare-earth orthochromites; magnetization reversal; magnetocaloric effect

---

**Citation:** Panwar, N.; Singh, K.; Kanwar, K.; Bitla, Y.; Kumar, S.; Puli, V.S. Low-Temperature Magnetic and Magnetocaloric Properties of Manganese-Substituted Gd$_{0.5}$Er$_{0.5}$CrO$_3$ Orthochromites. *Crystals* **2022**, *12*, 263. https://doi.org/10.3390/cryst12020263

Academic Editors: Alessandra Toncelli and Željka Antić

Received: 29 January 2022
Accepted: 12 February 2022
Published: 15 February 2022

**Publisher's Note:** MDPI stays neutral with regard to jurisdictional claims in published maps and institutional affiliations.

**Copyright:** © 2022 by the authors. Licensee MDPI, Basel, Switzerland. This article is an open access article distributed under the terms and conditions of the Creative Commons Attribution (CC BY) license (https://creativecommons.org/licenses/by/4.0/).

## 1. Introduction

Perovskite materials have been widely recognized for their diverse and intriguing properties, such as magnetic, electrical and photocatalytic, etc. [1–4]. As far as the magnetic properties of various perovskite materials are concerned, rare-earth orthochromites $RCrO_3$ ($R$ = rare-earth ion) exhibit interesting magnetic properties viz. exchange bias [5], spin reorientation [6], magnetization reversal [7–9] and magnetocaloric effect (MCE) [10–13]. It is now a well-established fact that there are three types of magnetic interactions, namely Cr$^{3+}$-Cr$^{3+}$, Cr$^{3+}$-$R^{3+}$ and $R^{3+}$-$R^{3+}$, in $RCrO_3$ materials, which govern their distinct magnetic properties in different temperature ranges [5]. The interaction Cr$^{3+}$-Cr$^{3+}$ is responsible for the canted antiferromagnetic behavior below the Néel transition temperature, whereas the Cr$^{3+}$-$R^{3+}$ interaction gives rise to negative magnetization or magnetization reversal below a certain temperature, known as the compensation temperature, provided that the $R^{3+}$ ion is magnetic. The negative magnetization occurs due to the antiparallel coupling between the Cr$^{3+}$ moments and rare-earth ion moments [14]. Since the last decade, researchers have also been exploring these materials for their usefulness as a low-temperature refrigerant employing the magnetocaloric effect, which is measured in terms of magnetic entropy change and relative cooling power [10–13,15–20]. Magnetic refrigeration involving solid materials is a cooling technology that is more efficient, inexpensive, safer and more environmentally friendly than the conventional vapor-compression-based technology. It was reported that Gadolinium and Gadolinium-based compounds are important working materials for magnetic refrigeration. For example, Yin et al. reported 31.6 J/kg-K magnetic entropy change in GdCrO$_3$ single crystal at 3 K under a 4.4 T applied magnetic field [18]. Shi et al. also reported a magnetic entropy change of 27.6 J/kg-K in Er$_{0.33}$Gd$_{0.67}$CrO$_3$ at 5 K and 7 T [19]. The effect of Mn substitution at the Cr site in Gd$_{0.5}$Er$_{0.5}$CrO$_3$ has not been explored yet. Therefore, it is worthwhile to investigate the magnetic and magnetocaloric properties of Mn-substituted Gd$_{0.5}$Er$_{0.5}$CrO$_3$ compounds.

In the present work, we synthesized Gd$_{0.5}$Er$_{0.5}$Cr$_{1-x}$Mn$_x$O$_3$ ($x$ = 0, 0.1, 0.2, 0.3, 0.4, 0.5) compounds and investigated the effect of Mn substitution on the magnetic and magnetocaloric properties. We also compared the experimental results with theoretical models.

## 2. Experimental Details

Gd$_{0.5}$Er$_{0.5}$Cr$_{1-x}$Mn$_x$O$_3$ ($x$ = 0, 0.1, 0.2, 0.3, 0.4, 0.5) (abbreviated as GECMO) orthochromites were prepared by the solid-state reaction method. Stoichiometric amounts of high-purity Gd$_2$O$_3$ (99.99%, Sigma Aldrich, St. Louis, MO, USA), Er$_2$O$_3$ (99.99%, Sigma Aldrich, St. Louis, MO, USA), Cr$_2$O$_3$ (99.99%, Sigma Aldrich, St. Louis, MO, USA) and MnO$_2$ (99.99%, Sigma Aldrich, St. Louis, MO, USA) were mixed and grounded. The mixed powders were calcined for 24 h at 800 °C and 1100 °C. Finally, the calcined powders were reground and pressed into uniform cylindrical pellets, which were sintered at 1450 °C for 48 h under ambient conditions. The samples were characterized for phase identification using an X-ray diffractometer (XRD; PANalytical-Empyrean, Almelo, The Netherlands) at room temperature using Cu-K$_\alpha$ radiation with wavelength 1.5404 Å at a step of 0.01° per second between 20° and 80°. Magnetic measurements were performed using a physical properties measurement system (PPMS DynaCool). Magnetization temperature data were collected between 2 K and 300 K in field-cooled (FC) and zero-field-cooled (ZFC) modes. The ramp rate was 4K/minute for both the modes. The magnetization–applied field ($M - H$) isotherms were measured under $H = \pm 8$ T.

## 3. Results and Discussion

### 3.1. Structural Study

Figure 1 displays the Rietveld-refined XRD patterns of the sintered GECMO orthochromites. It is clear from the XRD patterns that there are no impurity peaks in these series of samples and the diffraction peaks are well matched with the pristine GdCrO$_3$ compound (JCPDS Card No. 25-1056). The refinement was carried out by using Fullprof software with orthorhombic structure and *Pnma* space group. The calculated lattice parameters, unit cell volume (V), reliability factors, octahedral bond length (Cr–O), octahedral tilt angles, strain factor and Goldschmidt tolerance factor ($t$) of all compounds are presented in Table 1. It can be seen that there was an increase in the lattice constants $a$ and $c$, while the $b$ value was reduced with increasing Mn$^{3+}$ ion content. Their variation is shown in Figure 2a. Further, the lattice volume was also increased because of the larger ionic radius of the Mn$^{3+}$ ion (0.645 Å) compared to that of the Cr$^{3+}$ ion (0.615 Å) (Figure 2a). When the Mn$^{3+}$ ion was substituted at the Cr$^{3+}$ ion site, a strain developed in the lattice because of the different ionic radii, and, to ensure the stability of the crystal structure, the CrO$_6$ octahedra rotated or tilted. The strain factor was calculated by $S = 2(a - c)/(a + c)$, which arises as a result of octahedral rotation/tilting [21]. The distortion in the perovskite structure is also represented by the octahedral bond length and tilt angles $\theta_{[101]}$ and $\phi_{[010]}$. In the present study, the octahedral bond length Cr–O and tilt angles $\theta_{[101]}$ and $\phi_{[010]}$ were calculated by the Zhao formalism [22] as follows:

$$\text{Cr-O} = \left(\frac{ab}{4c}\right), \quad \theta_{[101]} = \cos^{-1}\left(\frac{c}{a}\right) \text{ and } \phi_{[010]} = \cos^{-1}\left(\frac{\sqrt{2}c}{b}\right)$$

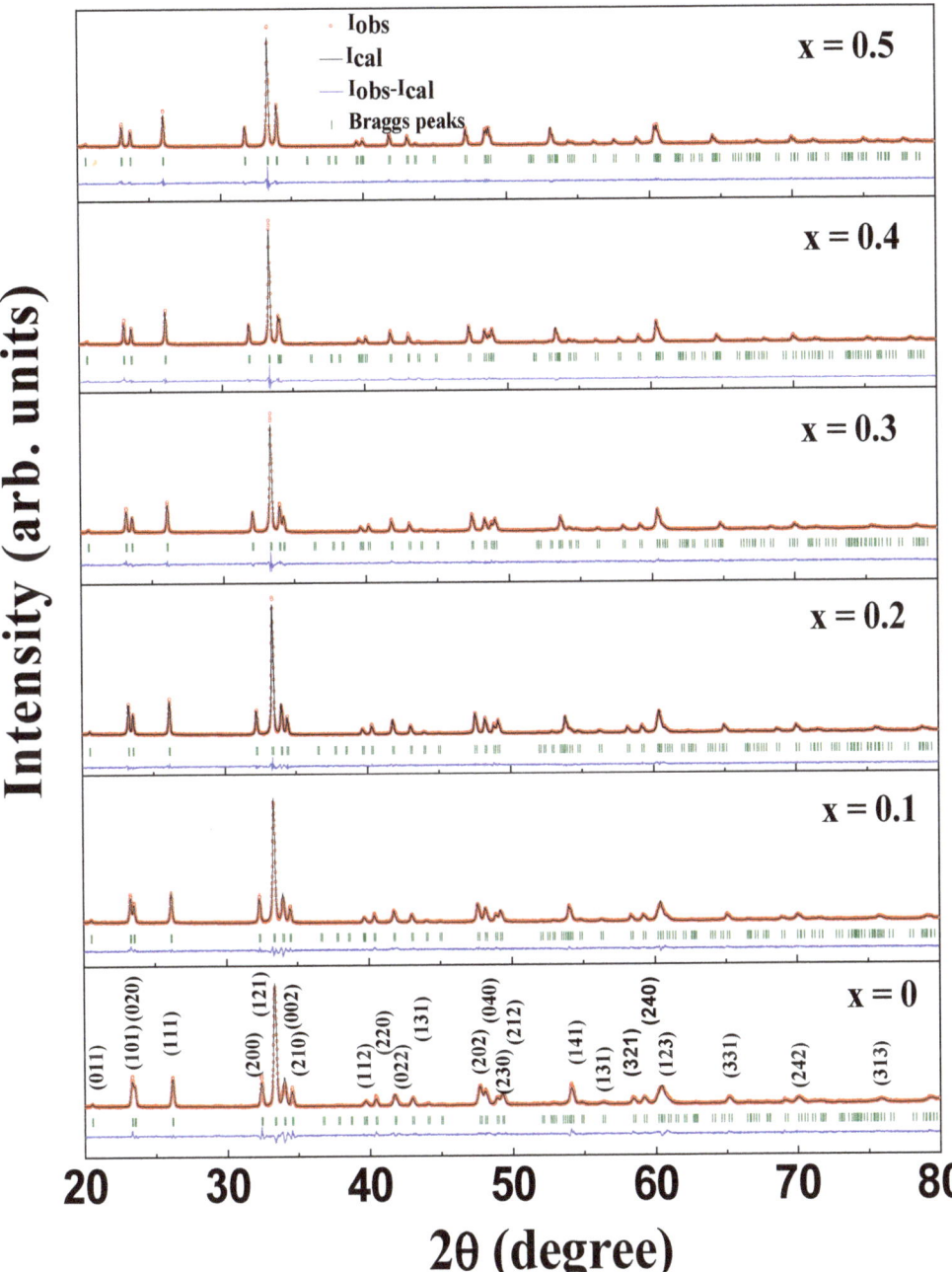

**Figure 1.** Rietveld-refined XRD patterns of the GECMO series.

Table 1. Lattice parameters, unit cell volume (V), reliability factors, octahedral bond length, tilt angles, strain factor and Goldschmidt tolerance factor (t) of GECMO series.

| | $x=0$ | $x=0.1$ | $x=0.2$ | $x=0.3$ | $x=0.4$ | $x=0.5$ |
|---|---|---|---|---|---|---|
| $a$ (Å) | 5.5220 | 5.5335 | 5.5530 | 5.5835 | 5.6072 | 5.6372 |
| $b$ (Å) | 7.5648 | 7.5541 | 7.5439 | 7.5376 | 7.5183 | 7.5029 |
| $c$ (Å) | 5.2691 | 5.2709 | 5.2721 | 5.2767 | 5.2740 | 5.2744 |
| $V$ (Å$^3$) | 220.109 | 220.330 | 220.860 | 222.081 | 222.340 | 223.085 |
| $\chi^2$ (%) | 3.29 | 2.48 | 2.08 | 1.94 | 2.62 | 2.03 |
| $R_P$ (%) | 19 | 16.7 | 16.1 | 16.7 | 19.5 | 19.2 |
| $R_{WP}$ (%) | 14.9 | 12.6 | 11.4 | 11.7 | 13.6 | 12.8 |
| Cr–O (Å) | 1.982 | 1.9825 | 1.9864 | 1.9925 | 1.9983 | 2.0047 |
| $\theta$ (°) | 17.40 | 17.71 | 18.29 | 19.06 | 19.84 | 20.65 |
| $\phi$ (°) | 9.922 | 9.32 | 8.75 | 8.06 | 7.22 | 6.19 |
| $S$ | 0.0468 | 0.0486 | 0.0518 | 0.0564 | 0.0612 | 0.0665 |
| $t$ | 0.8523 | 0.8510 | 0.8497 | 0.8484 | 0.8472 | 0.8459 |

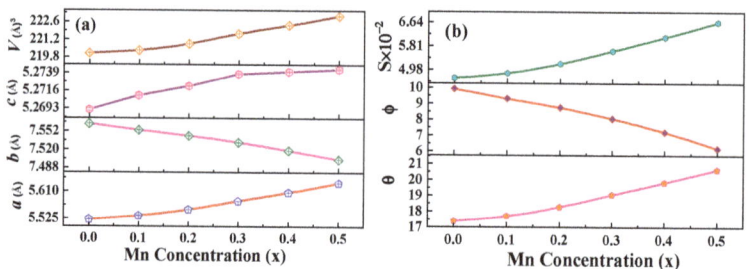

Figure 2. (a) Lattice parameters, volume, (b) strain parameter and tilt angle variation as a function of Mn concentration.

The octahedral bond length Cr–O increased in Mn-substituted compounds because the Mn$^{3+}$ ion had a larger ionic radius than the Cr$^{3+}$ ion. The rotation of the CrO$_6$ octahedron occurred in both planes, i.e., in-plane (along $a$ and $c$ axis) and out-of-plane (along $b$ axis). The tilt angle $\theta_{[101]}$ was found to increase, while the rotational angle $\phi_{[010]}$ decreased with an increase in the Mn$^{3+}$ ion concentration (Figure 2b). In determining the stability of the perovskite structure, the Goldschmidt tolerance factor ($t$) plays a key role. In order to calculate the tolerance factor ($t$) for the present compounds, we used the formula $t = <$Gd/Er–O$>/\sqrt{2} <$Cr/Mn–O$>$. The unity value of $t$ corresponds to an ideal perovskite structure, while $0.75 \leq t \leq 0.9$ favors a distorted orthorhombic structure [23,24]. Further, it should also be noted that the classical consideration of ionic size alone cannot explain the structural characteristics. To obtain a more accurate understanding of the structural characteristics when Mn$^{3+}$ ions replace Cr$^{3+}$ ions, we performed a detailed structural analysis. Figure 3a presents the crystal structure of the GECO compound generated using VESTA software. It can be noticed that the structure is distorted from the ideal cubic structure (distortion shown in Figure 3b). This results in variation in the bond lengths. Figure 3c shows the Cr–O bond length variation in the Cr–O$_6$ octahedron as a function of the Mn concentration. The difference among the three Cr–O bond lengths in all compounds can be correlated with the Jahn–Teller (JT) distortion of Mn$^{3+}$ ions, along with a contribution from intrinsic structural distortion. The in-plane (orthorhombic-like) and out-of-plane (tetragonal-like) distortions evaluated by $Q_2 [= l_y - l_x]$, and $Q_3 [= (2 \times l_z - l_x - l_y)/\sqrt{3}]$, respectively, are shown in Figure 3d. Here, $l_x$, $l_y$ and $l_z$ denote the long and short Cr–O$_2$ bond lengths mainly in the $ac$ plane and the middle bond length Cr–O$_1$ mainly along the $b$ axis (out of plane), respectively. O$_1$ and O$_2$ stand for the apical and equatorial oxygen atoms. Here, the large and positive magnitude of $Q_2$ is associated with the Jahn–Teller distortion and the low and negative value of $Q_3$ indicates that an out-of-plane distortion along the $c$

axis is competing with the JT distortion [25,26]. This implies that the lattice deformation is primarily confined to the *ac* plane. A higher value of $Q_2$ in comparison to $Q_3$ indicates a decrease in the *b* lattice parameter value and an increase in the *a* and *c* parameters, as shown in Figure 2a. The continuous variation in the in-plane and out-of-plane distortions with Mn substitution should have an impact on the magnetic properties.

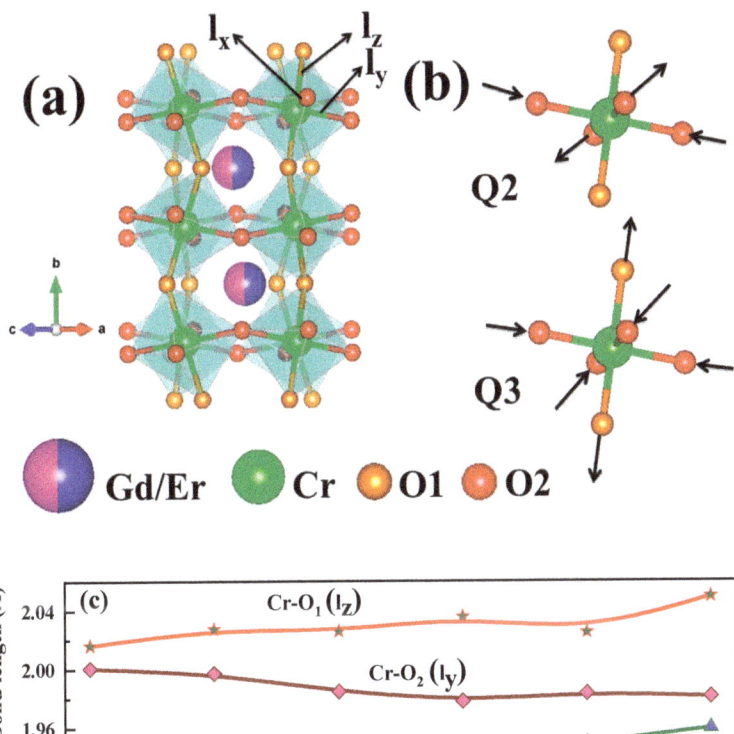

**Figure 3.** (a) Crystal structure of $Gd_{0.5}Er_{0.5}CrO_3$ compound obtained from the VESTA software; $l_x$, $l_y$ and $l_z$ denote the long and short Cr–$O_2$ bond lengths mainly in the *ac* plane and the middle length Cr–$O_1$ mainly along the *b* axis (out of plane), respectively. (b) $Q_2$ and $Q_3$ represent the orthogonal-like and tetragonal-like distortions, respectively. (c) Bond lengths and (d) octahedral distortion variation as a function of Mn concentration.

### 3.2. Magnetic Study

The zero-field-cooled (ZFC) and field-cooled (FC) magnetization plots of the GECMO series as a function of temperature from 2 K to 300 K at 100 Oe applied magnetic field are shown in Figure 4a–f. Above $T_N$ (i.e., the Néel transition temperature), both the FC and ZFC curves possess almost identical magnetization, whereas the curves become bifurcated below

$T_N$. The canted antiferromagnetism of $Cr^{3+}$ ions may be responsible for this bifurcation. The transition temperature for the pristine ($x = 0$) sample at 100 Oe magnetic field is ~144 K, which was evaluated by the first-order derivative of the magnetization temperature curve. The transition temperature decreased with an increase in $Mn^{3+}$ ion content. The decrease in $T_N$ was due to the weakening of the AFM interaction between $Cr^{3+}$ ions and the development of ferromagnetic double-exchange interaction $Cr^{3+}$-$Mn^{3+}$ via the $O^{2-}$ ion. It can be seen that, below $T_N$, magnetization in both modes (ZFC and FC) increased to a certain value with the lowering of temperature. It attained the maximum value below 10 K, corresponding to the ordering of rare-earth $Gd^{3+}$ and $Er^{3+}$ ions, and after this, it started decreasing. The magnetization increased with $Mn^{3+}$ ion doping for the samples with $x = 0, 0.1, 0.2, 0.3$ in FC mode, but it decreased in the case of the samples with higher Mn content. The magnetization in both modes (FC and ZFC) passed the zero magnetization and became negative for the two samples with $x = 0.4$ and $0.5$. The temperature where the magnetization crosses the temperature axis ($M = 0$) is known as the compensation temperature ($T_{comp}$). Further, in the case of the compound with $x = 0.4$, magnetization again started increasing below 10 K. This may have been due to the rotation of the net magnetic moment from one easy direction to the other. In the present study, it is likely that a higher concentration of manganese was responsible for the creation of a negative internal field that rotated the magnetization of $Gd^{3+}$ and $Er^{3+}$ ions in the opposite direction of the magnetization of $Cr^{3+}$ ions. As we shall see later, the calculated value of the internal field was negative (i.e., opposite to magnetization of $Cr^{3+}$ ions) for compounds with $x = 0.4$ and 0.5.

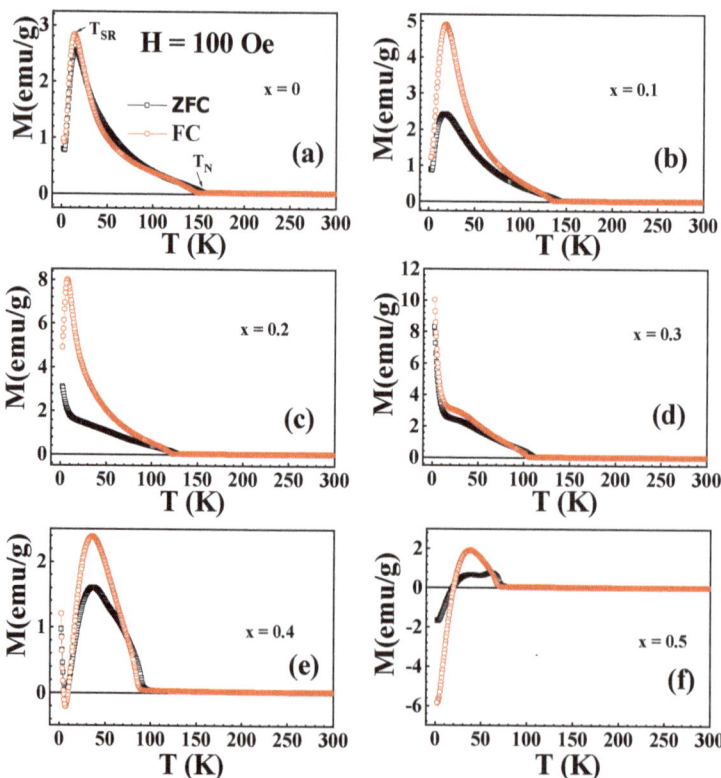

**Figure 4.** (a–f): ZFC and FC magnetization curves of GECMO series measured under 100 Oe magnetic field.

Figure 5a–d demonstrate the zero-field-cooled and field-cooled magnetization as a function of temperature from 2 K to 300 K for the compound with $x = 0.5$ at 500 Oe, 1 kOe, 2.5 kOe and 5 kOe applied magnetic field. Here, we observed that, for a higher magnetic field, negative magnetization disappeared. This means that the applied field of these values dominated the internal field, so the overall magnetization became positive because of the flipping of the magnetization of $Gd^{3+}/Er^{3+}$ ions in the direction of the applied field.

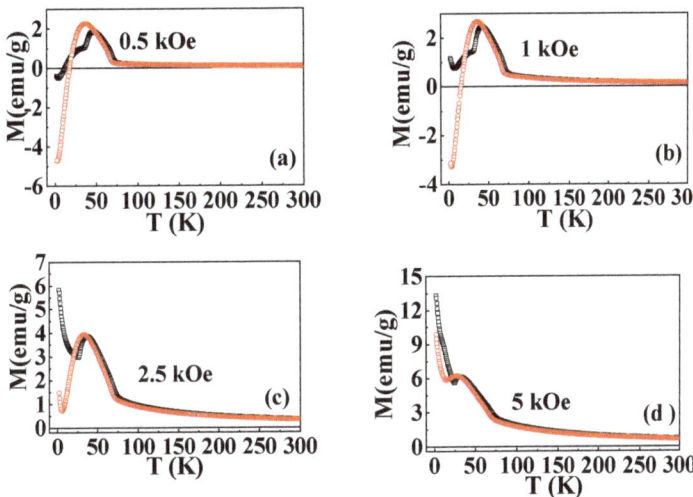

**Figure 5.** (a–d): ZFC and FC magnetization curves of $x = 0.5$ compound measured under 500 Oe, 1 kOe, 2.5 kOe and 5 kOe magnetic field.

Further, the FC magnetization data of the compounds, in which negative magnetization appeared, were fitted using the following equation [27]:

$$M = M_{Cr} + \frac{C(H + H_I)}{(T - \theta)} \quad (1)$$

In Equation (1), $M$ is the total magnetization, $M_{Cr}$ is the magnetization of canted $Cr^{3+}$ ions, $C$ is the Curie constant, $H_I$ is the internal field due to $Cr^{3+}$ ions, $H$ is the applied field and $\theta$ is the Weiss temperature. The fitting is shown by a solid red line in Figure 6a,b. The obtained parameters are listed in Table 2. We found negative values of internal field ($H_I$) for these compounds, which support the notion that induced magnetization occurred in the opposite direction to the applied field.

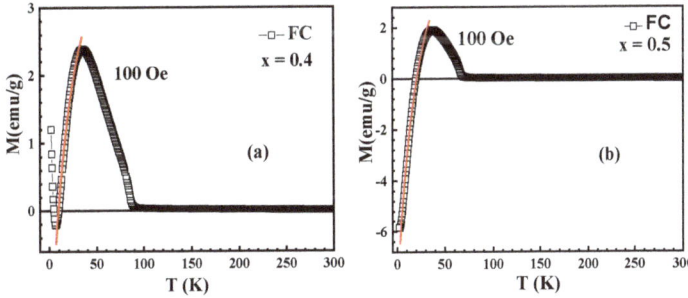

**Figure 6.** (a,b): The FC magnetization curves fitted to Equation (1) for compounds with $x = 0.4$ and $x = 0.5$ measured under 100 Oe magnetic field.

**Table 2.** List of fitting parameters for negative magnetization of M–T curves in FC mode as shown in Figure 6a,b.

| Compound | External Field (Oe) | $H_I$ (Oe) | $\theta$ (K) |
|---|---|---|---|
| x = 0.4 | 100 | −301.81 | −24.43 |
| x = 0.5 | 100 | −360.70 | −24.59 |

Figure 7a–f express the inverse of magnetic susceptibility ($\chi^{-1}$) versus temperature (T) plots of GECMO compounds measured at 100 Oe applied magnetic field. In the high-temperature region, the inverse of magnetic susceptibility obeys the Curie–Weiss law $\chi = C/(T-\theta)$, where C is the Curie constant and $\theta$ is the Weiss temperature. The solid red line shows the fitted curve using the Curie–Weiss law in the paramagnetic region above the transition temperature. The experimental value of the effective magnetic moment was calculated in the unit of Bohr magneton ($\mu_B$) using the fitted Curie constant (C) value by the following equation [28]:

$$\mu_{eff}^{exp.} = \sqrt{\frac{3K_B C}{N}} = \sqrt{8c} \quad (2)$$

where N is the Avogadro number and $k_B$ is the Boltzmann constant. The theoretical effective magnetic moment value was also estimated using the free ionic moments of $Gd^{3+}$ (7.9 $\mu_B$), $Er^{3+}$ (9.59 $\mu_B$), $Cr^{3+}$ (3.87 $\mu_B$) and $Mn^{3+}$ (4.90 $\mu_B$) by the following equation:

$$\mu_{eff}^{Th.} = \sqrt{(0.5)\left(\mu_{eff}^{Gd}\right)^2 + (0.5)\left(\mu_{eff}^{Er}\right)^2 + (1-x)\left(\mu_{eff}^{Cr}\right)^2 + x\left(\mu_{eff}^{Mn}\right)^2} \quad (3)$$

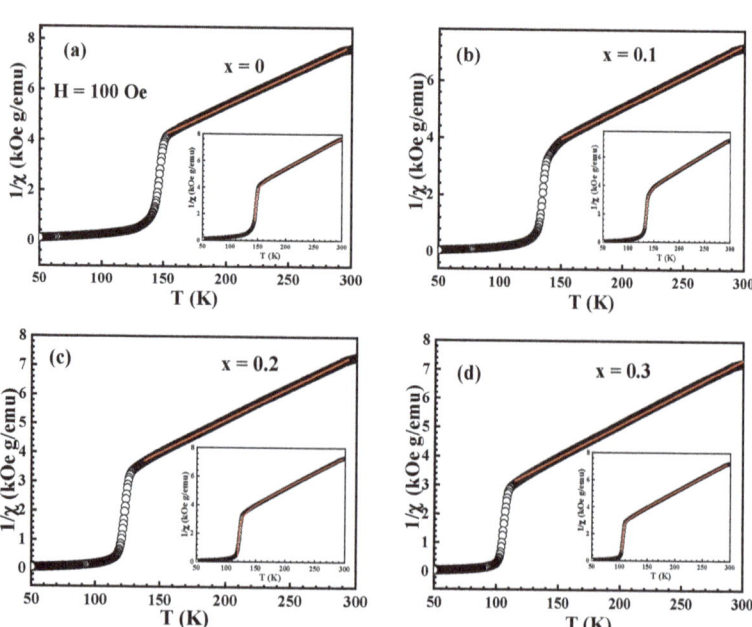

**Figure 7.** Cont.

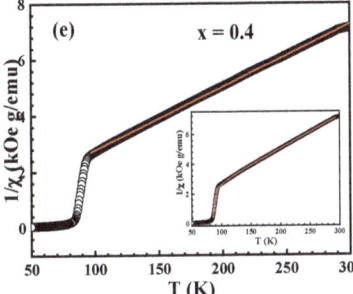
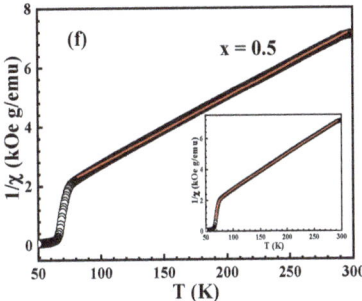

**Figure 7.** (a–f): Inverse magnetic susceptibility versus temperature plots of GECMO series measured at 100 Oe applied magnetic field. Open circles represent the experimental data points whereas solid lines of plot display the fitting to the Curie–Weiss law. Insets represent the fitting of inverse magnetic susceptibility versus temperature using Equation (4).

The value of fitted parameters ($C$ and $\theta$) and effective magnetic moment (calculated and experimental) are summarized in Table 3 for all compounds. Here, a negative value of $\theta$ suggests that the interaction is antiferromagnetic [29]. It is quite satisfactory that both the values of effective magnetic moment (calculated and experimental) match fairly well with each other.

**Table 3.** Magnetic parameters: Curie constant C (emu-K/Oe-mole), Weiss temperature $\theta$ (K), experimental and theoretical effective magnetic moment obtained from Curie–Weiss fitting, and $T_N^{Cr'}$ (K), $\theta'$ (K), C' (emu-K/Oe-mole), fitting parameter $T_0$ (K) and symmetric and antisymmetric exchange constant obtained from modified Curie–Weiss fitting of inverse susceptibility vs. temperature data.

| Compound | x = 0 | x = 0.1 | x = 0.2 | x = 0.3 | x = 0.4 | x = 0.5 |
|---|---|---|---|---|---|---|
| $\theta$ (K) | −26.59 | −28.57 | −26.29 | −24.39 | −23.73 | −21.48 |
| C(emu-K/Oe-mole) | 10.99 | 11.80 | 11.56 | 11.59 | 11.76 | 11.70 |
| $\mu_{eff}^{exp.}$ ($\mu_B$) | 9.38 | 9.72 | 9.62 | 9.63 | 9.70 | 9.68 |
| $\mu_{eff}^{th.}$ ($\mu_B$) | 9.60 | 9.65 | 9.69 | 9.74 | 9.79 | 9.83 |
| $T_N^{Cr'}$ (K) | 144.28 | 132.77 | 119.10 | 103.91 | 86.02 | 68.01 |
| $T_0$ (K) | 143.03 | 131.33 | 117.13 | 102.68 | 84.23 | 66.71 |
| $\theta'$ (K) | −53.42 | −48.98 | −51.07 | −36.70 | −36.8 | −28.34 |
| C'(emu-K/Oe-mole) | 11.91 | 12.50 | 12.41 | 12.00 | 12.19 | 11.91 |
| $J_e/k_B$ | 9.54 | 8.76 | 7.81 | 6.85 | 5.62 | 4.53 |
| $D_e/k_B$ | 2.53 | 2.60 | 2.88 | 2.13 | 2.33 | 1.80 |

As is apparent from Figure 7a–f, the $1/\chi$ versus T data fit well with the Curie–Weiss law above the transition temperature ($T_N$) and it deviates close to $T_N$ as there is an abrupt drop in $1/\chi$ with temperature reduction. The reason for this is that our samples were in a canted antiferromagnetic state instead of simple antiferromagnetic. Such a deviation from the Curie–Weiss behavior can be accounted for by the antisymmetric exchange (DM) interaction [30]. Thus, the susceptibility was fitted by the modified Curie–Weiss law obtained from the DM interaction:

$$\chi = \frac{C'}{(T-\theta')} \frac{(T-T_0)}{\left(T-T_N^{Cr'}\right)}, \qquad (4)$$

where $T_0$ and $T_N^{Cr'}$ are the fitted parameters, which can be calculated by the formula given by Moriya [30]:

$$T_0 = \frac{2J_e\ ZS(S+1)}{3k_B}, \qquad (5)$$

$$T_N^{Cr} = T_0 \left[1 + \left(\frac{D_e}{2 J_e}\right)^2\right]^{1/2} \qquad (6)$$

where $Z$ is the coordination number of $Cr^{3+}$ ions, $S$ is the spin quantum number of $Cr^{3+}$ ions, and $J_e$ and $D_e$ are the strength of symmetric and antisymmetric exchange interactions between $Cr^{3+}$ and $Cr^{3+}$ ions, respectively. The data fitted using Equation (4) are shown as the insets of Figure 7. The fitted parameters $C'$, $\theta'$, $T_0$ and $T_N^{Cr}$ are summarized in Table 3. The non-zero values of $D_e$ imply the presence of a DM interaction in these compounds.

### 3.3. Magnetocaloric Study

With the aim of exploring the efficiency of the materials for magnetic refrigeration applications, the isothermal magnetization $M - H$ curves of the GECMO samples were obtained in the first quadrant with an applied field of up to 8 T. Figure 8a–f show the $M - H$ curves obtained from 5 K to 170 K with an interval of 5 K for GECMO samples. There are two criteria that are used to characterize the magnetocaloric effect: the first is the magnetic entropy change ($\Delta S$) and the other is the relative cooling power (RCP). The magnetic entropy change was calculated by using the following Maxwell's relation [31]:

$$\Delta S(T)_{\Delta H} = \int_{H_I}^{H_F} \left(\frac{\partial M(T, H)}{H}\right) dH \qquad (7)$$

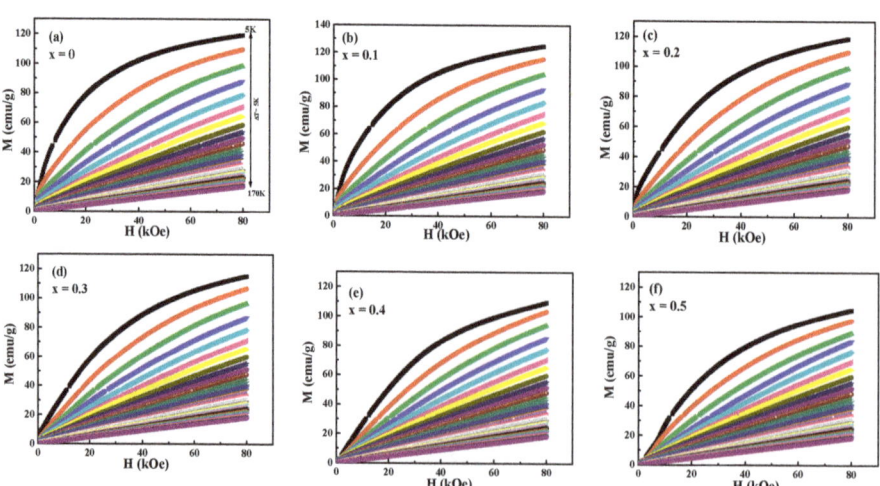

Figure 8. (a–f): Isothermal magnetization M–H curves of the GECMO series.

Figure 9a–f display the magnetic entropy change calculated from Equation (7) as a function of temperature under various applied field strengths. It can be noticed that the value of $-\Delta S$ increased rapidly with a decrease in temperature. The maximum value of $-\Delta S$ is 26.78 J Kg$^{-1}$K$^{-1}$ at 7.5 K and at 8 T applied field for the $x = 0$ compound. This is comparable to the $Er_{0.33}Gd_{0.67}CrO_3$ compound [19] and higher than other compounds such as $DyCrO_3$, $HoCrO_3$, $ErCrO_3$, etc., and dysprosium acetate tetrahydrate [32]. Furthermore, with the Mn doping in the pristine compound, the value of $-\Delta S$ decreased. RCP was evaluated by the following equation

$$RCP = \int_{T_1}^{T_2} \Delta S(T)_{\Delta H} dT \qquad (8)$$

where $T_1$ and $T_2$ are the lower and higher temperature of the refrigeration cycle ($T_1$ = 7.5 K and $T_2$ = 172.5 K in this study), respectively. A greater RCP value generally denotes a better magnetocaloric compound because of its high cooling efficiency. As displayed in the inset of Figure 9a–f, the value of RCP increased with the increase in applied field. The values of $-\Delta S$ and RCP for the GECMO series are listed in Table 4 for different applied magnetic field values.

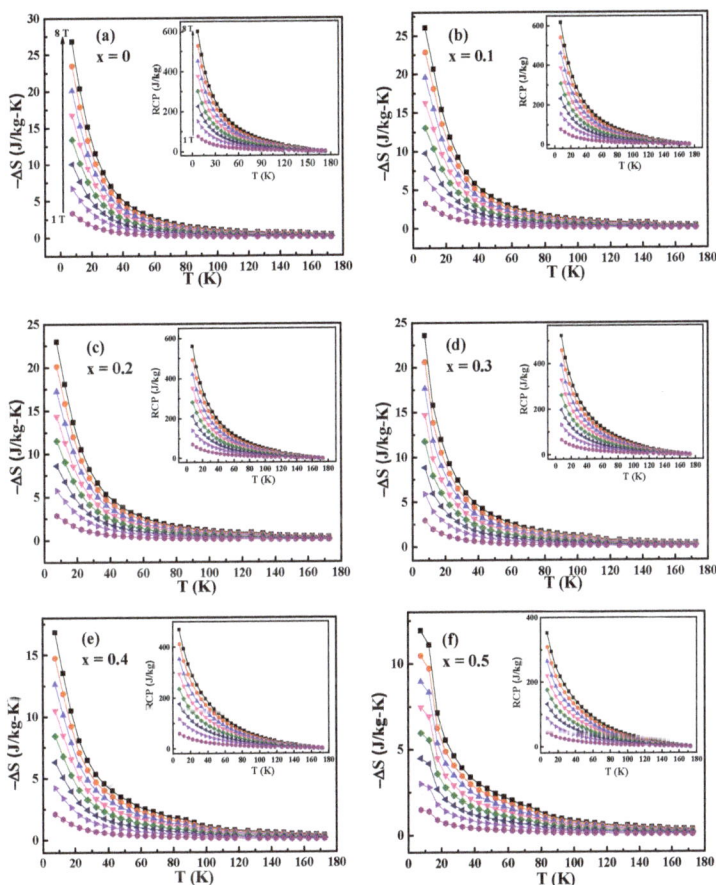

**Figure 9.** (a–f): The temperature dependence of magnetic entropy change ($-\Delta S$) and relative cooling power (inset) of GECMO series in magnetic fields up to 8 T.

**Table 4.** Magnetic entropy change ($-\Delta S$) and relative cooling power (RCP) at different applied magnetic fields for the $Gd_{0.5}Er_{0.5}Cr_{1-x}Mn_xO_3$ series.

| Compound | x = 0 | | x = 0.1 | | x = 0.2 | | x = 0.3 | | x = 0.4 | | x = 0.5 | |
|---|---|---|---|---|---|---|---|---|---|---|---|---|
| $H_{Max.}$ (T) | $-\Delta S_{Max.}$ (J/kg-K) | RCP (J/kg) | $-\Delta S_{Max.}$ (J/kg-K) | RCP (J/kg) | $-\Delta S_{Max.}$ (J/kg-K) | RCP (J/kg) | $-\Delta S_{Max.}$ (J/kg-K) | RCP (J/kg) | $-\Delta S_{Max.}$ (J/kg-K) | RCP (J/kg) | $-\Delta S_{Max.}$ (J/kg-K) | RCP (J/kg) |
| 1 | 3.34 | 75.18 | 3.25 | 77.19 | 2.87 | 70.15 | 2.94 | 65.52 | 2.10 | 58.73 | 1.49 | 44.04 |
| 2 | 6.69 | 150.37 | 6.51 | 154.38 | 5.74 | 140.30 | 5.88 | 131.04 | 4.20 | 117.46 | 2.98 | 88.09 |
| 3 | 10.04 | 225.56 | 9.77 | 231.57 | 8.62 | 210.46 | 8.82 | 196.56 | 6.31 | 176.20 | 4.47 | 132.13 |
| 4 | 13.39 | 300.75 | 13.02 | 308.76 | 11.49 | 280.61 | 11.76 | 262.08 | 8.41 | 234.93 | 5.96 | 176.18 |
| 5 | 16.74 | 375.94 | 16.28 | 385.95 | 14.36 | 350.76 | 14.70 | 327.61 | 10.52 | 293.67 | 7.46 | 220.22 |
| 6 | 20.09 | 451.13 | 19.54 | 463.14 | 17.24 | 420.92 | 17.64 | 393.13 | 12.62 | 352.40 | 8.95 | 264.27 |
| 7 | 23.44 | 526.32 | 22.79 | 540.33 | 20.11 | 491.07 | 20.58 | 458.65 | 14.72 | 411.13 | 10.44 | 308.32 |
| 8 | 26.78 | 601.51 | 26.05 | 617.52 | 22.99 | 561.23 | 23.53 | 524.17 | 16.83 | 469.87 | 11.93 | 352.36 |

To gain a better understanding of the magnetocaloric effect, we further explored the magnetic entropy change ($-\Delta S$) in view of the magneto-elastic coupling and electron interaction according to the Landau theory of phase transition [33,34]. According to this, the Gibbs free energy can be defined as:

$$G(T, M) = \frac{1}{2}AM^2 + \frac{1}{4}BM^4 + \frac{1}{6}CM^6 - MH \qquad (9)$$

where $A$, $B$ and $C$ are the temperature-dependent parameters (Landau coefficients). At equilibrium, energy should be minimized, i.e., $\frac{\partial G}{\partial M} = 0$, and Equation (9) gives

$$\frac{H}{M} = A + BM^2 + CM^4 \qquad (10)$$

The Landau coefficients were calculated from the Arrott's plots ($H/M$ vs. $M^2$) by the best fit method. Magnetic entropy change was obtained by differentiating Gibbs free energy with respect to temperature:

$$\Delta S_M(T, H) = -\frac{1}{2}\frac{\partial A}{\partial T}M^2 - \frac{1}{4}\frac{\partial B}{\partial T}M^4 - \frac{1}{6}\frac{\partial C}{\partial T}M^6 \qquad (11)$$

The magnetic entropy change calculated by Landau theory was compared with that calculated by Maxwell's thermodynamic relations under 5 T magnetic field and the comparison is shown in Figure 10a–f. The variation in the Landau coefficients is also shown in the inset of each plot. It can be seen from the plots that the magnetic entropy change versus temperature behavior is the same according to different models.

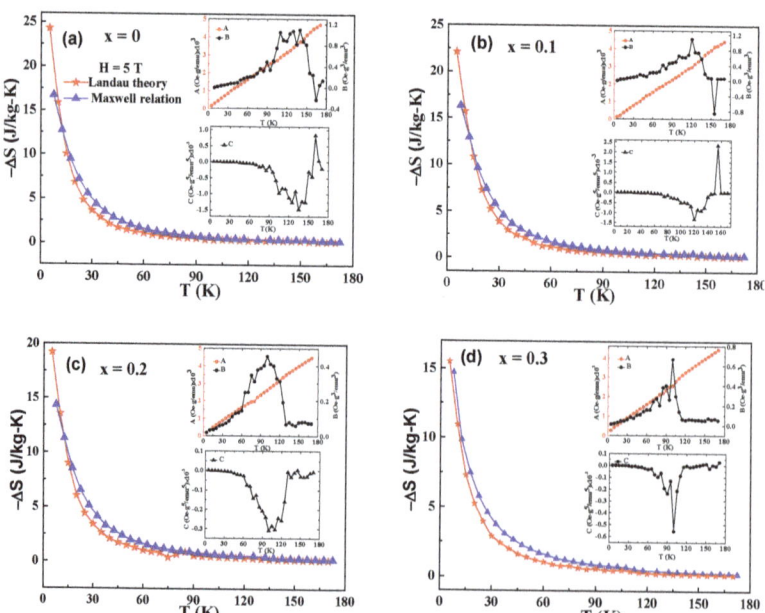

Figure 10. Cont.

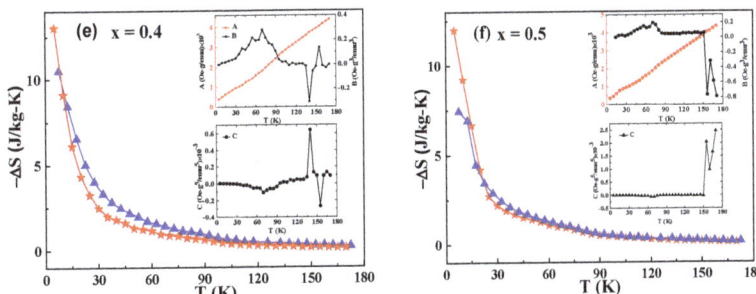

**Figure 10.** (a–f): Temperature dependence of magnetic entropy change (−ΔS) of GECMO series calculated using Landau theory and Maxwell's relation. Insets show the variation in Landau coefficients.

## 4. Conclusions

In this paper, we studied $Gd_{0.5}Er_{0.5}Cr_{1-x}Mn_xO_3$ ($x$ = 0, 0.1, 0.2, 0.3, 0.4, 0.5) compounds for their low-temperature magnetic and magnetocaloric properties. The $M$–$T$ measurements indicate that the transition temperature for the pristine sample was ~144 K, which decreased with Mn substitution. We also observed negative magnetization at lower magnetic fields in compounds with $x$ = 0.4 and 0.5 Mn concentrations. The existence of the magnetization reversal offers the characteristic switching of magnetization. The magnetocaloric parameters such as magnetic entropy change (−ΔS) varied from 16.74 J/kg-K to 7.46 J/kg-K and relative cooling power (RCP) from 375.94 J/kg to 220.22 J/kg with Mn concentration under a 5 T field change. These values are significant for these materials to be useful as low-temperature magnetic refrigerants, especially below 10 K.

**Author Contributions:** Conceptualization, N.P.; methodology, N.P.; validation, N.P.; formal analysis, K.S.; investigation, K.S., N.P., Y.B.; resources, S.K.; writing—original draft preparation K.S., K.K., V.S.P.; original draft correct N.P., Y.B. All authors have read and agreed to the published version of the manuscript.

**Funding:** The research was funded by DST SERB, New Delhi, India through grant No. ECR/2017/002681.

**Institutional Review Board Statement:** Not applicable.

**Informed Consent Statement:** Not applicable.

**Data Availability Statement:** Not applicable.

**Acknowledgments:** The authors are thankful to the low-temperature high-magnetic-field facility of CURAJ for the magnetic measurements. Neeraj Panwar would like to thank DST SERB, New Delhi, UGC DAE CSR Mumbai and IUAC New Delhi for the grants through Projects ECR/2017/002681, CRS-M-298 and UFR-62317, respectively. V.S.P. acknowledges the National Research Council (NRC) senior research associate fellowship program.

**Conflicts of Interest:** The authors declare no conflict of interest.

## References

1. Ateia, E.E.; Ismail, H.; Elshimy, H.; Abdelmaksoud, M.K. Structural and Magnetic Tuning of LaFeO3 Orthoferrite Substituted Different Rare Earth Elements to Optimize Their Technological Applications. *J. Inorg. Organomet. Polym. Mater.* **2021**, *31*, 1713–1725. [CrossRef]
2. Chakraborty, P.; Basu, S. Structural, electrical and magnetic properties of Er doped YCrO3 nanoparticles. *Mater. Chem. Phys.* **2021**, *259*, 124053. [CrossRef]
3. Yoshii, K. Magnetic properties of perovskite GdCrO3. *J. Solid State Chem.* **2001**, *159*, 204–208. [CrossRef]
4. Aamir, M.; Bibi, I.; Ata, S.; Majid, F.; Kamal, S.; Alwadai, N.; Sultan, M.; Iqbal, S.; Aadil, M.; Iqbal, M. Graphene oxide nanocomposite with Co and Fe doped LaCrO3 perovskite active under solar light irradiation for the enhanced degradation of crystal violet dye. *J. Mol. Liq.* **2021**, *322*, 114895. [CrossRef]
5. Wang, L.; Wang, S.W.; Zhang, X.; Zhang, L.L.; Yao, R.; Rao, G.H. Reversals of magnetization and exchange-bias in perovskite chromite YbCrO3. *J. Alloys Compd.* **2016**, *662*, 268–271. [CrossRef]

6. Hornreich, R.M. Magnetic interactions and weak ferromagnetism in the rare-earth orthochromites. *J. Magn. Magn. Mater.* **1978**, *7*, 280–285. [CrossRef]
7. Su, Y.; Zhang, J.; Feng, Z.; Li, L.; Li, B.; Zhou, Y.; Chen, Z.; Cao, S. Magnetization reversal and $Yb^{3+}/Cr^{3+}$ spin ordering at low temperature for perovskite $YbCrO_3$ chromites. *J. Appl. Phys.* **2010**, *108*, 013905. [CrossRef]
8. Cao, Y.; Cao, S.; Ren, W.; Feng, Z.; Yuan, S.; Kang, B.; Lu, B.; Zhang, J. Magnetization switching of rare earth orthochromite $CeCrO_3$. *Appl. Phys. Lett.* **2014**, *104*, 232405. [CrossRef]
9. Panwar, N.; Joby, J.P.; Kumar, S.; Coondoo, I.; Vashundhara, M.; Kumar, N.; Palai, R.; Singhal, R.; Katiyar, R.S. Observation of magnetization reversal behavior in $Sm_{0.9}Gd_{0.1}Cr_{0.85}Mn_{0.15}O_3$ orthochromites. *AIP Adv.* **2018**, *8*, 055818. [CrossRef]
10. Yin, S.; Sharma, V.; McDannald, A.; Reboredo, F.A.; Jain, M. Magnetic and magnetocaloric properties of iron substituted holmium chromite and dysprosium chromite. *RSC Adv.* **2016**, *6*, 9475–9483. [CrossRef]
11. Sharma, M.K.; Mukherjee, K. Magnetic and universal magnetocaloric behavior of rare-earth substituted $DyFe_{0.5}Cr_{0.5}O_3$. *J. Magn. Magn. Mater.* **2017**, *444*, 178–183. [CrossRef]
12. Yin, S.; Seehra, M.S.; Guild, C.J.; Suib, S.L.; Poudel, N.; Lorenz, B.; Jain, M. Magnetic and magnetocaloric properties of $HoCrO_3$ tuned by selective rare-earth doping. *Phys. Rev. B* **2017**, *95*, 184421. [CrossRef]
13. Zhang, H.; Qian, H.; Xie, L.; Guo, Y.; Liu, Y.; He, X. The spin reorientation and improvement of magnetocaloric effect in $HoCr_{1-x}Ga_xO_3$ ($0 \leq x \leq 0.5$). *J. Alloys Compd.* **2021**, *885*, 160863. [CrossRef]
14. Yoshii, K. Positive exchange bias from magnetization reversal in $La_{1-x}Pr_xCrO_3$ ($x \sim 0.7$–$0.85$). *Appl. Phys. Lett.* **2011**, *99*, 142501. [CrossRef]
15. Oliveira, G.N.P.; Machado, P.; Pires, A.L.; Pereira, A.M.; Araújo, J.P.; Lopes, A.M.L. Magnetocaloric effect and refrigerant capacity in polycrystalline $YCrO_3$. *J. Phys. Chem. Solids* **2016**, *91*, 182–188. [CrossRef]
16. Oliveira, G.N.P.; Pires, A.L.; Machado, P.; Pereira, A.M.; Araújo, J.P.; Lopes, A.M.L. Effect of chemical pressure on the magnetocaloric effect of perovskite-like $RCrO_3$ (R-Yb, Er, Sm and Y). *J. Alloys Compd.* **2019**, *797*, 269–276. [CrossRef]
17. Kumar, S.; Coondoo, I.; Vasundhara, M.; Kumar, S.; Kholkin, A.L.; Panwar, N. Structural, magnetic, magnetocaloric and specific heat investigations on Mn doped $PrCrO_3$ orthochromites. *J. Phys. Condens. Matter* **2017**, *29*, 195802. [CrossRef]
18. Yin, L.H.; Yang, J.; Kan, X.C.; Song, W.H.; Dai, J.M.; Sun, Y.P. Giant magnetocaloric effect and temperature induced magnetization jump in $GdCrO_3$ single crystal. *J. Appl. Phys.* **2015**, *117*, 133901. [CrossRef]
19. Shi, J.; Yin, S.; Seehra, M.S.; Jain, M. Enhancement in magnetocaloric properties of $ErCrO_3$ via A-site Gd substitution. *J. Appl. Phys.* **2018**, *123*, 193901. [CrossRef]
20. Kanwar, K.; Coondoo, I.; Anas, M.; Malik, V.K.; Kumar, P.; Kumar, S.; Kulriya, P.K.; Kaushik, S.D.; Panwar, N. A comparative study of the structural, optical, magnetic and magnetocaloric properties of $HoCrO_3$ and $HoCr_{0.85}Mn_{0.15}O_3$ orthochromites. *Ceram. Int.* **2021**, *47*, 7386–7397. [CrossRef]
21. Chan, T.S.; Liu, R.S.; Yang, C.C.; Li, W.-H.; Lien, Y.H.; Huang, C.Y.; Lee, J.F. Chemical Size Effect on the Magnetic and Electrical Properties in the $(Tb_{1-x}Eu_x)MnO_3$ ($0 \leq x \leq 1.0$) System. *J. Phys. Chem. B* **2007**, *111*, 2262–2267. [CrossRef] [PubMed]
22. Zhao, Y.; Weidner, D.J.; Parise, J.B.; Cox, D.E. Thermal expansion and structural distortion of perovskite-data for $NaMgF_3$ perovskite. Part I. *Phys. Earth Planet. Inter.* **1993**, *76*, 1–16. [CrossRef]
23. Søndenå, R.; Stølen, S.; Ravindran, P.; Grande, T.; Allan, N.L. Corner-versus face-sharing octahedra in $AMnO_3$ perovskites (A = Ca, Sr, and Ba). *Phys. Rev. B* **2007**, *75*, 184105. [CrossRef]
24. Kim, K.; Siegel, D.J. Correlating lattice distortions, ion migration barriers, and stability in solid electrolytes. *J. Mater. Chem. A* **2019**, *7*, 3216–3227. [CrossRef]
25. Mahana, S.; Pandey, S.K.; Rakshit, B.; Nandi, P.; Basu, R.; Dhara, S.; Turuchini, S.; Zema, N.; Manju, U.; Mahanti, S.D.; et al. Site substitution in $GdMnO_3$: Effects on structural, electronic, and magnetic properties. *Phys. Rev. B* **2020**, *102*, 245120. [CrossRef]
26. Chiang, F.K.; Chu, M.W.; Chou, F.C.; Jeng, H.T.; Sheu, H.S.; Chen, F.R.; Chen, C.H. Effect of Jahn-Teller distortion on magnetic ordering in Dy(Fe,Mn)$O_3$ perovskites. *Phys. Rev. B* **2011**, *83*, 245105. [CrossRef]
27. Cooke, A.H.; Martin, D.M.; Wells, M.R. Magnetic interactions in gadolinium orthochromite, $GdCrO_3$. *J. Phys. C Solid State Phys.* **1974**, *7*, 3133. [CrossRef]
28. Kumar, D.; Jena, P.; Singh, A.K. Structural, magnetic and dielectric studies on half-doped $Nd_{0.5}Ba_{0.5}CoO_3$ perovskite. *J. Magn. Magn. Mater.* **2020**, *516*, 167330. [CrossRef]
29. Deng, D.; Wang, X.; Zheng, J.; Qian, X.; Yu, D.; Sun, D.; Jing, C.; Lu, B.; Kang, B.; Cao, S.; et al. Phase separation and exchange bias effect in Ca doped $EuCrO_3$. *J. Magn. Magn. Mater.* **2015**, *395*, 283–288. [CrossRef]
30. Moriya, T. Anisotropic Superexchange Interaction and Weak Ferromagnetism. *Phys. Rev.* **1960**, *120*, 91–98. [CrossRef]
31. Tishin, A.M. Magnetocaloric Effect in the Vicinity of Phase Transitions. *Handb. Magn. Mater.* **1999**, *12*, 395–524. [CrossRef]
32. Lorusso, G.; Roubeau, O.; Evangelisti, M. Rotating Magnetocaloric Effect in an Anisotropic Molecular Dimer. *Angew. Chem. Int. Ed.* **2016**, *55*, 3360–3363. [CrossRef] [PubMed]
33. Landau, L.D.; Lifshitz, E.M. *Statistical Physics*; Pergamon: New York, NY, USA, 1958.
34. Amaral, V.S.; Amaral, J.S. Magnetoelastic coupling influence on the magnetocaloric effect in ferromagnetic materials. *J. Magn. Magn. Mater.* **2004**, *276*, 2104–2105. [CrossRef]

Article

# Photoluminescence of the Eu$^{3+}$-Activated Y$_x$Lu$_{1-x}$NbO$_4$ (x = 0, 0.25, 0.5, 0.75, 1) Solid-Solution Phosphors

Milica Sekulić, Tatjana Dramićanin, Aleksandar Ćirić, Ljubica Đačanin Far, Miroslav D. Dramićanin and Vesna Đorđević *

Center of Excellence for Photoconversion, Vinča Institute of Nuclear Sciences—National Institute of the Republic of Serbia, University of Belgrade, P.O. Box 522 Belgrade, Serbia; milicasekulic88@gmail.com (M.S.); tatjana@vin.bg.ac.rs (T.D.); aleksandar.ciric@ff.bg.ac.rs (A.Ć.); ljubica.far@vin.bg.ac.rs (L.Đ.F.); dramican@vinca.rs (M.D.D.)
* Correspondence: vesipka@vinca.rs; Tel.: +381-11-340-8191

**Abstract:** Eu$^{3+}$-doped Y$_x$Lu$_{1-x}$NbO$_4$ (x = 0, 0.25, 0.5, 0.75, 1) were prepared by the solid-state reaction method. YNbO$_4$:Eu$^{3+}$ and LuNbO$_4$:Eu$^{3+}$ crystallize as beta-Fergusonite (SG no. 15) in 1–10 μm diameter particles. Photoluminescence emission spectra show a slight linear variation of emission energies and intensities with the solid-solution composition in terms of Y/Lu content. The energy difference between Stark sublevels of $^5D_0 \rightarrow ^7F_1$ emission increases, while the asymmetry ratio decreases with the composition. From the dispersion relations of pure YNbO$_4$ and LuNbO$_4$, the refractive index values for each concentration and emission wavelength are estimated. The $\Omega_2$ Judd–Ofelt parameter shows a linear increase from 6.75 to 7.48 × 10$^{-20}$ cm$^2$ from x = 0 to 1, respectively, and $\Omega_4$ from 2.69 to 2.95 × 10$^{-20}$ cm$^2$. The lowest non-radiative deexcitation rate was observed with x = 1, and thus LuNbO$_4$:Eu$^{3+}$ is more efficient phosphor than YNbO$_4$:Eu$^{3+}$.

**Keywords:** phosphor; Eu$^{3+}$; niobates; Judd-Ofelt; quantum efficiency; refractive index

Citation: Sekulić, M.; Dramićanin, T.; Ćirić, A.; Đačanin Far, L.; Dramićanin, M.D.; Đorđević, V. Photoluminescence of the Eu$^{3+}$-Activated Y$_x$Lu$_{1-x}$NbO$_4$ (x = 0, 0.25, 0.5, 0.75, 1) Solid-Solution Phosphors. *Crystals* 2022, 12, 427. https://doi.org/10.3390/cryst12030427

Academic Editors: Alessandra Toncelli and Željka Antić

Received: 18 February 2022
Accepted: 15 March 2022
Published: 19 March 2022

**Publisher's Note:** MDPI stays neutral with regard to jurisdictional claims in published maps and institutional affiliations.

**Copyright:** © 2022 by the authors. Licensee MDPI, Basel, Switzerland. This article is an open access article distributed under the terms and conditions of the Creative Commons Attribution (CC BY) license (https:// creativecommons.org/licenses/by/ 4.0/).

## 1. Introduction

Phosphors designed with rare earth (RE)-activated compounds are a continuously rising research topic in both basic and applied science. Materials containing the trivalent europium ion (Eu$^{3+}$) are well known for their strong luminescence in the orange/red spectral region, making them suitable for use in artificial lights, display devices, luminescent sensing, and biomedical research, among other applications [1–4]. In addition, the Eu$^{3+}$ ion is a truly unique spectroscopic probe from a theoretical point of view. Because the 4f shell of Eu$^{3+}$ has an even number of electrons ([Xe]4f$^6$), the ion exhibits a non-degenerated ground ($^7F_0$) and excited ($^5D_0$) energy states, as well as non-overlapping $^{2S+1}L_J$ multiplets, resulting in emission spectra that are predictably dependent on the host material site symmetry. The energy level structure, the intensities of the f-f transitions (including the Judd–Ofelt theory), and the decay times of the excited states allow the Eu$^{3+}$ ion to be used as a spectroscopic probe [5–8].

Compounds with the ABO$_4$ composition (A = RE; B = P, V, Nb) are suggested to be excellent hosts for luminescent materials in the vast field of phosphors [9–11] due to the coupling between rare-earth (RE$^{3+}$) ions and the BO$_4$$^{3-}$ group. RE-Niobates (RENbO$_4$) have long been known, but have received far less attention than vanadates or phosphates. They are chemically stable and have good dielectric properties, as well as ion and proton conductivities [12–15]. They have been prepared as single crystals [16,17], thin films [18], and in crystalline power form [19,20] for a variety of potential applications, primarily as optical hosts. Under UV or x-ray excitation, YNbO$_4$ is a self-activated phosphor with a broad and strong emission band in the blue spectral region around 400 nm (associated with NbO$_4$$^{3-}$ groups from the host crystalline lattice) [21,22]. RE-Niobates' luminescent properties can be altered by doping with various rare-earth ions. Emission has been observed from: $Tm^{3+}$, $Tb^{3+}$, $Dy^{3+}$, $Sm^{3+}$, $Er^{3+}$, $Nd^{3+}$ [23–28], and $Eu^{3+}$ [20,29]. Recent applications of RE-doped niobates involved luminescent temperature sensors [27,30].

The $RENbO_4$ structure type has a more complex profile than the $REVO_4$ materials, which always occurs in the typical tetragonal zircon structure type. $YNbO_4$ was originally thought to be the parent material of the $RENbO_4$ group, like the natural mineral fergusonite, leading to the classification of the material as a tetragonal phase with a scheelite (I41/a) structure [31]. However, there are two major crystalline forms of $RENbO_4$. One is the low-temperature M-phase isostructure with a monoclinic form of the fergusonite, and the other is the high-temperature T-phase corresponding to the tetragonal scheelite. Between 500 and 850 °C, the reversible transition between two phases, monoclinic and tetragonal, occurs [32]. Subsequent research revealed that the low-temperature phase, monoclinic beta-Fergusonite structure can be described by the space group I-centered (I12/a1) or C-centered (C12/c1), as the I2/a space group is a non-standard setting of C2/c (SG no. 15) [33–35]. In both settings, Y is surrounded by 8 oxygens in a large, low-symmetry $YO_8$ octahedron [34]. When other rare earth ions, such as $Eu^{3+}$, occupy the Y site, the tilting of adjacent $NbO_4$ and $NbO_6$ polyhedra is expected to influence the luminescence properties of the dopant.

In this study, we wanted to investigate the effect of different ionic radii of $RE^{3+}$ ions on the photoluminescence of the $Eu^{3+}$-activated niobate host. We prepared a set of five Eu-doped $Y_xLu_{1-x}NbO_4$:Eu samples (x = 0, 0.25, 0.5, 0.75, and 1) with a fixed Eu concentration (5%) to investigate the influence of the Y to Lu ratio in the host niobate material on $Eu^{3+}$ luminescence features. In that sense, the crystal field splitting of $^7F_1$ manifold, R intensity ratio and Judd–Ofelt parameters were determined. The application of Judd–Ofelt theory was explained, and the difference in the refractive index of the materials was considered.

## 2. Materials and Methods

The set of five samples, Eu-doped $Y_xLu_{1-x}NbO_4$:$Eu^{3+}$ (x = 0, 0.25, 0.5, 0.75, and 1), were prepared by the solid-state reaction method. In the stoichiometric amounts of starting materials, Yttirium(III) oxide ($Y_2O_3$ Alfa Aesar, 99.9%), Lutetium(III) oxide ($Lu_2O_3$, Alfa Aeser 99.9%), and Niobium(V) oxide ($Nb_2O_5$ Alfa Aesar, 99.5%) were mixed with Europium oxide ($Eu_2O_3$ Alfa Aesar, 99.9%) added in order for Y or Lu to reach 95% (i.e., $Y_{0.95}Eu_{0.05}NbO_4$). With the addition of sodium sulphate decahydrate ($Na_2SO_4$ x $10H_2O$, Alfa Aesar, 99%) as flux and small amount of ethanol, the mixtures were homogenized in a ball mill (BM500, Anton Paar) for several hours. The dried mixture was pre-sintered at 800 °C for 2 h, then sintered at 1450 °C for 8 h and allowed to cool to room temperature. Such obtained powders or pellets prepared from the powder under a load of $5 \times 10^8$ Pa were used for measurements.

X-ray diffraction measurements were performed with a Rigaku SmartLab diffractometer (Tokyo, Japan) using Cu Kα radiation (30 mA, 40 kV) measured in the $2\theta$ range from 10° to 90°. The built-in package software was used to obtain relevant structural analysis results (crystal coherence size, microstrain values, unit cell parameters, and data fit parameters). A Mira3 Tescan field emission scanning electron microscope (FE-SEM) (Brno, Czech Republic) was used for microstructural characterization, operated at 20 keV and 5.00 kx magnification. Photoluminescence excitation spectra were measured by a Horiba Jobin Yvon Fluorolog FL3-221 spectrofluorometer (Palaiseau, France) equipped with a 450 W Xenon lamp, TBX-04-D PMT detector and a double-grating monochromator with 1200 g/mm. The excitations were performed by monitoring the emission maxima at 612 nm. Lifetime measurements were performed using the same instrument equipped with a xenon–mercury pulsed lamp. Photoluminescence emission spectra were recorded by a high-resolution spectrograph (FHR 1000) (Palaiseau, France) equipped with a Horiba Jobin–Yvon Intensified Charge Coupled Device–ICCD detector and selectable diffraction gratings of 300 and 1800 g/mm. Samples were excited by a 365 nm LED (Ocean Optics) and driven by an Ocean Insight LDC-1C controller.

## 3. Results

### 3.1. Crystal Structure and Morphology

X-ray diffraction measurements confirmed that YNbO$_4$:Eu crystallizes in a monoclinic Fergusonite-beta-(Y) structure that can be best fitted with the C2/c space group corresponding to pure YNbO$_4$ ICDD No. 01-083-1319, as can be seen in Figure 1 [36]. The isostructural incorporation of a bigger Eu$^{3+}$ ($r_{VIII}$ = 1.066 Å) ion instead of a Y$^{3+}$ ($r_{VIII}$ = 1.019 Å) ion of YO$_8$ dodecahedra results in a small maximum shift of approximately 0.03° to the lower $2\theta$ values [37]. X-ray diffraction measurements confirmed that LuNbO$_4$:Eu$^{3+}$ crystallizes in a monoclinic Fergusonite-beta-(Lu) structure that can be best fitted with the I2/a space group corresponding to pure LuNbO$_4$ ICDD No. 01-074-6538, as can be seen in Figure 1 [35]. A slight peak position shift of around 0.1° to the lower $2\theta$ values is also observed, resulting from the isostructural incorporation of a larger Eu$^{3+}$ ($r_{VIII}$ = 1.066 Å) ion into the Lu$^{3+}$ ($r_{VIII}$ = 0.977 Å) position of LuO$_8$ dodecahedra [36].

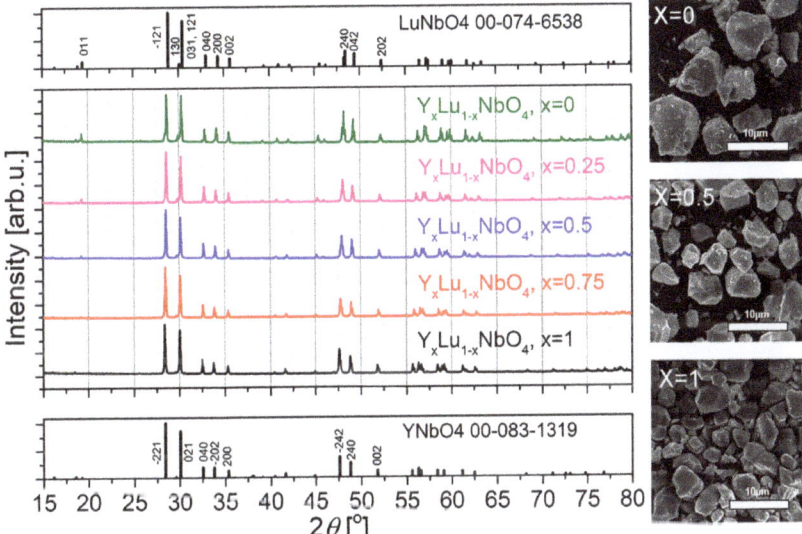

**Figure 1.** Comparison of the powder x-ray diffraction patterns of the Y$_x$Lu$_{1-x}$NbO$_4$:5%Eu (x = 0, 0.25, 0.5, 0.75, 1) samples with representative SEM images of Y$_x$Lu$_{1-x}$NbO$_4$:5%Eu (x = 0, 0.5, 1) with a 10-micron bar.

Table 1 shows the relevant structural characteristics for both final compositions of Y$_x$Lu$_{1-x}$NbO$_4$:Eu$^{3+}$ (x = 0 and 1), determined by Rietveld refinement of the experimental data derived from Rigaku SmartLab built-in package software. The YNbO$_4$:Eu$^{3+}$ (x = 1) parameters are refined to the C2/c space group, with a higher value of parameter a of the unit cell around 7.63 Å and a higher β of 138.44°, whereas LuNbO$_4$:Eu$^{3+}$ (x = 0) parameters are refined to the I2/a space group, with a lower value of parameter a of the unit cell of around 5.24 Å and a lower β of 94.43°. Both samples have crystallite sizes of roughly 50 ± 3 nm and low strain values. The SEM images of Y$_x$Lu$_{1-x}$NbO$_4$:5%Eu (x = 0, 0.5, 1) materials presented in Figure 1 show irregularly shaped particles ranging in size from 1 to 10 microns. The YNbO$_4$:5%Eu material comprises a higher proportion of smaller, densely packed particles, whereas the particles in the LuNbO$_4$:5%Eu material are larger and less densely packed.

**Table 1.** Structural parameters obtained by Rietveld refinement of XRD data for YNbO$_4$:Eu$^{3+}$ according to ICDD 01-083-1319, and for LuNbO$_4$:Eu$^{3+}$ according to ICDD 01-074-6538.

| Ref. Parameters | YNbO$_4$:Eu$^{3+}$ (C2/c) | LuNbO$_4$:Eu$^{3+}$ (I2/a) |
|---|---|---|
| Crystallite size (Å) | 533(6) | 479(6) |
| Strain (%) | 0.08(2) | 0.11(2) |
| a (Å) | 7.62775(17) | 5.23879(13) |
| b (Å) | 10.9636(3) | 10.8447(3) |
| c (Å) | 5.30476(15) | 5.04724(12) |
| α (°) | 90 | 90 |
| β (°) | 138.4389(11) | 94.4267(13) |
| γ (°) | 90 | 90 |
| Rwp [1] | 10.28% | 10.25% |
| Rp [2] | 6.30% | 6.55% |
| Re [3] | 3.23% | 3.32% |
| GOF [4] | 3.1833 | 3.0895 |

[1] Rwp—regression sum of weighted squared errors of fit; [2] Rp—regression sum of relative squared errors of fit; [3] Re—regression sum of relative errors of fit; [4] GOF—goodness of fit parameter.

### 3.2. Photoluminescence

The luminescence excitation and emission spectra of the Eu$^{3+}$ (4$f^6$) ion in Y$_x$Lu$_{1-x}$NbO$_4$ (x = 0, 0.25, 0.5, 0.75, 1) hosts clearly show characteristic $f$-$f$ transitions, as shown in Figure 2. The excitation spectra of all samples, observed in Figure 2a, as recorded at λem = 612 nm, show several dominant excitation bands centered at around 540 nm ($^7F_1 \rightarrow {}^5D_1$), 529 nm ($^7F_0 \rightarrow {}^5D_1$), 466 nm ($^7F_0 \rightarrow {}^5D_0$), 393 nm ($^7F_0 \rightarrow 5L_6$), and 365 nm ($^7F_0 \rightarrow {}^5D_4$). The emission spectra of all samples recorded with λexc = 365 nm excitation are presented in Figure 2b. Following the $^7F_0 \rightarrow {}^5D_4$ excitation, the electrons non-radiatively deexcite to the long-lived $^5D_0$ level, from where the radiative relaxation occurs to the $^7F_J$ ground multiplet. In the recorded region, five dominant emission bands from the $^5D_0$ level are centered at 595 nm ($^5D_0 \rightarrow {}^7F_1$), 612 nm ($^5D_0 \rightarrow {}^7F_2$), 655 nm ($^5D_0 \rightarrow {}^7F_3$), 710 nm ($^5D_0 \rightarrow {}^7F_4$), and 745 nm ($^5D_0 \rightarrow {}^7F_5$). The possible $^5D_0 \rightarrow {}^7F_6$ cannot be recorded by the used experimental setup. As is obvious from Figure 2, the luminescences in all the samples show overlapping bands characteristic for the Eu$^{3+}$ ion. The Stark level energies calculated from the recorded excitation and emission spectra are presented in Table 2.

The intensity of the Eu$^{3+}$ $^5D_0 \rightarrow {}^7F_2$ (ΔJ = 2) forced electric dipole transition is hypersensitive since it is strongly dependent on the coordinating polyhedron local site symmetry. Given that the environment of Eu$^{3+}$ ion is a (Y,Lu)O$_8$ dodecahedron in all samples, small changes can be expected because of the variation of ion positions governed by the different ionic radii of Y and Lu ions [36].

The intensity of the parity-allowed Eu$^{3+}$ $^5D_0 \rightarrow {}^7F_1$ (ΔJ = 1) magnetic dipole transition is considered as independent to the local symmetry; however, maximum splitting of the $^7F_1$ manifold of the Eu$^{3+}$ ion is a function of the host [38]. To observe the influence of the host on the $^7F_1$ manifold, additional measurements were performed using a 1800 g/mm diffraction grating monochromator. To precisely determine the positions of all three maxima, a deconvolution is applied, as presented in Figure 2c.

The positions (Peak energies [cm$^{-1}$]) of the characteristic emission peak maxima of Y$_x$Lu$_{1-x}$NbO$_4$:Eu (x = 0, 0.25, 0.5, 0.75, 1) are presented in Figure 3a. As can be seen, only small differences of around 10–20 cm$^{-1}$ are determined, given the similarity of the Eu$^{3+}$ ion local environment. Trends of maximum ΔE splitting, the Stark splitting of $^7F_1$ manifold, and the asymmetry ratio R corresponding to the Y/Lu content are presented in Figure 3b. As is evident, ΔE increases linearly with the increase of the x value (increase in Y content). For the same site symmetry and coordination number, the positions of the Stark levels of emissions originating from the $^5D_0$ level of Eu$^{3+}$ depend on the covalency of the metal-ligand bond [39].

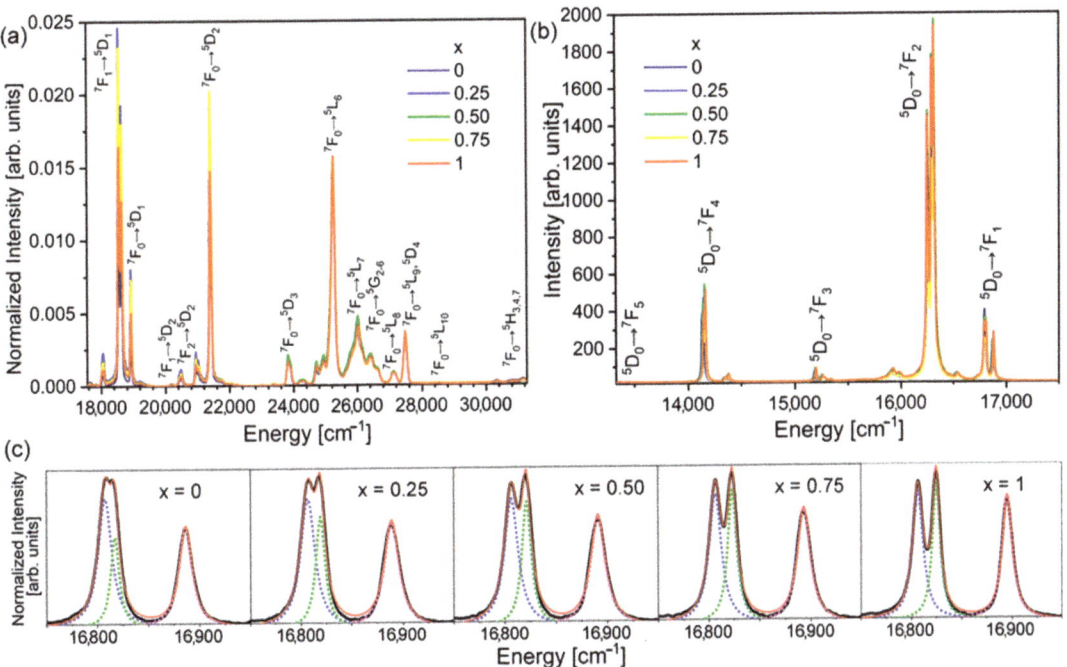

**Figure 2.** Luminescence spectra of $Y_xLu_{1-x}NbO_4:Eu^{3+}$ (x = 0, 0.25, 0.5, 0.75, 1): (**a**) Excitation recorded by monitoring 612 nm normalized to $^7F_0 \rightarrow {}^5L_6$; (**b**) Emission spectra of under 365 nm excitation and recorded with 300 g/mm, (**c**) Deconvolution of $^5D_0 \rightarrow {}^7F_1$ peaks recorded under 1800 g/mm diffraction grating.

**Table 2.** Calculated energy of Stark levels $Y_{0.5}Lu_{0.5}NbO_4$:5%Eu.

| Level | Observed Energy [cm$^{-1}$] |
|---|---|
| $^7F_0$ | 0 |
| $^7F_1$ | 369 |
| $^7F_2$ | 1200 |
| $^7F_3$ | 1960 |
| $^7F_4$ | 2999 |
| $^7F_5$ | 3799 |
| $^7F_6$ | N/A |
| $^5D_0$ | 17,204 |
| $^5D_1$ | 18,921 |
| $^5D_2$ | 21,402 |
| $^5D_3$ | 23,810 |
| $^5L_6$ | 25,253 |
| $^5L_7$ | 26,008 |
| $^5G_{2-6}$ | 26,420 |
| $^5L_8$ | 27,100 |
| $^5D_4 + {}^5L_9$ | 27,473 |
| $^5L_{10}$ | 28,531 |
| $^5H_{3,4,7}$ | 30,722 |

The asymmetry ratio R, which is the ratio of the two characteristic transitions, is frequently used to obtain information on the local site symmetry. As the ratio can be influenced by the refractive index ($n$) of the host, experimental R values can be corrected for $n$ using the following equation [40]:

$$R_{corr} = \left[\frac{9n^2}{(n^2+2)^2}\right]\frac{I(^5D_0 - {}^7F_2)}{I(^5D_0 - {}^7F_1)}. \quad (1)$$

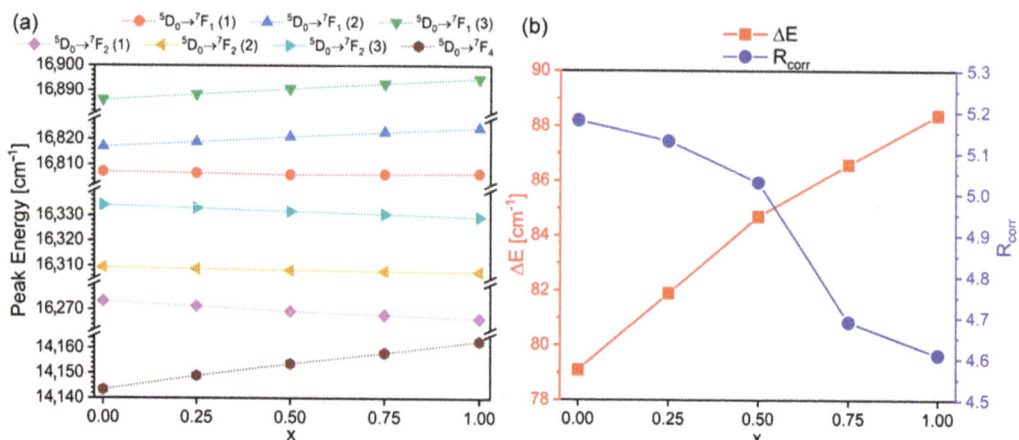

**Figure 3.** (a) Positions of the emission peak maxima of $Y_xLu_{1-x}NbO_4$:$Eu^{3+}$ (x = 0, 0.25, 0.5, 0.75, 1); (b) Energy difference ($\Delta E$) between $^5D_0 \to {}^7F_1$ (1) and (3), and corrected asymmetry ratio ($R_{corr}$).

Starting $n$ values for LuNbO$_4$ and YNbO$_4$ were obtained from Ref. [10]. The refractive index values for the other hosts were then calculated by approximation: $n(Y_xLu_{1-x}NbO_4) = x \cdot n(YNbO_4) + (1-x) \cdot n(LuNbO_4)$; and they are given in Figure 4. Characteristic $n$ values at different wavelengths are clearly marked in Figure 4, and they are used for further Judd–Ofelt calculations.

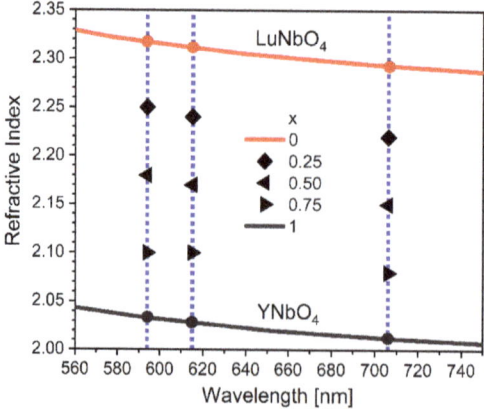

**Figure 4.** The calculated refractive index values for $Y_xLu_{1-x}NbO_4$ (x = 0, 0.25, 0.5, 0.75, 1).

### 3.3. Judd–Ofelt Analysis

$Eu^{3+}$ is the only ion with pure magnetic dipole transitions, $^5D_0 \to {}^7F_1$ and $^5D_1 \to {}^7F_0$, allowing for the self-referenced Judd–Ofelt analysis from the single emission spectrum [5,6,41]. As the magnetic dipole transition strengths are independent of the host matrix, the radiative transition probability of the magnetic dipole transition (MD) can be used for calibration of the spectrum [42]. Then, the Judd–Ofelt parameters can be estimated from the single emission spectrum (without the traditionally used fitting procedure [43,44]), directly from the relation [45,46]:

$$\Omega_\lambda \left[cm^2\right] = \frac{4.135 \times 10^{-23}}{U^\lambda} \left(\frac{\nu_{MD}}{\nu_\lambda}\right)^3 \frac{9n_{MD}^3}{n_\lambda(n_\lambda^2+2)^2} \frac{I_\lambda}{I_{MD}}, \quad (2)$$

where $\lambda = 2,4,6$ abbreviates $^5D_0 \rightarrow ^7F_\lambda$ transitions, and $U^\lambda$ are the squared reduced matrix elements with values of 0.0032, 0.0023, and 0.0002 for $\lambda = 2,4,6$, respectively. $\nu$ is the emission barycenter energy, $n$ is the refractive index, and $I$ are the integrated emission intensities (the employed MD transition here is $^5D_0 \rightarrow ^7F_1$). Equation (2) is given for intensities measured in counts. In the case where intensities are given in power units, the power to the ratio of barycenter energies should be equal to 4 [47].

From the Judd–Ofelt parameters, many derived quantities can be estimated directly: radiative transition probabilities, branching ratios, radiative lifetime, and emission cross-sections [48]. The radiative transition probabilities for a given $\lambda$ and MD transition are given by Equations (3) and (4), respectively [49]:

$$A_\lambda\left[s^{-1}\right] = 8.034 \times 10^9 \nu_\lambda^3 n_\lambda \left(n_\lambda^2+2\right)^2 U^\lambda \Omega_\lambda, \quad (3)$$

$$A_{MD}\left[s^{-1}\right] = 3 \times 10^{-12} \nu_{MD}^3 n_{MD}^3, \quad (4)$$

The radiative lifetime, branching ratios, and emission cross-sections are estimated by Equations (5)–(7), respectively [50,51]:

$$\tau_{rad}[ms] = \frac{10^3}{\sum_{\lambda=2,4,6} A_\lambda[s^{-1}] + A_{MD}[s^{-1}]}, \quad (5)$$

$$\beta_{\lambda,MD} = 10^{-3} A_{\lambda,MD}\left[s^{-1}\right] \tau_{rad}[ms], \quad (6)$$

$$\sigma_{\lambda,MD}\left[cm^2\right] = \frac{1.33 \times 10^{-5}}{\nu_{\lambda,MD}^4 n_{\lambda,MD}^2} \frac{\max i_{\lambda,MD}}{I_{\lambda,MD}} A_{\lambda,MD}, \quad (7)$$

where max($i$) is the intensity at the transition maximum

The experimentally measured lifetime values ($\tau_{obs}$) of the emissions from the $^5D_0$ level are given in Table 3, allowing for the calculation of the non-radiative lifetime ($\tau_{NR}$) by [52]:

$$\tau_{NR}[ms] = \frac{\tau_{rad}\tau_{obs}}{\tau_{rad} - \tau_{obs}}. \quad (8)$$

The Judd–Ofelt parameters and derived quantities were calculated by the JOES application software and are given in Table 3 [51]. With the increase of Y content in the hosts (increasing x), certain trends can be observed: the $\Omega_2$ parameter of the hypersensitive transition $^5D_0 \rightarrow ^7F_2$ increases with x, indicating an increasing degree of covalency and decrease of $Eu^{3+}$ site symmetry, and the $\Omega_4$ parameter, which relates to the viscosity and rigidity of the host matrix, shows no significant change [53]. The $\Omega_6$ parameter could not be estimated, as the emission $^5D_0 \rightarrow ^7F_6$ was not observed.

According to the changes in the refractive index and Judd–Ofelt parameters, the radiative transition probabilities and cross-sections show a monotonic increase with x, while the radiative lifetime decreases. As the observable lifetime also increases with x, the non-radiative lifetime is at a minimum in the pure YNbO$_4$, indicating that LuNbO$_4$ is a more efficient host matrix for the $Eu^{3+}$ emission.

As the Judd–Ofelt parameters depend on the environment of the ion, the parameters obtained by the semi-empirical method are a statistical average of the Judd–Ofelt parameters for each ion. Thus, if there is a mix of non-equivalent sites in the host matrix, the Judd–Ofelt parameters represent an average value for each site, weighted by their contribution [49]. Thus, the Judd–Ofelt parameters of the mixture of two hosts can be predicted from the Judd–Ofelt parameters of the two pure compounds (a,b):

$$\Omega_\lambda = x\Omega_\lambda^{(a)} + (1-x)\Omega_\lambda^{(b)}. \tag{9}$$

**Table 3.** Judd–Ofelt parameters of $Y_xLu_{1-x}NbO_4:Eu^{3+}$, branching ratios ($\beta$), radiative lifetimes ($\tau_{rad}$), emission cross-sections ($\sigma$), radiative transition probabilities (A), total radiative transition probabilities ($A_R$), observed lifetime ($\tau_{obs}$), non-radiative lifetime ($\tau_{NR}$), and intrinsic quantum yield ($\eta_{int}$). Transitions $^5D_0 \rightarrow \,^7F_{1,2,4}$ are abbreviated with MD, 2, and 4 in subscripts, respectively.

| x | 0 | 0.25 | 0.50 | 0.75 | 1 |
|---|---|---|---|---|---|
| $\Omega_2 \cdot 10^{20}$ [cm$^2$] | 6.75 | 6.88 | 7.37 | 7.41 | 7.48 |
| $\Omega_4 \cdot 10^{20}$ [cm$^2$] | 2.69 | 2.91 | 2.97 | 3.01 | 2.95 |
| $\beta_{MD}$ [%] | 14 | 14 | 14 | 14 | 14 |
| $\beta_2$ [%] | 73 | 73 | 73 | 73 | 73 |
| $\beta_4$ [%] | 13 | 14 | 14 | 14 | 13 |
| $\tau_{rad}$ [ms] | 0.782 | 0.859 | 0.916 | 1.027 | 1.134 |
| $\sigma_{MD} \cdot 10^{21}$ [cm$^2$] | 2.12 | 1.81 | 1.64 | 1.43 | 1.49 |
| $\sigma_2 \cdot 10^{21}$ [cm$^2$] | 9.96 | 9.40 | 9.71 | 9.01 | 8.87 |
| $\sigma_4 \cdot 10^{21}$ [cm$^2$] | 3.63 | 3.50 | 3.64 | 3.13 | 3.18 |
| $A_{MD}$ [s$^{-1}$] | 179 | 163 | 148 | 133 | 120 |
| $A_2$ [s$^{-1}$] | 928 | 838 | 795 | 705 | 628 |
| $A_4$ [s$^{-1}$] | 171 | 164 | 148 | 136 | 115 |
| $A_R$ [s$^{-1}$] | 1279 | 1164 | 1092 | 974 | 882 |
| $\tau_{obs}$ [ms] | 0.619 | 0.643 | 0.650 | 0.650 | 0.652 |
| $\tau_{NR}$ [ms] | 2.970 | 2.557 | 2.238 | 1.771 | 1.534 |
| $\eta_{int}$ [%] | 79 | 75 | 71 | 63 | 57 |

The estimation of the Judd–Ofelt parameters from the emission spectra inherently brings an error of about 10%. By fitting the Judd–Ofelt parameters to the linear relations for each x, the Judd–Ofelt parameters can be refined by the values on the fitted curve, as the estimate reduces the error of estimation for each by $\sqrt{k}$, where k is the number of measurements. Thus, in the case of the mixture of YNbO$_4$ and LuNbO$_4$ in five ratios, the error of each estimated Judd–Ofelt parameter is reduced about 2.2 times, and the more correct values are on the fitted curve for each concentration (Figure 5).

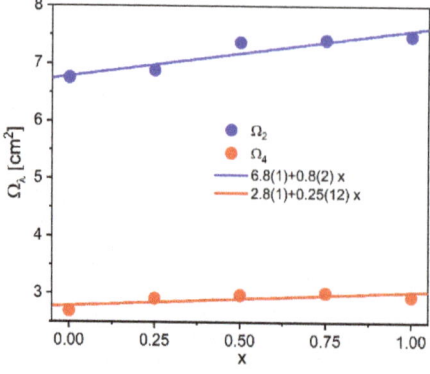

**Figure 5.** Experimentally obtained Judd–Ofelt parameters and corresponding linear fits.

## 4. Conclusions

In this study, we demonstrated how the Y-to-Lu ratio in $Y_xLu_{1-x}NbO_4$:$Eu^{3+}$ powder material influenced the $Eu^{3+}$ luminescence features. The materials were synthesized using the solid state reaction method. All of the structures crystallized as beta-Ferguosonite, in which the Eu ion replaced the Y or Lu ion in a large, low-symmetry octahedron. In all composition hosts, the luminescence excitation and emission spectra of the $Eu^{3+}$ ($4f^6$) ion showed characteristic $f$-$f$ transitions from which the Stark energy levels were calculated.

The specific features and energy positions of the characteristic $^5D_0 \rightarrow {^7F_1}$ magnetic dipole transition were determined when measured with a higher resolution and when spectra deconvolution was used. The maximum $\Delta E$ splitting of the Stark splitting of $^7F_1$ manifold and the asymmetry ratio R all exhibit Y/Lu content-dependent trends. Calculations based on the Judd–Ofelt theory were used to estimate specific quantities, concluding that $LuNbO_4$ is a more efficient host matrix for the $Eu^{3+}$ emission.

**Author Contributions:** Conceptualization, V.Đ. and M.D.D.; methodology, M.D.D.; validation, A.Ć.; formal analysis, A.Ć. and V.Đ.; investigation, M.S. and T.D.; data curation, L.Đ.F. and A.Ć.; writing—original draft preparation, V.Đ.; writing—review and editing, V.Đ., A.Ć. and M.D.D. All authors have read and agreed to the published version of the manuscript.

**Funding:** This research was funded by the NATO Science for Peace and Security Programme under grant id. [G5751]. The authors acknowledge funding from the Ministry of Education, Science and Technological Development of the Republic of Serbia.

**Institutional Review Board Statement:** Not applicable.

**Informed Consent Statement:** Not applicable.

**Data Availability Statement:** Not applicable.

**Acknowledgments:** The authors acknowledge Đorđe Veljović for SEM measurements.

**Conflicts of Interest:** The authors declare no conflict of interest.

## References

1. Yen, W.M.; Shionoya, S.; Yamamoto, H. (Eds.) *Practical Applications of Phosphors*; CRC Press: Boca Raton, FL, USA, 2007; ISBN 9781315219974. [CrossRef]
2. Atuchin, V.V.; Subanakov, A.K., Aleksandrovsky, A.S.; Bazarov, B.G.; Bazarova, J.G.; Gavrilova, T.A.; Krylov, A.S.; Molokeev, M.S.; Oreshonkov, A.S.; Stefanovich, S.Y. Structural and spectroscopic properties of new noncentrosymmetric self-activated borate $Rb_3EuB_6O_{12}$ with $B_5O_{10}$ units. *Mater. Des.* **2018**, *140*, 488–494. [CrossRef]
3. Atuchin, V.V.; Aleksandrovsky, A.S.; Chimitova, O.D.; Gavrilova, T.A.; Krylov, A.S.; Molokeev, M.S.; Oreshonkov, A.S.; Bazarov, B.G.; Bazarova, J.G. Synthesis and spectroscopic properties of monoclinic α-$Eu_2(MoO_4)_3$. *J. Phys. Chem. C* **2014**, *118*, 15404–15411. [CrossRef]
4. Shi, P.; Xia, Z.; Molokeev, M.S.; Atuchin, V.V. Crystal chemistry and luminescence properties of red-emitting $CsGd_{1-x}Eu_x(MoO_4)_2$ solid-solution phosphors. *Dalton Trans.* **2014**, *43*, 9669–9676. [CrossRef] [PubMed]
5. Binnemans, K. Interpretation of europium(III) spectra. *Coord. Chem. Rev.* **2015**, *295*, 1–45. [CrossRef]
6. Tanner, P.A. Some misconceptions concerning the electronic spectra of tri-positive europium and cerium. *Chem. Soc. Rev.* **2013**, *42*, 5090. [CrossRef]
7. Đorđević, V.; Antić, Ž.; Lojpur, V.; Dramićanin, M.D. Europium-doped nanocrystalline $Y_2O_3$–$La_2O_3$ solid solutions with bixbyite structure. *J. Phys. Chem. Solids* **2014**, *75*, 1152–1159. [CrossRef]
8. Đorđević, V.; Antić, Ž.; Nikolić, M.G.; Dramićanin, M.D. Comparative structural and photoluminescent study of $Eu^{3+}$-doped $La_2O_3$ and $La(OH)_3$ nanocrystalline powders. *J. Phys. Chem. Solids* **2014**, *75*, 276–282. [CrossRef]
9. Blasse, G.; Bril, A. Luminescence of Phosphors Based on Host Lattices $ABO_4$ (A is Sc, In; B is P, V, Nb). *J. Chem. Phys.* **2003**, *50*, 2974. [CrossRef]
10. Ding, S.; Zhang, H.; Liu, W.; Sun, D.; Zhang, Q. Experimental and first principle investigation the electronic and optical properties of $YNbO_4$ and $LuNbO_4$ phosphors. *J. Mater. Sci. Mater. Electron.* **2018**, *29*, 11878–11885. [CrossRef]
11. Rubin, J.J.; Van Uitert, L.G. Growth of Large Yttrium Vanadate Single Crystals for Optical Maser Studies. *J. Appl. Phys.* **1966**, *37*, 2920. [CrossRef]
12. Erdei, S. Growth of oxygen deficiency-free $YVO_4$ single crystal by top-seeded solution growth technique. *J. Cryst. Growth* **1993**, *134*, 1–13. [CrossRef]

13. Packer, R.J.; Tsipis, E.V.; Munnings, C.N.; Kharton, V.V.; Skinner, S.J.; Frade, J.R. Diffusion and conductivity properties of cerium niobate. *Solid State Ion.* **2006**, *177*, 2059–2064. [CrossRef]
14. Haugsrud, R.; Norby, T. Proton conduction in rare-earth ortho-niobates and ortho-tantalates. *Nat. Mater.* **2006**, *5*, 193–196. [CrossRef]
15. Wu, M.; Liu, X.; Gu, M.; Ni, C.; Liu, B.; Huang, S. Characterization and luminescence properties of sol–gel derived M′-type LuTaO$_4$:Ln$^{3+}$ (Ln = Pr, Sm, Dy) phosphors. *Mater. Res. Bull.* **2014**, *60*, 652–658. [CrossRef]
16. Takei, H.; Tsunekawa, S. Growth and properties of LaNbO$_4$ and NdNbO$_4$ single crystals. *J. Cryst. Growth* **1977**, *38*, 55–60. [CrossRef]
17. Fulle, K.; McMillen, C.D.; Sanjeewa, L.D.; Kolis, J.W. Hydrothermal Chemistry and Growth of Fergusonite-type RENbO$_4$ (RE = La-Lu, Y) Single Crystals and New Niobate Hydroxides. *Cryst. Growth Des.* **2016**, *16*, 4910–4917. [CrossRef]
18. Brunckova, H.; Mudra, E.; Medvecky, L.; Kovalcikova, A.; Durisin, J.; Sebek, M.; Girman, V. Effect of lanthanides on phase transformation and structural properties of LnNbO$_4$ and LnTaO$_4$ thin films. *Mater. Des.* **2017**, *134*, 455–468. [CrossRef]
19. Zhang, P.; Wang, T.; Xia, W.; Li, L. Microwave dielectric properties of a new ceramic system NdNbO$_4$ with CaF$_2$ addition. *J. Alloys Compd.* **2012**, *535*, 1–4. [CrossRef]
20. Đačanin, L.R.; Dramićanin, M.D.; Lukić-Petrović, S.R.; Petrović, D.M.; Nikolić, M.G.; Ivetić, T.B.; Gúth, I.O. Mechanochemical synthesis of YNbO$_4$:Eu nanocrystalline powder and its structural, microstructural and photoluminescence properties. *Ceram. Int.* **2014**, *40*, 8281–8286. [CrossRef]
21. Blasse, G.; Bril, A. Photoluminescent Efficiency of Phosphors with Electronic Transitions in Localized Centers. *J. Electrochem. Soc.* **1968**, *115*, 1067. [CrossRef]
22. Lee, S.K.; Chang, H.; Han, C.H.; Kim, H.J.; Jang, H.G.; Park, H.D. Electronic Structures and Luminescence Properties of YNbO$^4$ and YNbO$^4$:Bi. *J. Solid State Chem.* **2001**, *156*, 267–273. [CrossRef]
23. Dwivedi, A.; Mishra, K.; Rai, S.B. Tm$^{3+}$, Yb$^{3+}$ activated ANbO$_4$ (A = Y, Gd, La) phosphors: A comparative study of optical properties (downshifting and upconversion emission) and laser induced heating effect. *J. Phys. D Appl. Phys.* **2016**, *50*, 045602. [CrossRef]
24. Xiao, X.; Yan, B. REMO$_4$ (RE = Y, Gd; M = Nb, Ta) phosphors from hybrid precursors: Microstructure and luminescence. *J. Mater. Res.* **2008**, *23*, 679–687. [CrossRef]
25. Liu, C.; Zhou, W.; Shi, R.; Lin, L.; Zhou, R.; Chen, J.; Li, Z.; Liang, H. Host-sensitized luminescence of Dy$^{3+}$ in LuNbO$_4$ under ultraviolet light and low-voltage electron beam excitation: Energy transfer and white emission. *J. Mater. Chem. C* **2017**, *5*, 9012–9020. [CrossRef]
26. Dačanin, L.R.; Lukić-Petrović, S.R.; Petrović, D.M.; Nikolić, M.G.; Dramićanin, M.D. Temperature quenching of luminescence emission in Eu$^{3+}$- and Sm$^{3+}$-doped YNbO$_4$ powders. *J. Lumin.* **2014**, *151*, 82–87. [CrossRef]
27. Wang, X.; Li, X.; Xu, S.; Cheng, L.; Sun, J.; Zhang, J.; Li, L.; Chen, B. A comparative study of spectral and temperature sensing properties of Er$^{3+}$ mono-doped LnNbO$_4$ (Ln = Lu, Y, Gd) phosphors under 980 and 1500 nm excitations. *Mater. Res. Bull.* **2019**, *111*, 177–182. [CrossRef]
28. Ding, S.; Peng, F.; Zhang, Q.; Luo, J.; Liu, W.; Sun, D.; Dou, R.; Sun, G. Structure, spectroscopic properties and laser performance of Nd:YNbO$_4$ at 1066 nm. *Opt. Mater.* **2016**, *62*, 7–11. [CrossRef]
29. Massabni, A.M.G.; Montandon, G.J.M.; Couto, M.A.; Santos, D. Synthesis and luminescence spectroscopy of YNbO$_4$ doped with Eu(III). *Mater. Res.* **1998**, *1*, 01–04. [CrossRef]
30. Đačanin Far, L.; Lukić-Petrović, S.R.; Đorđević, V.; Vuković, K.; Glais, E.; Viana, B.; Dramićanin, M.D. Luminescence temperature sensing in visible and NIR spectral range using Dy$^{3+}$ and Nd$^{3+}$ doped YNbO$_4$. *Sens. Actuators A Phys.* **2018**, *270*, 89–96. [CrossRef]
31. Refguson, R.B. The crystallography of synthetic YTaO$_4$ and fused fergusonite | The Canadian Mineralogist | GeoScienceWorld. *Can. Mineral.* **1957**, *6*, 72–77.
32. Brixner, L.H.; Whitney, J.F.; Zumsteg, F.C.; Jones, G.A. Ferroelasticity in the LnNbO$_4$-type rare earth niobates. *Mater. Res. Bull.* **1977**, *12*, 17–24. [CrossRef]
33. Bayliss, R.D.; Pramana, S.S.; An, T.; Wei, F.; Kloc, C.L.; White, A.J.P.; Skinner, S.J.; White, T.J.; Baikie, T. Fergusonite-type CeNbO$_{4+\delta}$: Single crystal growth, symmetry revision and conductivity. *J. Solid State Chem.* **2013**, *204*, 291–297. [CrossRef]
34. Arulnesan, S.W.; Kayser, P.; Kimpton, J.A.; Kennedy, B.J. Studies of the fergusonite to scheelite phase transition in LnNbO$_4$ orthoniobates. *J. Solid State Chem.* **2019**, *277*, 229–239. [CrossRef]
35. Keller, C. Über ternäre Oxide des Niobs und Tantals vom Typ ABO$_4$. *Z. Anorg. Allg. Chem.* **1962**, *318*, 89–106. [CrossRef]
36. Weitzel, H. Kristallstrukturverfeinerungen von Euxenit, Y(Nb$_{0.5}$Ti$_{0.5}$)$_2$O$_6$, und M-Fergusonit, YNbO$_4$. *Z. Krist.-Cryst. Mater.* **1980**, *152*, 69–82. [CrossRef]
37. Shannon, R.D. Revised effective ionic radii and systematic studies of interatomic distances in halides and chalcogenides. *Acta Crystallogr. Sect. A* **1976**, *32*, 751–767. [CrossRef]
38. Malta, O.L.; Antic-Fidancev, E.; Lemaitre-Blaise, M.; Milicic-Tang, A.; Taibi, M. The crystal field strength parameter and the maximum splitting of the $^7F_1$ manifold of the Eu$^{3+}$ ion in oxides. *J. Alloys Compd.* **1995**, *228*, 41–44. [CrossRef]
39. Görller-Walrand, C.; Binnemans, K. Chapter 155 Rationalization of crystal-field parametrization. *Handb. Phys. Chem. Rare Earths* **1996**, *23*, 121–283. [CrossRef]

40. Srivastava, A.M.; Brik, M.G.; Beers, W.W.; Cohen, W. The influence of $nd^0$ transition metal cations on the $Eu^{3+}$ asymmetry ratio R = I($^5D_0-^7F_2$)/I($^5D_0-^7F_1$) and crystal field splitting of $^7F_1$ manifold in pyrochlore and zircon compounds. *Opt. Mater.* **2021**, *114*, 110931. [CrossRef]
41. Görller-Walrand, C.; Fluyt, L.; Ceulemans, A.; Carnall, W.T. Magnetic dipole transitions as standards for Judd-Ofelt parametrization in lanthanide spectra. *J. Chem. Phys.* **1991**, *95*, 3099–3106. [CrossRef]
42. Ćirić, A.; Stojadinović, S.; Brik, M.G.; Dramićanin, M.D. Judd-Ofelt parametrization from emission spectra: The case study of the $Eu^{3+}$ $^5D_1$ emitting level. *Chem. Phys.* **2020**, *528*, 110513. [CrossRef]
43. Judd, B.R. Optical Absorption Intensities of Rare-Earth Ions. *Phys. Rev.* **1962**, *127*, 750–761. [CrossRef]
44. Ofelt, G.S. Intensities of Crystal Spectra of Rare-Earth Ions. *J. Chem. Phys.* **1962**, *37*, 511–520. [CrossRef]
45. Ćirić, A.; Stojadinović, S.; Dramićanin, M.D. An extension of the Judd-Ofelt theory to the field of lanthanide thermometry. *J. Lumin.* **2019**, *216*. [CrossRef]
46. Babu, S.S.; Babu, P.; Jayasankar, C.K.; Sievers, W.; Tröster, T.; Wortmann, G. Optical absorption and photoluminescence studies of $Eu^{3+}$-doped phosphate and fluorophosphate glasses. *J. Lumin.* **2007**, *126*, 109–120. [CrossRef]
47. Suta, M.; Meijerink, A. A Theoretical Framework for Ratiometric Single Ion Luminescent Thermometers—Thermodynamic and Kinetic Guidelines for Optimized Performance. *Adv. Theory Simul.* **2020**, *3*, 2000176. [CrossRef]
48. Hehlen, M.P.; Brik, M.G.; Krämer, K.W. 50th anniversary of the Judd–Ofelt theory: An experimentalist's view of the formalism and its application. *J. Lumin.* **2013**, *136*, 221–239. [CrossRef]
49. Görller-Walrand, C.; Binnemans, K. Chapter 167 Spectral intensities of f-f transitions. *Handb. Phys. Chem. Rare Earths* **1998**, *25*, 101–264.
50. Babu, P.; Jayasankar, C.K. Optical spectroscopy of $Eu^{3+}$ ions in lithium borate and lithium fluoroborate glasses. *Phys. B Condens. Matter* **2000**, *279*, 262–281. [CrossRef]
51. Ćirić, A.; Stojadinović, S.; Sekulić, M.; Dramićanin, M.D. JOES: An application software for Judd-Ofelt analysis from $Eu^{3+}$ emission spectra. *J. Lumin.* **2019**, *205*, 351–356. [CrossRef]
52. Dramićanin, M.D. Sensing temperature via downshifting emissions of lanthanide-doped metal oxides and salts. A review. *Methods Appl. Fluoresc.* **2016**, *4*, 042001. [CrossRef] [PubMed]
53. Brik, M.G.; Antic, Ž.M.; Vuković, K.; Dramićanin, M.D. Judd-Ofelt Analysis of $Eu^{3+}$ Emission in $TiO_2$ Anatase Nanoparticles. *Mater. Trans.* **2015**, *56*, 1416–1418. [CrossRef]

Article

# Preparation of Cerium Oxide via Microwave Heating: Research on Effect of Temperature Field on Particles

Chao Lv [1,*], Hong-Xin Yin [1], Yan-Long Liu [1], Xu-Xin Chen [1], Ming-He Sun [1] and Hong-Liang Zhao [2,*]

[1] School of Control Engineering, Northeastern University, Qinhuangdao 066004, China; 2072123@stu.neu.edu.cn (H.-X.Y.); 2172254@stu.neu.edu.cn (Y.-L.L.); 2172245@stu.neu.edu.cn (X.-X.C.); 1971722@stu.neu.edu.cn (M.-H.S.)

[2] School of Metallurgical and Ecological Engineering, University of Science and Technology Beijing, Beijing 100083, China

* Correspondence: lvchao@neuq.edu.cn (C.L.); zhaohl@ustb.edu.cn (H.-L.Z.)

**Abstract:** Micro–nano cerium oxide particles with regular morphology and good dispersity have been widely used in polishing and other industrial fields. Microwave heating is an effective, controllable and green heating technology. The Venturi reactor used for microwave heating developed by our team was the core equipment utilized to study the effects of pyrolysis conditions on the purity and microstructure of cerium oxide particles. The experiments were carried out and the products were characterized using XRD, SEM and EDS. Microwave heating, fluid flow and chemical reaction were coupled using numerical simulation, the effects of microwave power, reactor location and waveguide arrangement on temperature fields were investigated. The results showed that with the microwave power increasing, the degree of crystallinity and purity of cerium oxide improved. The morphology gradually became sphere-like. Varied reactor locations and waveguide arrangements changed the gradient and dispersity of temperature fields. Bulk particles and agglomeration could be avoided, and cerium oxide particles with average size of 80 nm were produced when the reactor was located in the center of the cavity. Vertical arrangement of waveguides had the advantages of higher temperature value, gentle gradient and better dispersity.

**Keywords:** microwave heating; cerium oxide; temperature field; morphology

## 1. Introduction

Cerium oxide ($CeO_2$) is a cost-effective rare earth oxide that can be utilized in multiple applications. $CeO_2$ has been widely used in industrial applications, including chemical–mechanical polishing, fuel cells and catalyzers [1–4]. In recent years, the preparation method of $CeO_2$ has been extensively investigated. L. Xiang [5] prepared micro–nano $CeO_2$ by the static roasting of $CeCl_3$. The purity of $CeO_2$ increased when the roasting temperature increased. However, a serious agglomeration of particles appeared in this method. This was because the conventional heating method had a significant temperature gradient, which was not of benefit for obtaining a good morphology of particles [6]. Microwaves are electromagnetic waves of which wavelength ranges from 1 mm to 1 m. When an ion solution is heated, polar molecules and ions move intensely and crash into each other, which contributes to the heat release [7]. Microwave heating has the advantages of accelerating the chemical reaction rate [8,9], selective heating [10,11] and easy control. H. Chen [12] used zirconyl chloride octahydrate ($ZrOCl_2·8H_2O$) and ammonia water ($NH_3·H_2O$) as raw materials to prepare nanometer zirconium dioxide ($ZrO_2$) via the coprecipitation–conventional heating method and the coprecipitation–microwave pyrolysis method. It was found that the $ZrO_2$ prepared using the conventional heating method had the disadvantages of uneven particle size and agglomeration, the size of the $ZrO_2$ particles exceeded 60 nm. The particles prepared using microwave heating had better dispersity, of which size was less 40 nm.

During experiments of microwave heating, chemical reaction process cannot be observed because of the non-transparent equipment, therefore numerical simulation has become an essential tool to predict and visualize the temperature field [13–16]. S. Yu et al. [17] investigated the pyrolysis process of methyl ricinoleate using microwave-assisted heating. The effects of waveguide quantity and heating power in pyrolysis process were studied. It was found that higher energy efficiency was obtained at higher power. The waveguide quantity had much influence on the electric field distribution; however, no obvious difference on temperature distribution was observed. J. Zhu et al. [18] coupled electromagnetism and heat by computing the Maxwell's equations, wave equation and heat transfer equation. The effects of microwave power on the temperature field of oil shale were investigated using numerical simulation. The selective characteristic of microwave heating was beneficial to the porosity and permeability enhancement of the sample. Higher microwave power contributed to higher temperature, which increased the heating efficiency. The above studies demonstrated that both microwave powers and waveguides had significant effects on the intensity and distribution of electromagnetic field.

In this study, different temperature fields were obtained when microwave power, reactor location and waveguide distribution were changed. Then, both experiment and numerical simulation were carried out to investigate the influence of different temperature fields on purity and morphology of products.

## 2. Experimental

The schematic diagram of the experiment was shown in Figure 1. Oxygen bottles provided oxygen for the chemical reaction. Oxygen velocity was measured using the gas flow meter. $CeCl_3$ solution was prepared by the $CeCl_3 \cdot 7H_2O$ (99%, analytical grade, Sinopharm Chemical Reagent Co., Ltd., Shanghai, China). It was jetted into the Venturi reactor, which was developed by our team, the droplets were collided into smaller droplets by high-speed oxygen. These droplets mixed well with oxygen due to an increased contact area with oxygen, which accelerated the chemical reaction. In the industrial production process of pyrolysis method, HCl was used to dissolve or recycle rare earth ore in the process of extraction separation. However, there were no conditions for HCl recovery in the laboratory. Therefore, the gas produced by the reaction was absorbed by the $Na_2CO_3$ solution.

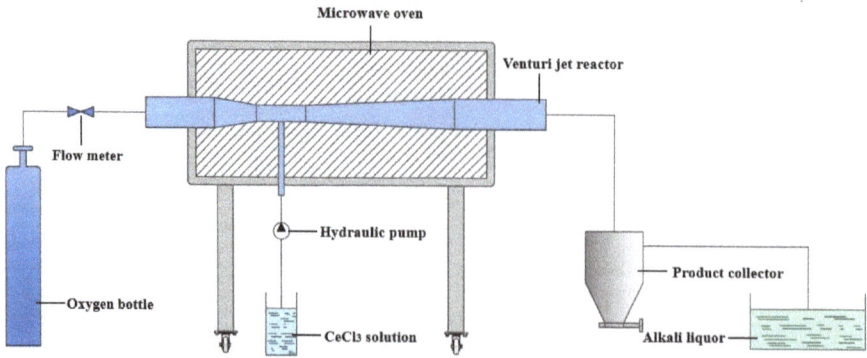

**Figure 1.** Schematic diagram of the experiment.

The phase composition of the products was determined using X-ray diffraction (XRD, Empyrean S3, PANalytical, Mumbai, India) at a scanning rate of 5°/min in the 2θ range from 10° to 90° with Cu $K_\alpha$ radiation (λ = 1.540598 nm). The microstructure of particles was observed using scanning electron microscopy (SEM, apreo2c, Thermo Fisher, Waltham, MA, USA). The average content of residual chlorine element was measured using energy dispersive spectroscopy (EDS, ultim max 40, Oxford, UK).

## 3. Numerical Simulation

### 3.1. Geometry

According to the microwave oven used in the experiment (custom-made by the Shan-Lang Experimental Material Management Department), the simplified three-dimensional model was shown in Figure 2. Three rectangular BJ26 waveguides were equidistantly distributed on the surface of the microwave cavity. Dimensions of the microwave oven and Venturi tube were shown in Table 1.

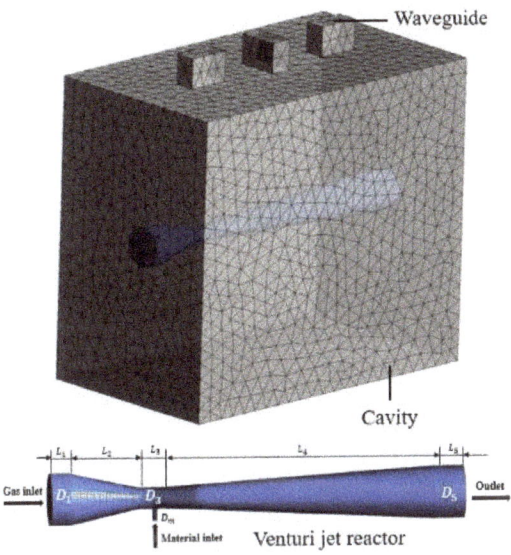

Figure 2. Simplified three-dimensional model of the equipment.

Table 1. Dimensions of models.

|  | Length/mm | Width/mm | Height/mm | Diameter/mm |
|---|---|---|---|---|
| Cavity | 690 | 400 | 600 | - |
| Waveguide | 86.36 | 43.18 | 60 | - |
| L1 | - | - | 30 | - |
| L2 | - | - | 110 | - |
| L3 | - | - | 40 | - |
| L4 | - | - | 419 | - |
| L5 | - | - | 50 | - |
| D1 | - | - | - | 80 |
| D3 | - | - | - | 32 |
| D5 | - | - | - | 80 |
| Dm | - | - | - | 7 |

### 3.2. Governing Equation

The electromagnetic field inside the Venturi reactor is determined using the Maxwell's equation. The time-varying form of the Maxwell's equations can be expressed in the form below:

$$\nabla \cdot D = \rho_e \quad (1)$$

$$\nabla \cdot B = 0 \quad (2)$$

$$\nabla \times E = -\frac{\partial B}{\partial t} \quad (3)$$

$$\nabla \times H = -\frac{\partial D}{\partial t} + J \quad (4)$$

where $B$ is the magnetic flux density (Wb/m$^3$), $D$ is the electric displacement or electric flux density (C/m$^2$), $E$ is the electric field intensity (V/m), $H$ is the magnetic field intensity (A/m), $J$ is the current density (A/m$^2$) and $\rho_e$ is the electric charge density (C/m$^3$).

The complex relative permittivity, ($\varepsilon_r$), is defined as:

$$\varepsilon_r = \varepsilon' - j\varepsilon'' \tag{5}$$

where $j = \sqrt{-1}$, $\varepsilon'$ is the dielectric constant and $\varepsilon''$ is the corresponding loss factor (that is, the imaginary part of the dielectric constant). $\varepsilon'$ represents the ability of microwaves to penetrate the material, while $\varepsilon''$ represents the ability of the material to store electricity.

The differential wave equation deduced using Maxwell's equations was used to describe the electromagnetic field distribution:

$$\nabla \times \mu_r^{-1}(\nabla \times E) - k_0^2\left(\varepsilon_r - \frac{j\sigma}{\omega\varepsilon_0}\right)E = 0 \tag{6}$$

$$k_0 = \omega/c_0 \tag{7}$$

where $\omega$ is the angular frequency (rad/s), $\varepsilon_0$ is the free space permittivity (8.85 × 10$^{-12}$ F/m), $\mu_r$ is the relative permeability, $\sigma$ represents the electrical conductivity (S/m), $k_0$ is the wave number in free space and $c_0$ represents the speed of light in vacuum (m/s).

When the electromagnetic field interacts with the solution, part of the electromagnetic energy transforms into thermal energy, which is the source item of the energy equation. The energy absorbed by the per unit volume solution can be expressed using the following equation:

$$Q = \frac{1}{2}\omega\varepsilon_0\varepsilon''|E|^2 \tag{8}$$

where $|E|$ is the electric field modulus.

The energy equation is given as:

$$\frac{\partial(\rho T)}{\partial t} + \nabla\cdot(v\rho T) = \nabla\cdot\left(\frac{k}{c_p}\nabla T\right) + Q \tag{9}$$

where $T$ represents the temperature (°C), $k$ is the thermal conductivity [W/(m·K)], $\rho$ is the density of the fluid, $c_p$ represents the specific heat capacity, $v$ is the velocity and $Q$ is the source item.

The chemical reaction occurring inside the Venturi reactor can be expressed as:

$$4CeCl_3(s) + 6H_2O(g) + O_2(g) = 4CeO_2(s) + 12HCl(g) \tag{10}$$

In the reactor, every relative species needs to observe the species mass conservation equation:

$$\frac{\partial(\rho Y_i)}{\partial t} + \nabla\cdot(\rho v Y_i) = -\nabla J_i + R_i + S_i \tag{11}$$

where $Y_i$ is the mass fraction of the $i$th substance, $R_i$ is the net produce rate of the chemical reaction of the ith substance, $S_i$ is the discrete phase of the $i$th substance that is responsible for the additional produce rate caused by the user-defined source item and $J_i$ is the diffusive flux of the $i$th substance produced by the concentration gradient.

### 3.3. Material and Boundary Conditions

The main physical parameters, including permittivity, conductivity, thermal conductivity, specific heat and density, were measured in this study. The measured results of the parameters were shown in Table 2. Enthalpy, entropy and other necessary parameters were determined from the Practical Inorganic Thermodynamics Data Manual [19].

Table 2. Main physical parameters measured at 363 K.

| Properties | Dielectric Constant | Loss Factor | Electrical Conductivity | Thermal Conductivity | Specific Heat | Density |
|---|---|---|---|---|---|---|
| Symbol | $\varepsilon'$ | $\varepsilon''$ | $\sigma$ | $k$ | $C_p$ | $\rho$ |
| Value | 80.49 | 0.785 | 0.1383 S/m | 0.53 W/(m·K) | 5.025 J/(g·K) | 1.023 g/cm$^3$ |
| Function | T | T | T | T | T | - |
| Source | This study | This study | This study | This study | This study | This study |

Where $\varepsilon' = 0.0003T^2 - 0.006T + 46.535$, $\varepsilon'' = -0.0011T^2 + 0.9937T - 151.307$, $\sigma = 3.29 \times 10^{-5}T^2 - 0.0197T + 2.9515$, $k = -1.06 \times 10^{-5}T^2 + 0.0081T - 1.0261$, $C_p = 2.87 \times 10^{-5}T^2 - 0.016T + 7.043$. The density did not change much when the temperature changed.

To decrease the difficulty associated with the simulation and the computing period, the following assumptions were proposed before solving the mathematical models:

1. Microwave frequency is constant (2.45 GHz).
2. The initial temperature inside the reactor is uniform (25 °C).
3. The fluid inside the reactor is incompressible.
4. Permeability is constant at 1.

The boundary conditions of the model were shown in Table 3. Three-dimensional transient algorithm was selected in the Fluent 19.0 software. The following models were used for the simulation: Eulerian multiphase, standard k-ε turbulence, energy equation and species transport models. Pressure and velocity were coupled using the SIMPLE algorithm. The computations were considered to converge when the residuals for all the quantities did not exceed more than $10^{-3}$, except for energy equation with the residual below $10^{-6}$.

Table 3. Boundary conditions.

| | Boundary Type | Value |
|---|---|---|
| Excitation | Wave port | - |
| Wall of waveguides and cavity | Perfect E boundary | - |
| Gas velocity | Velocity-inlet | 7 m/s |
| Material velocity | Velocity-inlet | 0.2 m/s |
| Outlet | Outflow | - |

### 3.4. Independence Verification of Grid

The whole model was composed of unstructured grids. A local coordinate system and an influence sphere were built at the Venturi tube, which contributed to the local encrypted grids. The average quality of the grids was 0.842. Independence verification of grids is shown in Figure 3. When the quantity of the grids reached 410 K, the error of the temperature value at outlet was less than 10 K, which indicated that the grid quantity did not cause too much fluctuation of the simulated results. To raise the simulated efficiency, grid quantity of 410 K was used in this study.

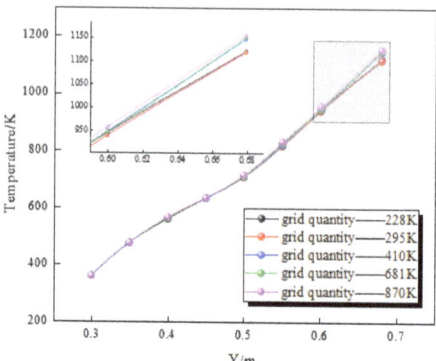

**Figure 3.** Independence verification of the grid.

## 4. Results and Discussion

### 4.1. Effect of Microwave Power

The experiment and simulation were both carried out when microwave power was 12, 13, 14, 15 and 16 kW, respectively. The comparation of $Cl^-$ content of the product between experiment and simulation was shown in Figure 4. The error was below 3% (the error bar in Figure 4 was 3%). It proved that the used model, parameters and boundary conditions were rational. $Cl^-$ content was less than 2% when the power reached 16 kW. The inserted XRD pattern showed that a single phase $CeO_2$ was obtained when the power reached 14 kW. The increased power caused higher strength of the diffraction peaks. The software, MDI Jade, and the method of multipeak separation were used to analyze the degree of crystallinity of products at each diffraction peak. As for the products prepared at 16 kW, the degree of crystallinity at the diffraction peak at $2\theta = 28.56°$, $56.383°$ and $59.097°$ was obviously higher than that prepared at 14 kW. The degree of crystallinity at the other diffraction peaks was almost equal under the two microwave powers. It indicated that higher microwave power was beneficial to the improvement of the degree of crystallinity.

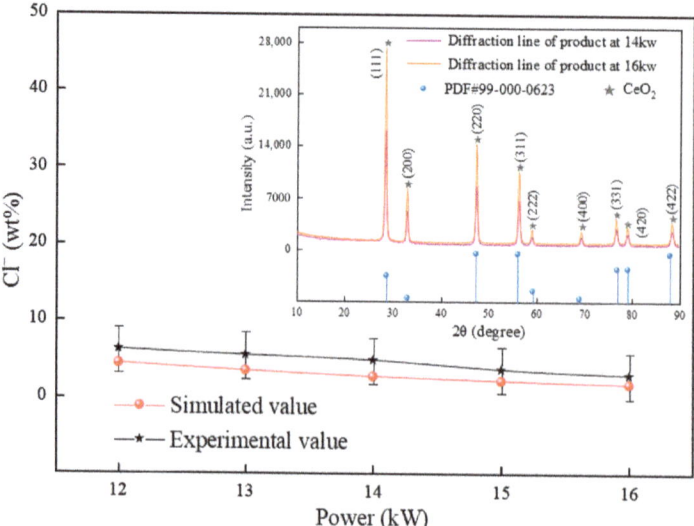

**Figure 4.** Comparation of $Cl^-$ content between experiment and simulation.

The comparison of SEM images was shown in Figure 5. $CeO_2$ particles had a sharp morphology when it was produced at 14 kW. The increased power led to higher temperature. Sphere-like particles were obtained when power reached 16 kW. It indicated that the higher microwave power was beneficial to the improvement of the particles' morphology.

**Figure 5.** SEM images. (**a**) 14 kW, (**b**) 16 kW.

### 4.2. Effects of Reactor Location

Different reactor locations led to variations in the electromagnetic and temperature fields inside the reactor. The schemes were shown in Figure 6a. The simulated temperature fields at 12 kW were shown in Figure 6b. Temperature value was higher in the center than it was when closer to the wall, which accorded with the features of the microwave heating model. With the reactor location changing, temperature fields varied considerably. However, the model of center heating was not changed. The temperature value was always 300 K in the front part of the reactor because only oxygen existed there, which is a substance that cannot absorb microwaves.

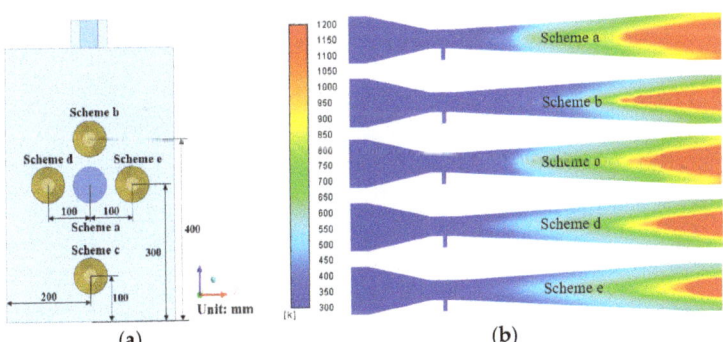

**Figure 6.** (**a**) Schemes of different reactor locations, (**b**) temperature field.

Scheme (a) shows the highest average temperature and the gentlest temperature gradient in the radial direction. Scheme (b) shows the worst temperature dispersity in the axial direction because the region of high temperature was concentrated at the end of the reactor. The temperature dispersity of Scheme (d) was better than Scheme (b), which benefited the reactant as it was heated earlier. Scheme (e) had the worst temperature field distribution because there was a significant radial temperature gradient and the lowest average temperature. It is obvious that the temperature field of Scheme (c) was similar to Scheme (a). Therefore, the products of Schemes (a), (b), (d) and (e) were characterized using SEM, as shown in Figure 7. The $CeO_2$ particles produced via Scheme (a) had a regular morphology and its average particle size was near 80 nm, the smallest among the four Schemes. However, each single particle produced via Scheme (b) had a crude morphology, which may be attributed to the shortened heating time of the solution caused by the concentrative temperature field. Additionally, it had a bad particle dispersity. $CeO_2$ produced via Scheme (d) had a relative regular morphology, even particle size and good dispersity. In Figure 7d, it is obvious that serious agglomeration and many bulk particles appeared. According to the above results, it can be concluded that a gentler temperature gradient was beneficial for eliminating the agglomeration and bulk particles. Comparing the simulated temperature fields and SEM images of Scheme (b) with (d), it was found that a better temperature dispersity contributed to a better dispersity of particles.

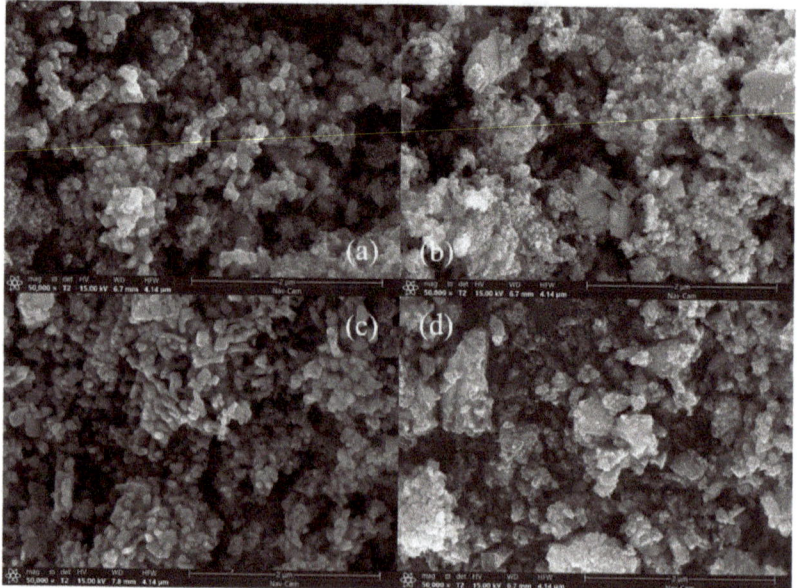

**Figure 7.** SEM images. (**a**) Scheme (a), (**b**) Scheme (b), (**c**) Scheme (d) and (**d**) Scheme (e).

### 4.3. Effects of Waveguide Distribution

The effect of various waveguide distributions on the temperature field was investigated. Numerical simulation was used to predict the morphology and purity of the products due to the high cost of experimental equipment. Differently to the experimental Scheme, three new Schemes of waveguide arrangements and their temperature fields were shown in Figure 8. All the Schemes had a good axial temperature dispersity. However, in Scheme (a), the waveguides were parallelly arranged and had an extremely large temperature gradient at the end of the reactor, and the temperature at most regions was below 600 K, which may not cause the chemical reaction. The vertically arranged waveguides of Schemes (b) and (c) led to higher than average temperature values and gentler temperature gradients. Based on the former deductions and the SEM images of the experiment, as seen in Scheme (a), it can be predicted that there is a high probability that the agglomeration and bulk particles will appear and, for each particle, that sharp morphology will appear. Comparing the temperature fields of Scheme (b) and (c) with the experimental field, it can be deduced that the phenomenon of agglomeration is unlikely to appear because of the gentle temperature gradient and that $CeO_2$ particles produced via Scheme (b) will have a more regular morphology than Scheme (c). Figure 9 showed the temperature and $Cl^-$ content at the outlet of the three new Schemes. $Cl^-$ content reached 16.85% due to the low overall temperature of Scheme (a). It can be predicted that the $Cl^-$ content of $CeO_2$ produced by Scheme (b) will be close to 2.07%.

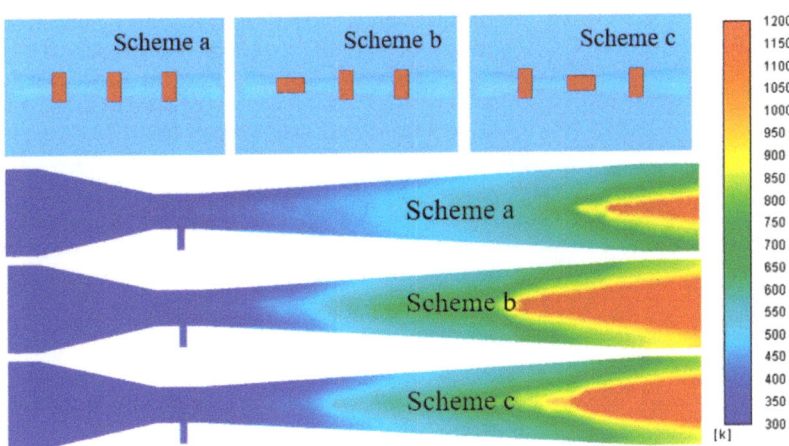

**Figure 8.** Schemes of waveguide distributions and their temperature fields.

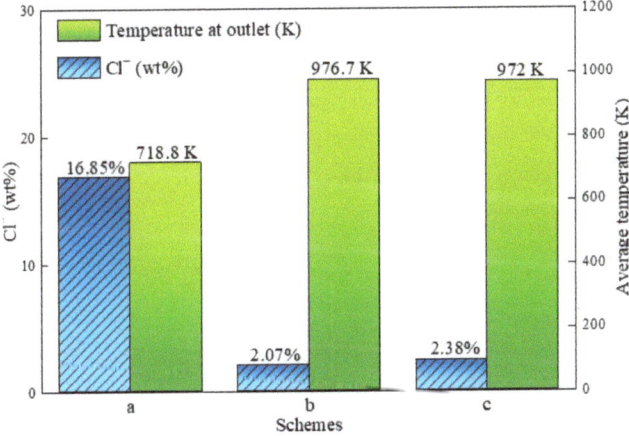

**Figure 9.** Comparation of temperature at outlet and Cl⁻ content.

## 5. Conclusions

In this study, the effects of microwave power, reactor location and waveguide distribution on the $CeO_2$ purity and microstructure were investigated. The following conclusions were obtained:

1. Microwave power had significant effects on the purity of the products. Higher power contributed to higher purity and an improvement of the degree of crystallinity of $CeO_2$. When microwave power reached 16 kW, sphere-like $CeO_2$ particles were obtained and the residual $Cl^-$ content was only 0.15%. Increasing microwave power also had benefits for obtaining a more regular morphology of particles.
2. Various reactor locations caused various temperature fields to present varied temperature gradients and dispersities, which had a remarkable effect on the microstructure of $CeO_2$ particles. The simulated nephogram of temperature fields combined with SEM images showed that large temperature gradients caused agglomeration and bulk particles to appear, and better temperature dispersity contributed to a better dispersity of particles. When the reactor was located in the center of the cavity, $CeO_2$ particles with a regular morphology were prepared, which had an average size of 80 nm.

3. Varied waveguide arrangements also caused various temperature fields. The vertical arrangement of waveguides had the advantages of higher temperature value, gentle gradient and better dispersity. It was predicted that the actual value of $Cl^-$ content was close to 2.07% when the waveguides were distributed in Scheme (b). Moreover, the morphology was similar to Scheme (c) and better than Scheme (a).

From the perspective of energy efficiency, at the current stage, there is a problem of high energy consumption. Reducing energy consumption and improving the energy utilization while ensuring the product quality will be the focus in the next stage of study. In addition, gas velocity determined the atomization effect, which can change the microstructure of the particles. In the next stage, the effect of gas velocity on particle microstructure will be worth studying.

**Author Contributions:** Conceptualization, C.L.; Project administration, C.L.; Funding acquisition, C.L.; Methodology, C.L. and H.-X.Y.; Software, H.-X.Y.; Validation, Y.-L.L. and X.-X.C.; Data Curation, M.-H.S.; Visualization, H.-L.Z. All authors have read and agreed to the published version of the manuscript.

**Funding:** This research was supported by the National Natural Science Foundation of China (51904069), the Natural Science Foundation of Hebei Province of China (E2019501085), the Fundamental Research Funds for the Central Universities (N2223026), the Scientific Research Fund Project of Northeastern University at Qinhuangdao (XNY201808).

**Institutional Review Board Statement:** Not applicable.

**Informed Consent Statement:** Not applicable.

**Data Availability Statement:** Not applicable.

**Conflicts of Interest:** The authors declare no conflict of interest.

## References

1. Fujiwara, A.; Yoshida, H. In-Situ diffuse reflectance spectroscopy analysis of $Pd/CeO_2$–$ZrO_2$ model three-way catalysts under Lean-Rich cycling condition. *Catal. Today* **2021**, *376*, 269–275. [CrossRef]
2. Chen, Y.S.; Fan, J. Synthesis of high stability nanosized $Rh/CeO_2$–$ZrO_2$ three-way automotive catalysts by Rh chemical state regulation. *J. Energy Inst.* **2020**, *93*, 2325–2333. [CrossRef]
3. Fan, J.; Chen, Y.S. A simple and effective method to synthesize $Pt/CeO_2$ three-way catalysts with high activity and hydrothermal stability. *J. Environ. Chem. Eng.* **2020**, *8*, 104236. [CrossRef]
4. Jiang, D.; Wan, G. Elucidation of the Active Sites in Single-Atom $Pd1/CeO_2$ Catalysts for Low-Temperature CO Oxidation. *ACS Catal.* **2020**, *10*, 11356–11364. [CrossRef]
5. Xiang, L. Study on the Preparation of Light Rare Earth Oxide from Rare Earth Chloride Roasted Statically. Master's Thesis, Northeastern University, Shenyang, China, 2010.
6. Li, H.; Xu, J.L. Food waste pyrolysis by traditional heating and microwave heating: A review. *Fuel* **2022**, *324*, 124574. [CrossRef]
7. Arballo, J.R.; Goni, S.M. Modeling of fluid dynamics and water vapor transport in microwave ovens—ScienceDirect. *Food Bioprod. Process.* **2020**, *119*, 75–87. [CrossRef]
8. Wang, B.; Zheng, H.B. Microwave-assisted fast pyrolysis of waste tires: Effect of microwave power on products composition and quality. *J. Anal. Appl. Pyrolysis* **2021**, *155*, 104979.
9. Zhou, N.; Dai, L.L. Catalytic pyrolysis of plastic wastes in a continuous microwave assisted pyrolysis system for fuel production. *Chem. Eng. J.* **2021**, *418*, 129412. [CrossRef]
10. Luo, W.; Huang, S.Y. Three-dimensional mesostructure model of coupled electromagnetic and heat transfer for microwave heating on steel slag asphalt mixtures. *Constr. Build. Mater.* **2022**, *330*, 127235. [CrossRef]
11. Shen, X.; Li, H. Imaging of liquid temperature distribution during microwave heating via thermochromic metal organic frameworks. *Int. J. Heat Mass Transfer.* **2022**, *189*, 122667. [CrossRef]
12. Chen, H. Study on Preparation and Mechanism of Yttria-Stabilized Zirconia Nanosized Powder by Microwave Pyrolysis. Master's Thesis, Zhengzhou University, Zhengzhou, China, 2014.
13. Sadeghi, A.; Hassanzadeh, H. Thermal analysis of high frequency electromagnetic heating of lossy porous media. *Chem. Eng. Sci.* **2017**, *172*, 13–22. [CrossRef]
14. Lv, C.; Sun, M.H. Pyrolysis Preparation Process of $CeO_2$ with the Addition of Citric Acid: A Fundamental Study. *Crystals* **2021**, *11*, 912. [CrossRef]
15. Xu, T.; Song, G. Visualization and simulation of steel metallurgy processes. *Int. J. Miner. Metall. Mater.* **2021**, *28*, 1387–1396. [CrossRef]

16. Yuan, F.; Xu, A.J. Development of an improved CBR model for predicting steel temperature in ladle furnace refining. *J. Miner. Metall. Mater.* **2021**, *28*, 1321–1331. [CrossRef]
17. Yu, S.Z.; Duan, Y. Three-dimensional simulation of a novel microwave-assisted heating device for methyl ricinoleate pyrolysis. *Appl. Therm. Eng.* **2019**, *153*, 341–351. [CrossRef]
18. Zhu, J.Y.; Yi, L.P. Three-dimensional numerical simulation on the thermal response of oil shale subjected to microwave heating. *Chem. Eng. J.* **2021**, *407*, 127197. [CrossRef]
19. Ye, D.L. *Practical Inorganic Thermodynamics Data Manual*; Cao, S.L., Ed.; Metallurgical Industry Press: Beijing, China, 1981; pp. 262–263, 265–266.

Article

# Room Temperature Synthesis of Various Color Emission Rare-Earth Doped Strontium Tungstate Phosphors Applicable to Fingerprint Identification

Soung-Soo Yi [1] and Jae-Yong Jung [2],*

[1] Division of Materials Science and Engineering, Silla University, Busan 45985, Korea; ssyi@silla.ac.kr
[2] Research and Business Development Foundation, Engineering Building, Silla University, Busan 45985, Korea
* Correspondence: eayoung21@naver.com; Tel.: +82-50-999-6441

**Abstract:** Crystalline $SrWO_4$ was synthesized at room temperature using a co-precipitation method. To use the $SrWO_4$ as a phosphor, green and red phosphors were synthesized by doping with $Tb^{3+}$ and $Eu^{3+}$ rare earth ions. The synthesized samples had a tetragonal structure, and the main peak (112) phase was clearly observed. When the sample was excited using the absorption peak observed in the ultraviolet region, $SrWO_4:Tb^{3+}$ showed an emission spectrum of 544 nm, and $SrWO_4:Eu^{3+}$ showed an emission spectrum of 614 nm. When $Tb^{3+}$ and $Eu^{3+}$ ions were co-doped to realize various colors, a yellow-emitting phosphor was realized as the doping concentration of $Eu^{3+}$ ions increased. When the synthesized phosphor was scattered on a glass substrate with fingerprints, as used in the field of fingerprint recognition, the fingerprint was revealed by green, red, and yellow emissions in response to a UV lamp.

**Keywords:** $SrWO_4$; phosphors; luminescence; fingerprint

## 1. Introduction

Crystalline tungsten has excellent thermal and chemical stability and has been applied in various fields. A material that is thermally and chemically stable has high energy transfer efficiency from tungsten ion to rare earth ion in the rare earth doped phosphor; thus, it is suitable for use as a host material [1–3]. Rare earth (RE) ions doped in the host lattice can generate high intensity emissions and various emission wavelengths, with a narrow bandgap due to energy transfer between the 4f-4f shells [4–6]. The type and site symmetry of the rare earth ions doped in a thermally and chemically stable host lattice are important factors in the performance of various types of lighting, laser, and display devices [7–9].

It has been reported that the emission wavelength of phosphors used in various types of light devices can vary depending on the type and concentration of the doped rare earth ions, the sintering temperature, crystal grain size, excitation wavelength, and synthesis conditions [10–13]. In particular, the main emission wavelength of the rare earth ions is determined by competition between electric dipole transitions and magnetic dipole transitions. If the electric dipole transition is strong, it reacts sensitively to the local environment around the rare earth ions located in the host lattice, but magnetic dipole transitions are hardly affected by external environmental factors [14–16]. For example, two types of emission wavelengths occur in a phosphor doped with europium ($Eu^{3+}$) ions. In one emission spectrum, an orange emission (~597 nm) spectrum is generated by the $^5D_0 \rightarrow {}^7F_1$ magnetic dipole transition, and the other is a rare red orange (~620 nm) emission signal from the $^5D_0 \rightarrow {}^7F_2$ electric dipole transition. It is known that either the magnetic dipole transition or the electric dipole transition will become the main transition depending on whether $Eu^{3+}$ rare earth ions located in the host lattice are in the inversion-doping region or not, and this determines the emission wavelength [17–19]. Yu et al. synthesized the $BaWO_4:Eu^{3+}$, $Bi^{3+}$ phosphor powder using a solid-state method and chemical immersion method and observed that the red emission at 613 nm increased as the doping concentration of $Bi^{3+}$ ions increased [20]. Jung et al. synthesized crystalline $BaWO_4$ by preparing a precursor by co-precipitation and heat-treating it at 800 °C. By doping $Dy^{3+}$, $Tb^{3+}$, and $Sm^{3+}$ rare earth ions, phosphors emitting yellow, green, and red were synthesized and applied to anti-counterfeiting [21]. Shinde et al. synthesized $NaCaPO_4$ phosphor doped with $Ce^{3+}$, $Eu^{3+}$, and $Dy^{3+}$ rare earth ions using the combustion method. In the case of the $Ce^{3+}$ ion-doped phosphor, an emission wavelength of 367 nm was obtained. Blue light emission at 482 nm ($^4F_{9/2} \rightarrow {}^6H_{15/2}$, magnetic dipole transition) and light emission at 576 nm ($^4F_{9/2} \rightarrow {}^6H_{13/2}$, electric dipole transition) were observed [22].

In this study, crystalline $SrWO_4$ was synthesized at room temperature by co-precipitation. Then, green and red phosphors were synthesized by doping with rare earth ions $Tb^{3+}$ and $Eu^{3+}$, respectively, and yellow phosphors were synthesized by co-doping the two rare earth ions. The structure of the synthesized phosphor, the size and shape of particles, and their luminescence characteristics were investigated. The synthesized phosphor was reacted with a UV lamp to visualize a fingerprint using the emission color, suggesting that it can be applied to the field of anti-counterfeiting.

## 2. Materials and Methods

### 2.1. Synthesis of $SrWO_4$:$RE^{3+}$ by Co-Precipitation at Room Temperature

Starting materials: Strontium acetate (($CH_3CO_2$)Sr, Sigma-Aldrich, reagent grade), Sodium tungstate ($Na_2WO_4 \cdot 2H_2O$, Sigma-Aldrich, $\geq$99%), Terbium nitrate ($Tb(NO_3)_3 \cdot xH_2O$, $Tb^{3+}$, Sigma-Aldrich, 99.999%), Europium nitrate ($Eu(NO_3)_3 \cdot 6H_2O$, $Eu^{3+}$, Sigma-Aldrich, 99.9%)

The synthesis process was as follows. First 1 mmol ($CH_3CO_2$)Sr was placed in beaker 'A' and stirred with 50 mL distilled water. Then, 1 mmol $Na_2WO_4 \cdot 2H_2O$ was placed in beaker 'B' and stir with 50 mL distilled water (Figure 1). When the solutions in the 'A' and 'B' beakers are completely dissolved and become transparent, pour the 'B' solution into the 'A' beaker, and stir at room temperature for about 20 min. The reacted solutions change to a white opaque color and a powder is formed. The formed powder is recovered by centrifugation at 4000 rpm for 10 min. The recovered powder is washed 3 times with distilled water to remove unreacted substances and then centrifuged again to recover the powder and dried at 80 °C for 16 h (Figure 1). To synthesize the phosphor, 0.25 mmol each of $Tb^{3+}$ and $Eu^{3+}$ was added to the 'A' beaker and processed in the same manner. White light phosphor was synthesized by co-doping by fixing the amount of $Tb^{3+}$ and controlling the amount of $Eu^{3+}$.

### 2.2. Fabricated Fingerprint Identification Application

To use the synthesized phosphor for fingerprint recognition, a thumb fingerprint was imprinted on a glass substrate. After spraying the synthesized phosphor on the glass substrate and removing the remaining powder with a brush, the fingerprint on the glass substrate could be visualized by illuminating it with a UV lamp, which revealed the unique luminous color of the phosphor.

### 2.3. Characterization

The crystal structure of the synthesized phosphor powder was measured using an X-ray diffraction apparatus (X'Pert PRO MPD, 40 kV, 30 mA) having Cu–K$\alpha$ radiation (wavelength: 1.5406 Å) at a scan rate of 4° per minute at a diffraction angle of 10° to 70°. The size and microscopic surface shape of the crystal grains were photographed with a scanning electron microscope (TESCAN MIRA 3 LMH FE-SEM, TESCAN, Brno, Czech Republic), and a fluorescence photometer (FS-2, Scinco) with a xenon lamp was used as a light source to obtain emission and absorption characteristics.

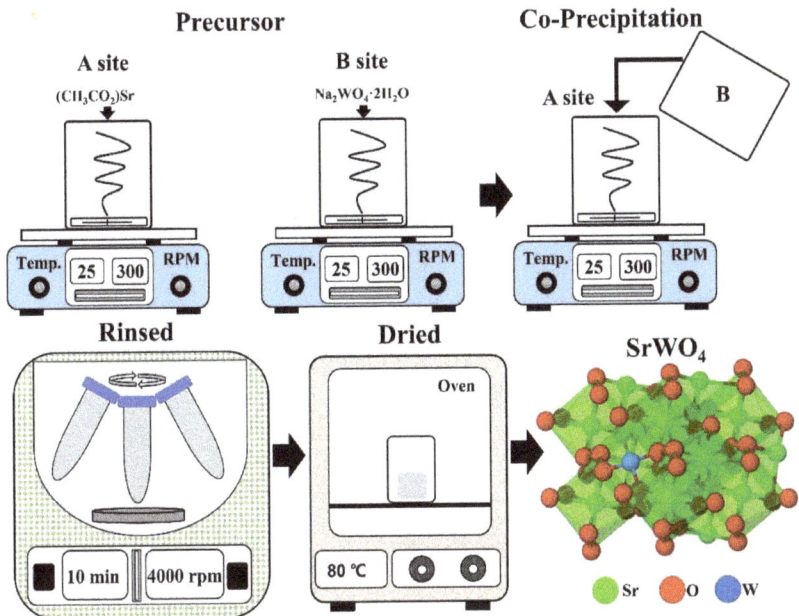

**Figure 1.** Procedure for co-precipitation.

## 3. Results & Discussion

### 3.1. Characteristics of SrWO$_4$ and SrWO$_4$:RE$^{3+}$

Figure 2a shows the XRD measurement results of SrWO$_4$, SrWO$_4$:Tb$^{3+}$, and SrWO$_4$:Eu$^{3+}$. SrWO$_4$ synthesized by co-precipitation showed a tetragonal (a = 5.400 Å, b = 5.400 Å, c = 11.910 Å) structure consistent with ICDD # 01-089-2568. The (112) peak, which is the main diffraction peak, was clearly observed, and the sample to which the rare earth was added also clearly exhibited the main peak.

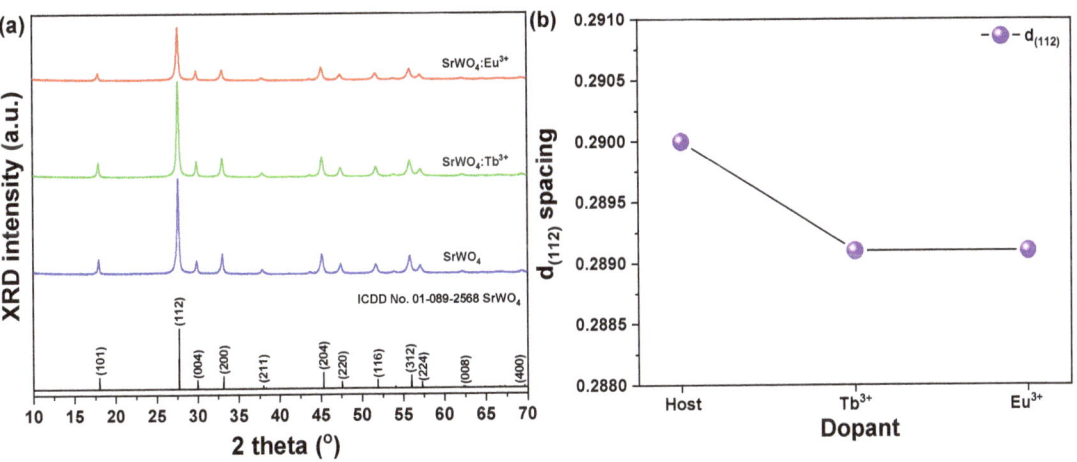

**Figure 2.** (a) XRD patterns, (b) d$_{(112)}$ spacing of SrWO$_4$ and SrWO$_4$:RE$^{3+}$.

Based on the 'Lewis's acid–base' reaction, the samples synthesized by the co-precipitation method showed an explosive reaction when the solution dissolved in beaker 'A' became

'Homo' and the solution dissolved in beaker 'B' became 'Lumo' [23,24], and crystalline SrWO$_4$ was easily synthesized at room temperature. Figure 2b shows the lattice constant change with and without rare earth doping with the (112) plane, which is the main peak of the sample, of the hosts SrWO$_4$, SrWO$_4$:Tb$^{3+}$ and SrWO$_4$:Eu$^{3+}$, respectively. The lattice constant of the (112) phase, which is the main peak of SrWO$_4$, was slightly changed by the rare earth doping (SrWO$_4$: 0.291 nm, SrWO$_4$:Tb$^{3+}$: 0.2892 nm, SrWO$_4$:Eu$^{3+}$: 0.2892 nm). It is considered that the change in the crystal lattice is due to the doping with rare earth ions, which have a relatively large ionic radius [25]. FE-SEM images of the synthesized samples are shown in Figure 3. The samples showed a long cylindrical shape with and without doping.

**Figure 3.** FE-SEM images of (**a**) SrWO$_4$, (**b**) SrWO$_4$:Tb$^{3+}$, and (**c**) SrWO$_4$:Eu$^{3+}$ samples.

The particle size of SrWO$_4$ was about 5.78 µm in the longitudinal direction and about 2.36 µm in the transverse direction (Figure 3a). Rare earth doped SrWO$_4$:Tb$^{3+}$ particles were about 3.57 µm in the longitudinal direction and about 2.29 µm in the transverse direction (Figure 3b), and SrWO$_4$:Eu$^{3+}$ particles had a size of about 4.82 µm in the longitudinal direction and about 2.31 µm in the transverse direction (Figure 3c). Krishna at al. reported BaMoO$_4$ is synthesized by reacting with MoO$_4{}^{-2}$, which is a monomer of oxyanion and grows in the vertical direction immediately after mixing the Ba aqueous solution and Mo aqueous solution. It is reported that the shape of the shuttle was clearly visible due to the larger rift in Oswald. It was synthesized using basic materials and explained by the action of the bases [26].

The host SrWO$_4$ showed absorption in a wide range, from 220 to 340 nm, and peaked at 277 nm. When the sample was excited at the highest peak of 277 nm, it was broad from 350 to 650 nm, and the peak at 492 nm showed blue–white emission spectrum (Figure 4a). Figure 4b shows the emission spectrum of SrWO$_4$:Tb$^{3+}$ phosphor synthesized by doping Tb$^{3+}$ rare earth ions into SrWO$_4$. The absorption spectrum of the phosphor powder under 544 nm showed that the band of charge-transfer transition (CTB) generated between the O$^{-2}$ and W$^{6+}$ of the WO$_4{}^{2-}$ groups was widely distributed in the 210~290 nm region with a peak at 254 nm [27]. When the phosphor powder was excited with 254 nm, peaks at 487, 544, 586, 620, and 649 nm were observed in the emission spectrum. Among these peaks, the intensity of the green emission spectrum produced by the magnetic dipole transition was the strongest. This emission intensity was 2.99 times stronger than the blue emission intensity produced by the electric dipole transition. The Tb$^{3+}$ ion in the SrWO$_4$ lattice is located at the inversion symmetric site because the emission intensity due to the magnetic dipole transition of green emission is strong [28].

Figure 4c shows the absorption and emission spectrum of the SrWO$_4$:Eu$^{3+}$ phosphor synthesized by doping with the rare earth ion Eu$^{3+}$. The absorption spectrum of the phosphor powder under 614 nm shows the absorption spectrum by CTB generated between O$^{2-}$ and Eu$^{3+}$ ions, which appear over the 230~310 nm region and have a peak at 277 nm, with Eu$^{3+}$ observed over the 310~400 nm region. Absorption signals due to the 4f-4f transition of ions were observed [29]. The emission spectrum of the synthesized phosphor was measured by excitation at 277 nm. The phosphor powder showed a red–orange emission spectrum with a peak emission intensity at a wavelength of 614 nm and a spectrum with peaks at 590, 650, and 700 nm. Among these peaks, the 614 nm peak due to the electric dipole transition and the 590 nm peak due to the magnetic dipole transition signal had an intensity difference of about 7.94 times, indicating that the $Eu^{3+}$ ions in the host are located in non-inversion symmetric sites [30].

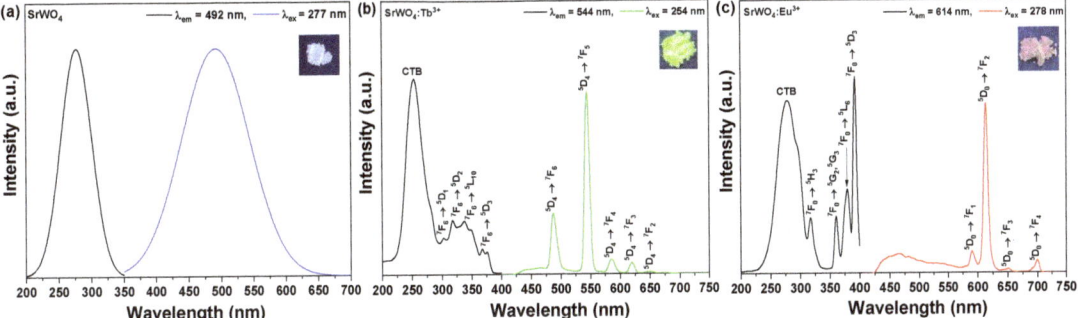

Figure 4. Photoluminescence spectra of (a) $SrWO_4$, (b) $SrWO_4:Tb^{3+}$, and (c) $SrWO_4:Eu^{3+}$ samples.

### 3.2. Characteristics of the $SrWO_4$: $[Eu^{3+}]$:$[Tb^{3+}]$ Phosphors

Figure 5a shows the X-ray diffraction peak of $SrWO_4$ co-doped with rare earth ions $Tb^{3+}$ and $Eu^{3+}$ as a white light-emitting phosphor. In the XRD pattern of the synthesized samples, a secondary phase caused by rare earth doping was not found, and the diffraction signal of the main peak (112) was clearly observed. Figure 5b shows the lattice constant change of the (112) phase, which is the main peak of the rare earth co-doped $SrWO_4:[Eu^{3+}]/[Tb^{3+}]$ samples. Previously, the lattice constants of the $SrWO_4:Tb^{3+}$ and $SrWO_4:Eu^{3+}$ samples doped with a single rare earth decreased, but the lattice constants of the samples doped with both increased. It is believed that the crystal lattice is distorted, or the structure is changed by the amount of added rare earth ions, which have a relatively large ionic radius.

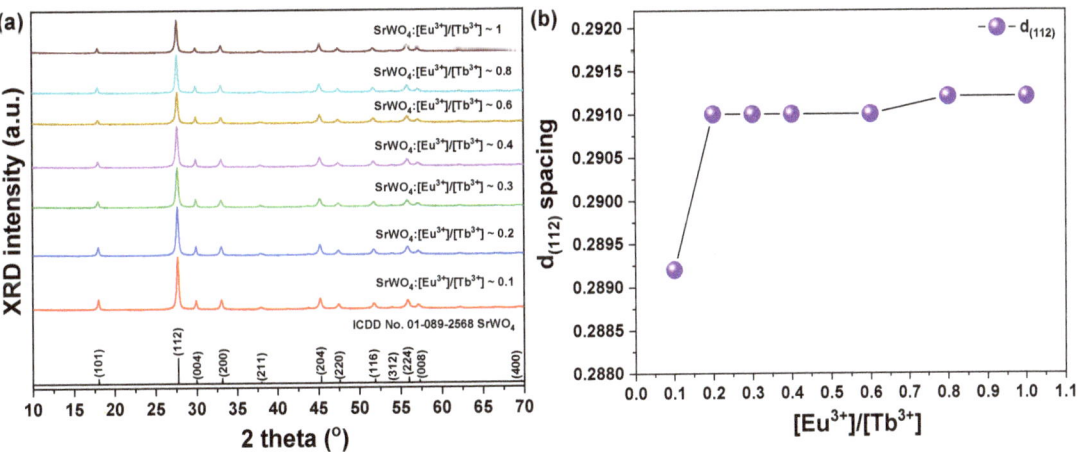

Figure 5. (a) XRD patterns of $SrWO_4:[Eu^{3+}]/[Tb^{3+}]$ and (b) change of $d_{(112)}$ spacing.

Figure 6 shows the FE-SEM images and energy dispersive X-ray spectroscopy (EDS) mapping component analysis results of the synthesized $SrWO_4:[Eu^{3+}][Tb^{3+}]$ phosphor. The shape of the particles grew in the longitudinal direction with a cylindrical shape close to the shape of a dumbbell. The particles were about 3.18 μm in the longitudinal direction and about 1.45 μm in the transverse direction. In the EDS component analysis, Sr, W, O, Tb, and Eu components were detected, which confirmed that rare earth ions had been successfully doped.

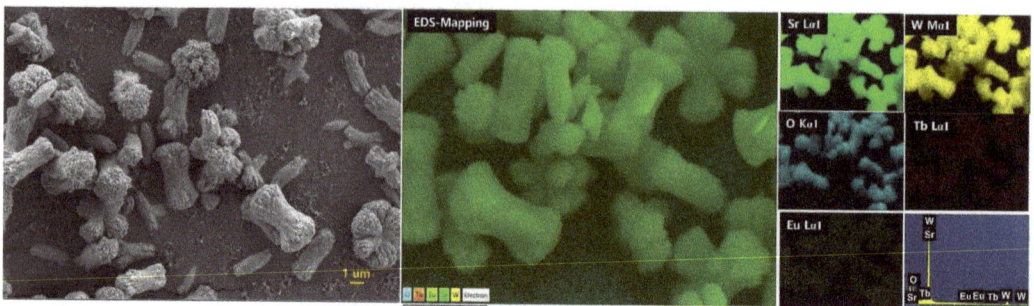

**Figure 6.** FE-SEM images and EDS mapping analysis of SrWO$_4$:[Eu$^{3+}$]/[Tb$^{3+}$].

Figure 7a shows the emission spectrum of the SrWO$_4$:[Eu$^{3+}$]/[Tb$^{3+}$] phosphor powder co-doped with changing Eu$^{3+}$ ion concentrations, while the Tb$^{3+}$ ion concentration remained fixed. The emission spectra of the two rare earth ions were shown when excited at a wavelength of 254 nm, as the doping concentration of Eu$^{3+}$ increased. Green at 544 nm and orange–red at 614 nm were simultaneously observed. As the concentration of Eu$^{3+}$ ions increased, the intensity of the green emission by Tb$^{3+}$ ions decreased, which means that the emission energy was converted from Tb$^{3+}$ ions in the host lattice to Eu$^{3+}$ ions (Figure 7b). The energy transfer efficiency from Tb$^{3+}$ to Eu$^{3+}$ ions can be expressed by Equation (1) [31].

$$\eta = 1 - I/I_0 \tag{1}$$

Here, $I$ is the emission intensity of the Tb$^{3+}$ ions in the SrWO$_4$:[Eu$^{3+}$]/[Tb$^{3+}$] phosphors, and $I_0$ is the emission intensity of Tb$^{3+}$ ions in the SrWO$_4$:Tb$^{3+}$ phosphors. As shown in Figure 7c, as the amount of added Eu$^{3+}$ ions increases, the energy transfer efficiency tends to increase. However, the emission intensity decreased, which is a concentration-quenching phenomenon due to excessive rare earth doping [23]. In the CIE color coordinates, as the doping concentration of Eu$^{3+}$ ions increased, the green coordinates moved to the yellow region (Figure 7d). According to Zhu et al. [32], among the Tb$^{3+}$ and Eu$^{3+}$ rare earth ions co-doped with the CaCO$_3$ cubic structure, the green emission of Tb$^{3+}$ decreases and the intensity of the red emission of Eu$^{3+}$ increases as the doping concentration of Eu$^{3+}$ increases. Regarding the energy from Tb$^{3+}$ to Eu$^{3+}$ ions, it was reported that a transfer occurred. In this study, as the doping concentration of Eu$^{3+}$ ions increased in the Tb$^{3+}$ and Eu$^{3+}$ ions co-doped with SrWO$_4$, the intensity of green emission decreased, and the intensity of red emission increased as the energy transfer occurred.

Figure 8 shows the schematic energy diagram of terbium and europium ions luminescence mechanisms in SrWO$_4$:[Eu$^{3+}$]/[Tb$^{3+}$] phosphors. The phosphors under 254 nm were excited, and $^5D_3$ states of Tb$^{3+}$ can luminesce non-radiatively to the energetically lower $^3D_4$ excited states. Since the $^5D_4$ states of the Tb$^{3+}$ and the $^5D_3$ states of Eu$^{3+}$ are energetically closed to each other, excitation energy availability transferred to the $^5D_3$ states from Eu$^{3+}$ by the path of resonance transmission. Thus, Tb$^{3+}$ mainly emitted green peaks due to $^5D_4 \rightarrow {}^7F_5$, and Eu$^{3+}$ emitted red due to $^5D_0 \rightarrow {}^7F_J$ [32].

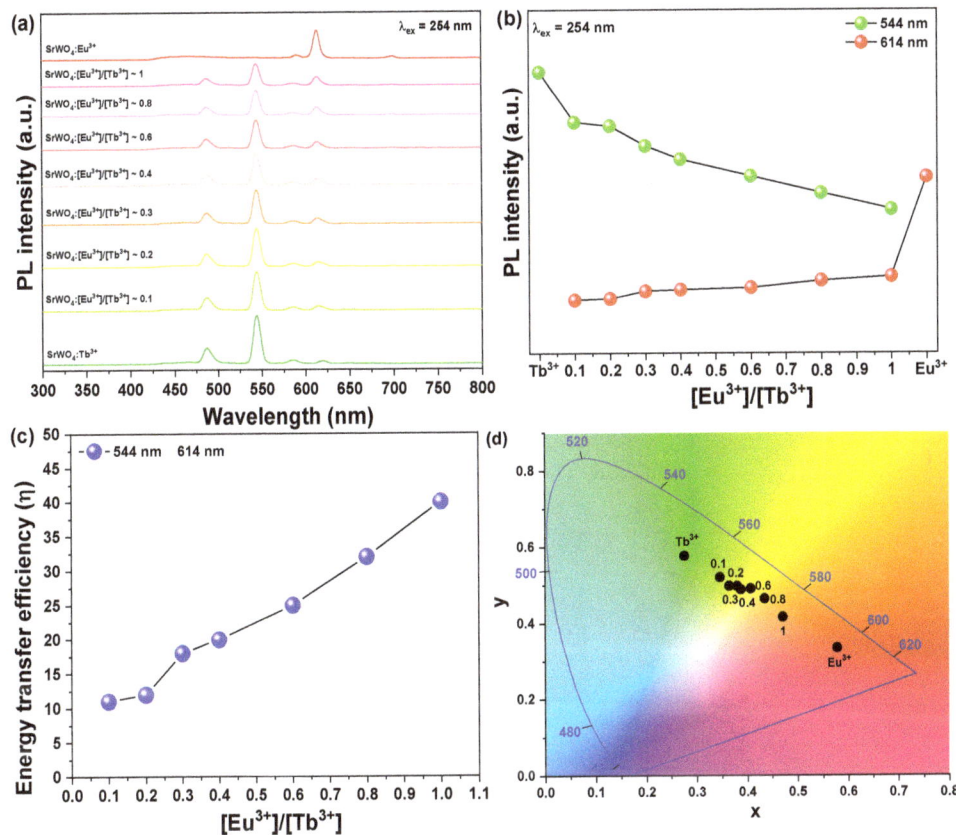

**Figure 7.** (a) Pl spectra under 254 nm, (b) change in PL intensity, (c) energy transfer efficiency, and (d) CIE coordination SrWO$_4$:[Eu$^{3+}$]/[Tb$^{3+}$].

Table 1 shows a reported phosphor synthesized by adding various rare earth ions, with tungsten oxide as a host. Several types of phosphors have been reported, such as up-conversion pre-conversion phosphors co-doped with Yb$^{3+}$ and Er$^{3+}$ and red phosphors co-doped with Eu$^{3+}$ and Sm$^{3+}$. A phosphor was synthesized, and a yellow-emitting phosphor was synthesized by co-doping with Tb$^{3+}$ and Eu$^{3+}$ ions to produce various light-emitting materials as in the previously reported research [33–37].

**Table 1.** Comparison of previous work in tungsten oxide phosphors.

| No. | Host | Rare Earth | Type | Wavalength (nm) |
|---|---|---|---|---|
| 1 [33] | SrWO$_4$ | Er$^{3+}$/Yb$^{3+}$ | Up conversion | 489, 525 |
| 2 [34] | SrWO$_4$ | Tm$^{3+}$/Yb$^{3+}$ | Up conversion | 684, 814 |
| 3 [35] | CaWO$_4$ | Sm$^{3+}$/Eu$^{3+}$ | Down conversion | 592, 615 |
| 4 [36] | CaWO$_4$ | Eu$^{3+}$/Sm$^{3+}$ | Down conversion | 622, 630 |
| 5 [37] | SrWO$_4$ | Eu$^{3+}$/Sm$^{3+}$ | Down conversion | 590, 613 |
| This work | SrWO$_4$ | Eu$^{3+}$/Tb$^{3+}$ | Down conversion | 544, 614 |

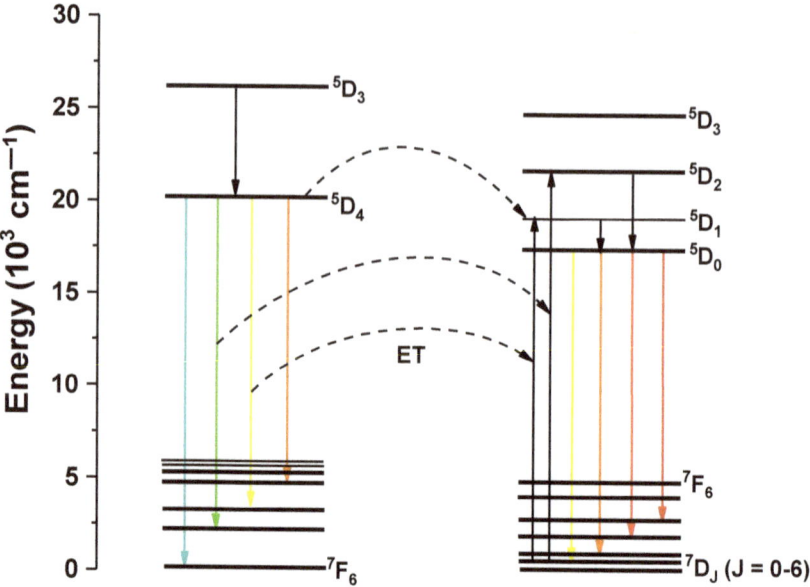

**Figure 8.** Schematic of $Tb^{3+}$ and $Eu^{3+}$ energy levels indicating the energy transfer processes in the $SrWO_4$ phosphors.

### 3.3. Applied for Fingerprint Identification

To clearly observe the fingerprint of the author's thumb on the glass substrate, the synthesized phosphor powder was scattered, and then the shape of the fingerprint was visualized by illuminating it with a UV lamp. The phosphor doped with rare earth ions revealed the fingerprints in emissions of green and red, which are their own colors, and the specimens co-doped with $Eu^{3+}$ and $Tb^{3+}$ revealed the fingerprints by emitting yellow light. The results suggest that the synthesized phosphor can be used for fingerprint identification (Figure 9).

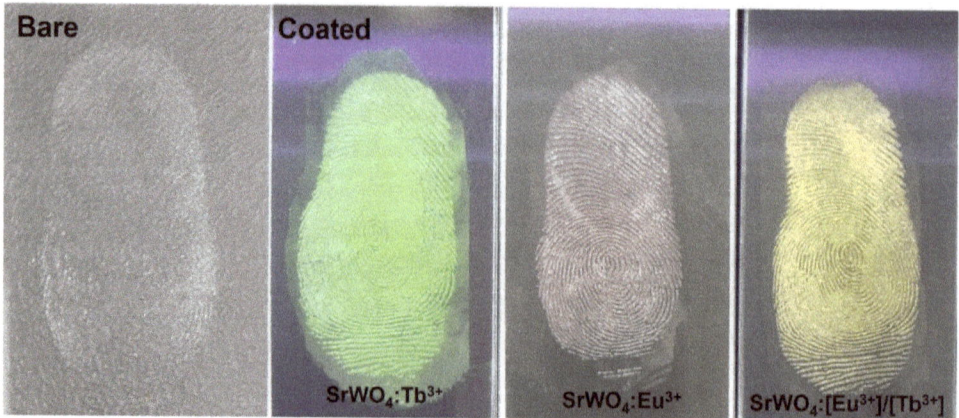

**Figure 9.** Images of fingerprint coated with phosphors under a UV lamp.

## 4. Conclusions

Crystalline SrWO$_4$ was synthesized at room temperature by co-precipitation. Then, green and red phosphors were synthesized by doping with rare earth ions, Tb$^{3+}$ and Eu$^{3+}$, respectively. The synthesized samples clearly exhibited the (112) phase, which was the main peak in the X-ray diffraction pattern, and the lattice constant was changed by doping with rare earth ions. The synthesized specimens had a size of several microns and a cylindrical shape. In addition, when each specimen was excited using an absorption peak in the ultraviolet region, SrWO$_4$:Tb$^{3+}$ exhibited green, and SrWO$_4$:Eu$^{3+}$ emitted red due to the doped rare earth. In the specimen co-doped with Tb$^{3+}$ and Eu$^{3+}$ to obtain various color emissions, the color coordinates shifted to the yellow region as the doping concentration of Eu$^{3+}$ ions increased. The synthesized phosphor was scattered on the glass substrate on which the fingerprint was printed, and when a UV lamp was lit, the green, red, and yellow emission colors were visualized so that the fingerprint could be clearly recognized.

**Author Contributions:** Conceptualization, J.-Y.J.; methodology, J.-Y.J.; software, J.-Y.J.; validation, J.-Y.J. and S.-S.Y.; formal analysis, J.-Y.J.; investigation, J.-Y.J.; resources, J.-Y.J.; data curation, J.-Y.J.; writing—original draft preparation, J.-Y.J. and S.-S.Y.; writing—review and editing, J.-Y.J. and S.-S.Y.; visualization, J.-Y.J.; supervision, J.-Y.J.; project administration, S.-S.Y. and J.-Y.J.; funding acquisition, S.-S.Y. All authors have read and agreed to the published version of the manuscript.

**Funding:** This research was supported by the Basic Research Program through the National Research Foundation of Korea (NRF) funded by the Ministry of Education (NRF-2020R1F1A1072676).

**Data Availability Statement:** The data presented in this study are available on request from the corresponding author.

**Conflicts of Interest:** The authors declare no conflict of interest.

## References

1. Do, Y.R.; Huh, Y.D. Optical Properties of Potassium Europium Tungstate Phosphors. *J. Electrochem. Soc.* **2000**, *147*, 4385–4388. [CrossRef]
2. Zhao, Y.; Wang, X.; Zhang, Y.; Li, Y.; Yao, X. Winning wide-temperature-range and high-sensitive thermometry by a multichannel strategy of dual-lanthanides in the new tungstate phosphors. *J. Alloys Compd.* **2020**, *834*, 154998. [CrossRef]
3. LEE, G.; KIM, T.; YOON, C.; KANG, S. Effect of local environment and Sm$^{3+}$-codoping on the luminescence properties in the Eu$^{3+}$-doped potassium tungstate phosphor for white LEDS. *J. Lumin.* **2008**, *128*, 1922–1926. [CrossRef]
4. Hua, Y.; Yu, J.S. Double-excited states of charge transfer band and 4f 4f in single-phase K$_3$Gd(VO$_4$)$_2$:Tb$^{3+}$/Sm$^{3+}$ phosphors with superior sensing sensitivity for potential luminescent thermometers. *J. Mater. Sci. Technol.* **2021**, *91*, 148–159. [CrossRef]
5. Qin, X.; Liu, X.; Huang, W.; Bettinelli, M.; Liu, X. Lanthanide-Activated Phosphors Based on 4f-5d Optical Transitions: Theoretical and Experimental Aspects. *Chem. Rev.* **2017**, *117*, 4488–4527. [CrossRef]
6. Li, L.; Yang, P.; Xia, W.; Wang, Y.; Ling, F.; Cao, Z.; Jiang, S.; Xiang, G.; Zhou, X.; Wang, Y. Luminescence and optical thermometry strategy based on emission and excitation spectra of Pr$^{3+}$ doped SrMoO$_4$ phosphors. *Ceram. Int.* **2021**, *47*, 769–775. [CrossRef]
7. Wu, H.; Niu, P.; Pei, R.; Zheng, Y.; Jin, W.; Li, X.; Jiang, R. Tb$^{3+}$ and Sm$^{3+}$ co-doped CaWO$_4$ white light phosphors for plant lamp synthesized via solid state method: Phase, photoluminescence and electronic structure. *J. Lumin.* **2021**, *236*, 118146. [CrossRef]
8. Pollnau, M.; Romanyuk, Y.E.; Gardillou, F.; Borca, C.N.; Griebner, U.; Rivier, S.; Petrov, V. Double Tungstate Lasers: From Bulk Toward On-Chip Integrated Waveguide Devices. *JSTQE* **2007**, *13*, 661–671. [CrossRef]
9. Semenov, P.A.; Meshchanin, A.P.; Davidenko, A.M.; Kormilitsin, V.A.; Batarin, V.A.; Goncharenko, Y.M.; Stone, S.; Kravtsov, V.I.; Matulenko, Y.A.; Semenov, V.K.; et al. Design and performance of LED calibration system prototype for the lead tungstate crystal calorimeter. *Nucl. Instrum. Methods Phys. Res. Sect. A Accel. Spectrometers Detect. Assoc. Equip.* **2005**, *556*, 94–99. [CrossRef]
10. Jung, J.-Y. Luminescent Color-Adjustable Europium and Terbium Co-Doped Strontium Molybdate Phosphors Synthesized at Room Temperature Applied to Flexible Composite for LED Filter. *Crystals* **2022**, *12*, 552. [CrossRef]
11. Feng, L.; Chen, X.; Mao, C. A facile synthesis of SrWO$_4$ nanobelts by the sonochemical method. *Mater. Lett.* **2010**, *64*, 2420–2423. [CrossRef]
12. Thongtem, T.; Phuruangrat, A.; Thongtem, S. Preparation and characterization of nanocrystalline SrWO$_4$ using cyclic microwave radiation. *Curr. Appl. Phys.* **2008**, *8*, 189–197. [CrossRef]
13. Gopal, R.; Kumar, A.; Manam, J. Enhanced photoluminescence and abnormal temperature dependent photoluminescence property of SrWO$_4$:Dy$^{3+}$ phosphor by the incorporation of Li$^+$ ion. *Mater. Chem. Phys.* **2021**, *272*, 124960. [CrossRef]
14. Long, Q.; Xia, Y.; Huang, Y.; Liao, Y.; Gao, Y.; Huang, J.; Liang, J.; Cai, J. Na+ induced electric-dipole dominated transition ($^5D_0 \rightarrow ^7F_2$) of Eu$^{3+}$ emission in AMgPO$_4$:Eu$^{3+}$ (A = Li$^+$, Na$^+$, K$^+$) phosphors. *Mater. Lett.* **2015**, *145*, 359–362. [CrossRef]

15. Roh, H.; Lee, S.; Caliskan, S.; Yoon, C.; Lee, J. Luminescence and electric dipole in Eu$^{3+}$ doped strontium phosphate: Effect of SiO$_4$. *J. Alloys Compd.* **2019**, *772*, 573–578. [CrossRef]
16. Blasse, G.; Bril, A. Luminescence of Phosphors Based on Host Lattices ABO$_4$ (A is Sc, In; B is P, V, Nb). *J. Chem. Phys.* **1969**, *50*, 2974–2980. [CrossRef]
17. Song, X.; Wang, X.; Xu, X.; Liu, X.; Ge, X.; Meng, F. Crystal structure and magnetic-dipole emissions of Sr$_2$CaWO$_6$: RE$^{3+}$ (RE = Dy, Sm and Eu) phosphors. *J. Alloys Compd.* **2018**, *739*, 660–668. [CrossRef]
18. Yu, R.; Wang, C.; Chen, J.; Wu, Y.; Li, H.; Ma, H. Photoluminescence Characteristics of Eu$^{3+}$-Doped Double-Perovskite Phosphors. *ECS J. Solid State Sci. Technol.* **2014**, *3*, R33–R37. [CrossRef]
19. Tian, L.; Yu, B.; Pyun, C.; Park, H.L.; Mho, S. New red phosphors BaZr(BO$_3$)$_2$ and SrAl$_2$B$_2$O$_7$ doped with Eu$^{3+}$ for PDP applications. *Solid State Commun.* **2004**, *129*, 43–46. [CrossRef]
20. Yu, P.; Su, L.; Xu, J. Synthesis and Luminescence Properties of Eu$^{3+}$, Bi$^{3+}$-Doped BaWO$_4$ Phosphors. *Opt. Rev.* **2014**, *21*, 455–460. [CrossRef]
21. Jung, J.; Kim, J.; Shim, Y.; Hwang, D.; Son, C.S. Structure and Photoluminescence Properties of Rare-Earth (Dy$^{3+}$, Tb$^{3+}$, Sm$^{3+}$)-Doped BaWO$_4$ Phosphors Synthesized via Co-Precipitation for Anti-Counterfeiting. *Materials* **2020**, *13*, 4165. [CrossRef]
22. Shinde, K.N.; Dhoble, S.J.; Kumar, A. Combustion synthesis of Ce$^{3+}$, Eu$^{3+}$ and Dy$^{3+}$ activated NaCaPO$_4$ phosphors. *J. Rare Earths* **2011**, *29*, 527–535. [CrossRef]
23. Qi, G.; Yang, R.T.; Chang, R. MnO$_x$-CeO$_2$ mixed oxides prepared by co-precipitation for selective catalytic reduction of NO with NH$_3$ at low temperatures. *Appl. Catal. B Environ.* **2004**, *51*, 93–106. [CrossRef]
24. Thongtem, T.; Kungwankunakorn, S.; Kuntalue, B.; Phuruangrat, A.; Thongtem, S. Luminescence and absorbance of highly crystalline CaMoO$_4$, SrMoO$_4$, CaWO$_4$ and SrWO$_4$ nanoparticles synthesized by co-precipitation method at room temperature. *J. Alloys Compd.* **2010**, *506*, 475–481. [CrossRef]
25. Shu, Y.; Travert, A.; Schiller, R.; Ziebarth, M.; Wormsbecher, R.; Cheng, W. Effect of Ionic Radius of Rare Earth on USY Zeolite in Fluid Catalytic Cracking: Fundamentals and Commercial Application. *Top. Catal.* **2015**, *58*, 334–342. [CrossRef]
26. Krishna Bharat, L.; Lee, S.H.; Yu, J.S. Synthesis, structural and optical properties of BaMoO4:Eu$^{3+}$ shuttle like phosphors. *Mater. Res. Bull.* **2014**, *53*, 49–53. [CrossRef]
27. Liao, J.; Qiu, B.; Wen, H.; Chen, J.; You, W. Hydrothermal synthesis and photoluminescence of SrWO$_4$:Tb$^{3+}$ novel green phosphor. *Mater. Res. Bull.* **2009**, *44*, 1863–1866. [CrossRef]
28. Barja, B.; Baggio, R.; Garland, M.T.; Aramendia, P.F.; Peña, O.; Perec, M. Crystal structures and luminescent properties of terbium(III) carboxylates. *Inorg. Chim. Acta* **2003**, *346*, 187–196. [CrossRef]
29. Tseng, T.; Choi, J.; Davidson, M.; Holloway, P.H. Synthesis and luminescent characteristics of europium dopants in SiO$_2$/Gd$_2$O$_3$ core/shell scintillating nanoparticles. *J. Mater. Chem.* **2010**, *20*, 6111–6115. [CrossRef]
30. Dhara, S.; Imakita, K.; Mizuhata, M.; Fujii, M. Europium doping induced symmetry deviation and its impact on the second harmonic generation of doped ZnO nanowires. *Nanotechnology* **2014**, *25*, 225202. [CrossRef]
31. Sun, S.; Guo, R.; Zhang, Q.; Lv, X.; Leng, P.; Wang, Y.; Huang, Z.; Wang, L. Efficient deep-blue thermally activated delayed fluorescence emitters based on diphenylsulfone-derivative acceptor. *Dye. Pigment.* **2020**, *178*, 108367. [CrossRef]
32. Zhu, H.; Qian, B.; Zhou, X.; Song, Y.; Zheng, K.; Sheng, Y.; Zou, H. Tunable luminescence and energy transfer of Tb$^{3+}$/Eu$^{3+}$ co-doped cubic CaCO$_3$ nanoparticles. *J. Lumin.* **2018**, *203*, 441–446. [CrossRef]
33. Pandey, A.; Rai, V.K.; Kumar, V.; Kumar, V.; Swart, H.C. Upconversion based temperature sensing ability of Er$^{3+}$–Yb$^{3+}$codoped SrWO$_4$: An optical heating phosphor. *Sens. Actuators B Chem.* **2015**, *209*, 352–358. [CrossRef]
34. Song, H.; Wang, C.; Han, Q.; Tang, X.; Yan, W.; Chen, Y.; Jiang, J.; Liu, T. Highly sensitive Tm$^{3+}$/Yb$^{3+}$ codoped SrWO$_4$ for optical thermometry. *Sens. Actuators A Phys.* **2018**, *271*, 278–282. [CrossRef]
35. Li, G.; Gao, S.; He, W. Preparation and photoluminescence properties of the Sm$^{3+}$, Eu$^{3+}$ co-doped CaWO$_4$ phosphors. *Optik* **2015**, *126*, 3272–3275. [CrossRef]
36. Kang, F.; Hu, Y.; Wu, H.; Mu, Z.; Ju, G.; Fu, C.; Li, N. Luminescence and red long afterglow investigation of Eu$^{3+}$–Sm$^{3+}$ CO-doped CaWO$_4$ phosphor. *J. Lumin.* **2012**, *132*, 887–894. [CrossRef]
37. Ren, Y.; Liu, Y.; Yang, R. A series of color tunable yellow–orange–red-emitting SrWO$_4$:RE (Sm$^{3+}$, Eu$^{3+}$–Sm$^{3+}$) phosphor for near ultraviolet and blue light-based warm white light emitting diodes. *Superlattices Microstruct.* **2016**, *91*, 138–147. [CrossRef]

Article

# Rapid Aqueous-Phase Synthesis and Photoluminescence Properties of $K_{0.3}Bi_{0.7}F_{2.4}$:$Ln^{3+}$ (Ln = Eu, Tb, Pr, Nd, Sm, Dy) Nanocrystalline Particles

Weili Wang, Shihai Miao *, Dongxun Chen * and Yanjie Liang *

Key Laboratory for Liquid-Solid Structure Evolution and Processing of Materials, Ministry of Education, Shandong University, Jinan 250061, China; wangweili@sdu.edu.cn
* Correspondence: 202020485@mail.sdu.edu.cn (S.M.); 202120530@mail.sdu.edu.cn (D.C.); yanjie.liang@sdu.edu.cn (Y.L.)

**Abstract:** Trivalent lanthanides ($Ln^{3+}$) doped bismuth-based inorganic compounds have attracted considerable interest as promising candidates for next-generation inorganic luminescent materials. Here, a series of $K_{0.3}Bi_{0.7}F_{2.4}$ (KBF) nanocrystalline particles with controlled morphology have been synthesized through a low-temperature aqueous-phase precipitation method. Using KBF as the host matrix, $Eu^{3+}$, $Tb^{3+}$, $Pr^{3+}$, $Nd^{3+}$, $Sm^{3+}$, and $Dy^{3+}$ ions are introduced to obtain $K_{0.3}Bi_{0.7}F_{2.4}$:$Ln^{3+}$ (KBF:Ln) nanophosphors. The as-prepared KBF:Ln nanophosphors exhibit commendable photoluminescence properties, in which multicolor emissions in a single host lattice can be obtained by doping different $Ln^{3+}$ ions when excited by ultraviolet light. Moreover, the morphology and photoluminescence performance of these nanophosphors remain unchanged under different soaking times in water, showing good stability in a humid environment. The proposed simple and rapid synthesis route, low-cost and nontoxic bismuth-based host matrix, and tunable luminescent colors will lead the way to access these KBF:Ln nanophosphors for appealing applications such as white LEDs and optical thermometry.

**Keywords:** $K_{0.3}Bi_{0.7}F_{2.4}$; nanophosphors; lanthanide; photoluminescence

Citation: Wang, W.; Miao, S.; Chen, D.; Liang, Y. Rapid Aqueous-Phase Synthesis and Photoluminescence Properties of $K_{0.3}Bi_{0.7}F_{2.4}$:$Ln^{3+}$ (Ln = Eu, Tb, Pr, Nd, Sm, Dy) Nanocrystalline Particles. *Crystals* **2022**, *12*, 963. https://doi.org/10.3390/cryst12070963

Academic Editors: Alessandra Toncelli and Željka Antić

Received: 17 June 2022
Accepted: 8 July 2022
Published: 10 July 2022

**Publisher's Note:** MDPI stays neutral with regard to jurisdictional claims in published maps and institutional affiliations.

**Copyright:** © 2022 by the authors. Licensee MDPI, Basel, Switzerland. This article is an open access article distributed under the terms and conditions of the Creative Commons Attribution (CC BY) license (https://creativecommons.org/licenses/by/4.0/).

## 1. Introduction

Inorganic luminescent nanoparticles have aroused tremendous attention compared with their bulk counterparts owing to their unique features such as tunable emission peaks, sharp emission bandwidth, large stokes shift, and excellent luminescence stability against heat and irradiation [1–4]. As a result, trivalent lanthanides doped inorganic nanoparticles show a rather wide range of applications in areas such as spectroscopic analysis, anti-counterfeiting, biomedicine, and light-emitting diodes (LEDs) [5–9]. In the past two decades, a large number of novel luminescent nanomaterials have been designed and developed to take advantage of their nanoscale size and distinct luminescence properties. Among them, rare earth fluorides are becoming a hot research topic in consideration of their high chemical stability, very low lattice phonon energy, and good accommodation for $Ln^{3+}$ [10–15]. Nevertheless, substantial rare earth salts are usually required to synthesize these rare earth fluoride nanomaterials, which greatly limits their further development and application. Therefore, it is urgent and necessary to further explore new and efficient host materials for luminescent lanthanides.

Bismuth-based inorganic materials are considered promising candidates for developing the next-generation inorganic phosphors because of the advantages of the relatively inexpensive, nontoxic, and unique electronic structure of bismuth elements [16–19]. Among these bismuth-based inorganic hosts, bismuth fluorides have attracted significant attention [20,21]. As a result of extensive research, lanthanide-doped bismuth fluorides have been found numerous applications in photocatalysis (e.g., BiOF:$Eu^{3+}$ and

$Bi_7O_5F_{11}:Eu^{3+}$) [22,23], white LEDs (e.g., $NaBiF_4:Eu^{3+}$ and $BiF_3:Eu^{3+}$) [24,25] and optical thermometry (e.g., $BiF_3:Yb^{3+},Er^{3+}$) [26]. Recently, $K_{0.3}Bi_{0.7}F_{2.4}$ is emerging as a more attractive host matrix for lanthanide doping to realize tunable emissions in a single phosphor. For instance, An et al. synthesized a series of $Yb^{3+}/Ln^{3+}$ (Ln = Er, Ho, Tm) co-doped $K_{0.3}Bi_{0.7}F_{2.4}$: upconverting nanoparticles for dual-modal in vivo imaging through a solvothermal method at 200 °C for 10 h in ethylene glycol [27]. Gao et al. reported a room-temperature chemical precipitation method to synthesize $K_{0.3}Bi_{0.7}F_{2.4}:Yb^{3+}$, $Er^{3+}$, and $K_{0.3}Bi_{0.7}F_{2.4}:Eu^{3+}$ nanoparticles by using ethylene glycol as the reaction solvent [28,29]. Du et al. have used a similar precipitation method to prepare $Tb^{3+}/Eu^{3+}$-codoped $K_{0.3}Bi_{0.7}F_{2.4}$ nanoparticles for white LED application [30]. In our previous research, we synthesized the $Yb^{3+}/Ln^{3+}$ (Ln = Er, Ho, Tm)-doped $K_{0.3}Bi_{0.7}F_{2.4}$ nanoparticles by using a very fast (1 min only) method in a water-based system at very low temperature (room temperature−90 °C) [31]. However, rapid synthesis of other lanthanide ions ($Pr^{3+}$, $Nd^{3+}$, $Sm^{3+}$, $Dy^{3+}$, $Eu^{3+}$, and $Tb^{3+}$) doped KBF nanocrystals in the water-based system and their photoluminescence properties are rarely reported.

Here we report the controlled synthesis and multicolor luminescence properties of KBF:Ln nanophosphors synthesized by a low-temperature water-based precipitation method. The composition and morphologies of the obtained KBF:Ln nanophosphors are investigated in detail by varying the synthesis conditions. The obtained KBF:Ln nanophosphors show tunable emission colors and good chemical stability against humidity. The results suggest that the as-prepared KBF:Ln nanophosphors are promising candidates for application in white LEDs and optical thermometry.

## 2. Experimental Process

$BiCl_3$ (Macklin, Shanghai, China), KF (99.9%, Macklin, Shanghai, China), HCl (37 wt%, Sinopharm, Shanghai, China), $PrCl_3 \cdot 6H_2O$ (99.99%, Aladdin, Shanghai, China), $NdCl_3 \cdot 6H_2O$ (99.9%, Aladdin, Shanghai, China), $SmCl_3 \cdot 6H_2O$ (99.99%, Aladdin, Shanghai, China), $DyCl_3 \cdot 6H_2O$ (99.99%, Aladdin, Shanghai, China), $EuCl_3 \cdot 6H_2O$ (99.99%, Aladdin, Shanghai, China) and $TbCl_3 \cdot 6H_2O$ (99.9%, Aladdin, Shanghai, China) are used as the starting materials. The pure and doped KBF samples were synthesized by a low-temperature water-based precipitation method in accordance with our earlier study [31]. The detailed synthesis procedure and sample characterization are presented in the Supporting Information.

## 3. Results and Discussion

### 3.1. Phase Composition of the Pure and Doped KBF Nanocrystals

The phase composition of the prepared pure KBF nanocrystals is first checked by XRD. XRD patterns of the as-prepared non-doped KBF hosts at room temperature are depicted in Figure 1. When the ratios of F/Bi are 4 and 8, the XRD patterns of the obtained samples can be well matched with the corresponding standard XRD data of the cubic $K_{0.3}Bi_{0.7}F_{2.4}$ (JCPDS No. 84–0534), as presented in Figure 1a. The narrow and strong XRD peaks reveal that the pure KBF nanocrystalline particles are highly crystalline. However, as the ratio of F/Bi reaches 12 or higher, it is found that the impurity phase starts to appear at around 27°, assigned to the hexagonal $KBiF_4$ (JCPDS No. 37–0972). Meanwhile, when the F/Bi ratio is set to 8, all the XRD results of the pure KBF nanocrystals obtained at different temperatures (room temperature, 50 °C, and 80 °C) are consistent with the standard pattern of $K_{0.3}Bi_{0.7}F_{2.4}$ crystal, as shown in Figure 1b, indicating that pure KBF can be synthesized in the water-based system at very low temperature. Furthermore, the full width at half-maximum (FWHM) with the dominant XRD peak at 26° presents a declining trend with the increase of reaction temperature, illuminating that the degree of crystallization becomes better for pure KBF nanocrystalline particles at a high synthetic temperature.

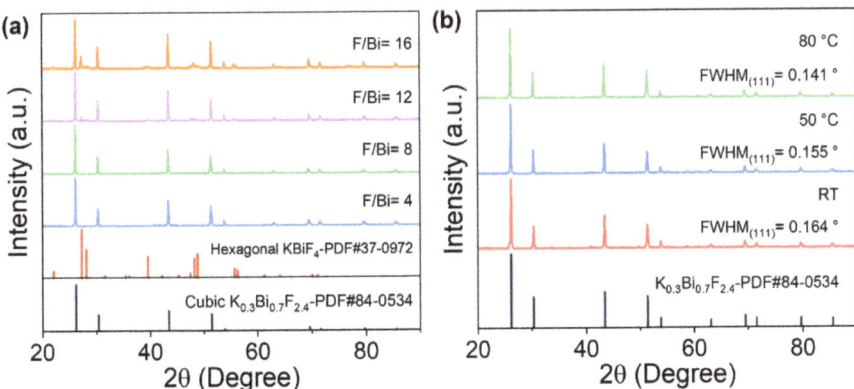

**Figure 1.** XRD patterns of pure KBF synthesized at different F/Bi ratios (a) and reaction temperatures (b).

Figure 2 presents the XRD results of the KBF:Ln$^{3+}$ (Ln = Pr, Nd, Sm, Dy, Eu, and Tb) nanocrystalline particles. It can be noted that the XRD patterns of the doped KBF nanocrystals are consistent with the cubic K$_{0.3}$Bi$_{0.7}$F$_{2.4}$ compound, which suggests that introducing different lanthanide ions will not lead to the formation of an impurity phase. Meanwhile, the KBF host matrix can accommodate a high concentration of Ln$^{3+}$ (e.g., 25%Eu$^{3+}$ and 30%Tb$^{3+}$) without changing its crystal structure, indicating that KBF is an attractive host for Ln$^{3+}$ doping to obtain tunable luminescence properties. Furthermore, the average size of KBF:Ln nanocrystalline particles can be estimated by using the Debye Scherrer equation. And the crystallite sizes of KBF:30%Tb, KBF:25%Eu, KBF:5%Dy, KBF:5%Sm, KBF:2%Nd and KBF:5%Pr are calculated to be 20.41, 28.31, 39.75, 45.94, 50.87 and 44.98 nm, respectively.

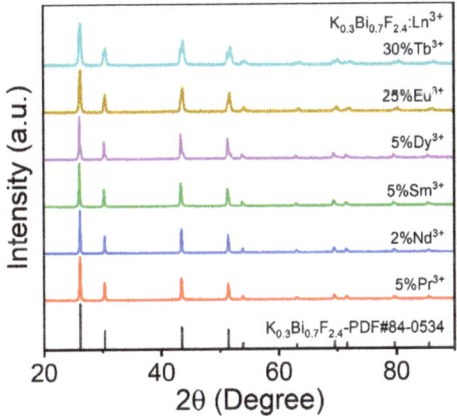

**Figure 2.** XRD patterns of the doped KBF nanocrystals.

### 3.2. Morphology Characterization of Pure and Doped KBF Nanocrystalline Particles

The SEM graphs can present the particle size and typical morphology of the as-prepared pure and doped KBF nanocrystals directly and vividly. Figure 3 depicts the representative morphologies of the non-doped KBF nanocrystals synthesized at different ratios of F/Bi. Note that the reaction temperature is room temperature. As presented in Figure 3a,b, when the ratios of F/Bi are determined to be 4 and 8, the obtained pure KBF presents a regular cube shape, and the average size of the cubes decreases as the F/Bi ratio increases. However, as the ratio value of F/Bi increases to 12 or 16, the morphology of the obtained nanocrystalline particles changes from a cubic shape to an irregular polyhedron. These results indicate that the F/Bi ratio significantly affects the morphology of the final products.

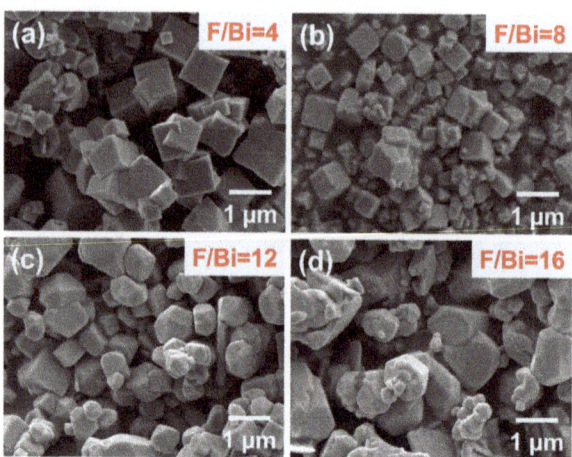

**Figure 3.** SEM photographs of the pure KBF nanocrystals synthesized at different ratios of F/Bi, (**a**) F/Bi = 4; (**b**) F/Bi = 8; (**c**) F/Bi = 12; (**d**) F/Bi = 16.

Figure 4 presents the SEM graphs of the KBF nanocrystals obtained at different synthetic temperatures. The F/Bi ratio is set to 8. As the reaction temperature increases from room temperature to 80 °C, the cubic morphology of the samples remains almost unchanged, but it is found that the average particle size of the cubic particles increases slightly with an increase in the synthetic temperature.

**Figure 4.** SEM photographs of pure KBF nanocrystals obtained at different synthetic temperatures, (**a**) room temperature; (**b**) T = 50 °C; (**c**) T = 80 °C.

The SEM images of the KBF:Ln (Ln = Pr, Nd, Sm, Dy, Eu, and Tb) nanocrystalline particles synthesized at room temperature are given in Figure 5. When the ratio of F/Bi is determined to be 8, all the as-prepared KBF:Ln samples consist of cubic particles and irregular nanoparticles attached to cubes after introducing different lanthanide ions into the KBF lattice. Compared with the non-doped KBF, the as-prepared KBF:Ln nanocrystalline particles significantly decrease in size. Moreover, the proportion of nanoparticles shows an increasing trend, and the content of cubic particles gradually decreases with an increase in the doping concentration of $Ln^{3+}$.

**Figure 5.** SEM images of the KBF:Ln (Ln = Pr, Nd, Sm, Dy, Eu, and Tb) nanocrystalline particles synthesized at room temperature, introducing (**a**) 5% $Pr^{3+}$; (**b**) 2% $Nd^{3+}$; (**c**) 5% $Sm^{3+}$; (**d**) 5% $Dy^{3+}$; (**e**) 25% $Eu^{3+}$; (**f**) 25% $Tb^{3+}$.

### 3.3. Photoluminescence Properties of KBF:Ln Nanophosphors

The concentration-dependent emission spectra of KBF:Eu and KBF:Tb nanocrystals are presented in Figure S1. And the optimum doping concentration of $Eu^{2+}$ and $Tb^{3+}$ is determined to be 25% and 30%, respectively. The photoluminescence excitation and emission spectra of KBF:25%Eu are given in Figure 6a. The emission spectrum of KBF:Eu consists of a group of sharp lines at 578, 592, 612, 651, and 700 nm upon 393 nm near-UV light excitation, which can be attributed to the characteristic $^5D_0 \rightarrow {}^7F_J$ ($J$ = 0, 1, 2, 3, 4) electron transitions of $Eu^{3+}$ ions, respectively [32]. Among these emission bands, the most intense one, peaking at 612 nm, originates from the $Eu^{3+}$ $^5D_0 \rightarrow {}^7F_2$ electric-dipole transition. The excitation spectrum monitored at 612 nm emission is made up of multiple excitation bands centered at 362, 376, 393, and 414 nm due to the electron transitions from the $^7F_0$ ground state to the $^5D_4$, $^5G_3$, $^5L_6$, and $^5D_3$ excited levels. Figure 6d shows the excitation and emission spectra of KBF:30%Tb nanocrystalline particles. The emission spectrum exhibits four distinct peaks at 487, 544, 582, and 622 nm when excited by 368 nm of UV light, resulting from the $^5D_4 \rightarrow {}^7F_J$ ($J$ = 6, 5, 4, 3) electron transitions of $Tb^{3+}$ ions, respectively [33]. And the most prominent emission peak is located at 544 nm, assigned to the $^5D_4 \rightarrow {}^7F_5$ transition of $Tb^{3+}$. The excitation spectrum monitored 544 nm emission comprises excitation bands at 318, 350, and 368 nm due to the $^7F_6 \rightarrow {}^5D_{1,0}$, $^7F_6 \rightarrow {}^5G_{6-2}$, and $^7F_6 \rightarrow {}^5D_{2,3}$ transitions of $Tb^{3+}$ ions. Furthermore, it is found that the photoluminescence intensities of KBF:Eu and KBF:Tb nanocrystals decrease gradually with increasing the test temperature from room temperature to 210 °C, which can be attributed to the increase in the non-radiative transition probability, as shown in Figure S2.

To further reveal the chemical stability of the prepared KBF:Ln nanocrystalline particles, the emission spectra of KBF:25%Eu and KBF:30%Tb phosphor powders immersed in deionized water for different days (0, 15, and 30 days) are shown in Figure 6b,e. It can be noticed that the photoluminescence intensity of both KBF:25%Eu and KBF:30%Tb nanophosphors remains almost unchanged under different soaking times in water, indicating that the as-prepared nanophosphors show excellent chemical stability against water.

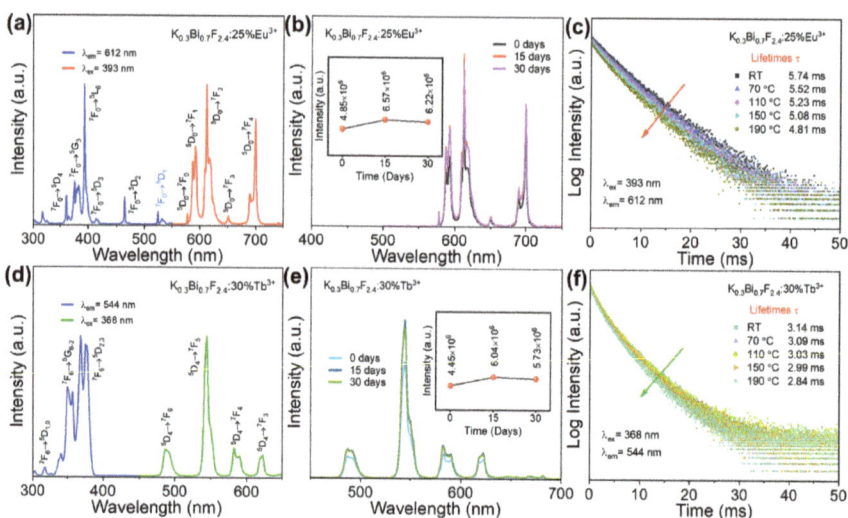

**Figure 6.** Photoluminescence properties of KBF:25%Eu (**a–c**) and KBF:30%Tb nanocrystalline particles (**d–f**).

Luminescence decay curves of KBF:25%Eu and KBF:30%Tb samples are depicted in Figure 6c,f. The decay curves can be fitted with a single exponential equation [34]:

$$I(t) = I_0 \exp(-t/\tau) \tag{1}$$

where $I(t)$ and $I_0$ represent the photoluminescence intensity at time $t$ and the background intensity for the dark current of the detector, and $\tau$ is the lifetime. Therefore, the average lifetimes of KBF:25%Eu were estimated to be 5.74, 5.52, 5.23, 5.08, and 4.81 ms at different testing temperatures, while the average lifetimes of KBF:30%Tb were determined to be 3.14, 3.09, 3.03, 2.99 and 2.84 ms at different temperatures. For both KBF:25%Eu and KBF:30%Tb samples, the lifetimes show a decreasing trend with the increase of the test temperature, which can be attributed to the increase in the probability of non-radiative transition at high temperature [35]. In addition, the internal quantum efficiencies of KBF:25%Eu and KBF:30%Tb nanocrystals are measured to be 42.39% and 13.75%, respectively (Figure S3). The obtained quantum efficiency values of KBF:25%Eu and KBF:30%Tb samples are comparable with that of the previous reported fluorides, such as $K_{0.3}Bi_{0.7}F_{2.4}$:40%Eu$^{3+}$ (14.9%) [29] and $K_{0.3}Bi_{0.7}F_{2.4}$:9%Tb$^{3+}$,0.6%Eu$^{3+}$ (50.8%) [30].

The photoluminescence emission and excitation spectra of the KBF:Ln$^{3+}$ (Ln = Pr, Nd, Sm, Dy) nanocrystals are depicted in Figure 7. As presented in Figure 7a, the excitation spectrum of KBF:5%Pr nanocrystal monitored at 606 nm emission is composed of three distinct excitation bands centered at 443, 467, and 481 nm, which are caused by the characteristic $^3H_4 \rightarrow {}^3P_J$ ($J$ = 0, 1, 2) electron transitions of Pr$^{3+}$ ions, respectively [36]. The emission spectrum with a dominant peak at 606 nm in a wavelength range over 500–750 nm can be assigned to the Pr$^{3+}$ $^1D_2 \rightarrow {}^3H_4$ transition [37]. The emission spectrum of KBF:2%Nd, upon 808 nm excitation, shows three emission bands at 870, 1054, and 1325 nm. This can be attributed to the characteristic $^4F_{3/2} \rightarrow {}^4I_{9/2}$, $^4F_{3/2} \rightarrow {}^4I_{11/2}$ and $^4F_{3/2} \rightarrow {}^4I_{13/2}$ transitions of Nd$^{3+}$, respectively, as shown in Figure 7b, while its excitation spectrum is made up of a group of sharp lines in the visible and NIR regions arising from intraconfigurational f-f transitions of Nd$^{3+}$ [38]. Figure 7c gives the photoluminescence properties of the yielded KBF:5%Sm nanocrystalline particles. The emission spectrum shows several distinct bands centered at 559, 595, 641, and 700 nm ascribed to the $^4G_{5/2} \rightarrow {}^6H_J$ ($J$ = 5/2, 7/2, 9/2, and 11/2) transitions of Sm$^{3+}$ ions upon 399 nm excitation, respectively [36]. The excitation spectrum of KBF:5%Sm monitored at 595 nm emission spans from 300 to 500 nm with a dominant excitation peak at 399 nm ascribed to the $^6H_{5/2} \rightarrow {}^4K_{11/2}$ electron transition of $Sm^{3+}$ [39]. Figure 7d depicts the photoluminescence properties of KBF:5%Dy nanocrystals. The excitation spectrum monitored at 572 nm emission is composed of a group of narrow excitation bands in the UV assigned to the electron transitions from $^6H_{15/2}$ to $^6P_{3/2}$, $^6P_{7/2}$, $^6P_{5/2}$, and $^4I_{13/2}$ excited levels of $Dy^{3+}$ ions, respectively [40]. Under the excitation of 364 nm, the emission spectrum exhibits four distinct bands with a maximum emission peak of 572 nm due to the $Dy^{3+}$ $^4F_{9/2} \rightarrow {}^6H_{13/2}$ transition. Overall, the emission colors of the as-prepared nanocrystals can be finely tuned by doping different $Ln^{3+}$ in a single KBF host.

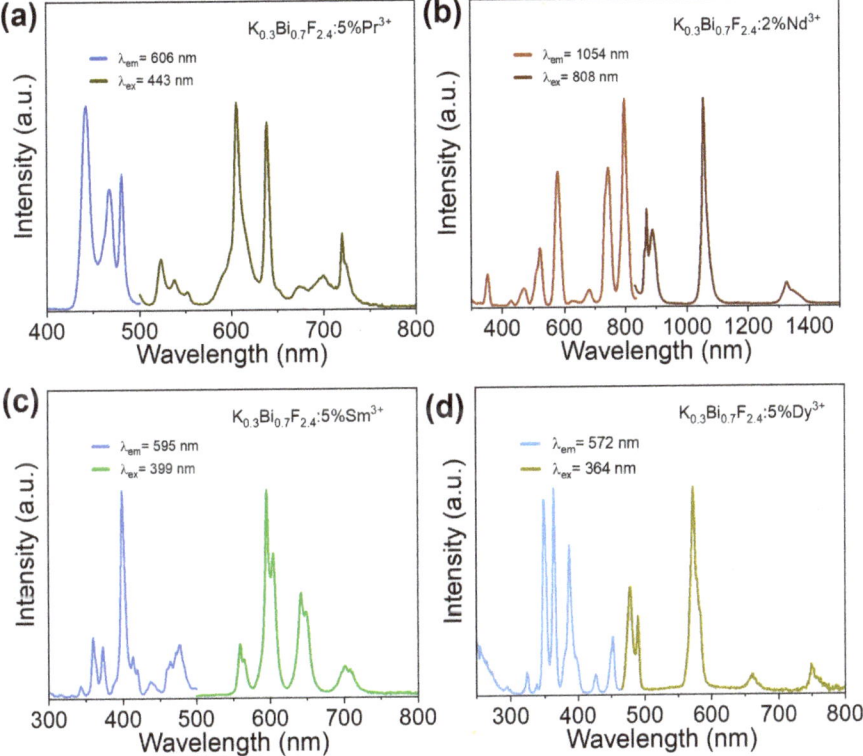

**Figure 7.** Photoluminescence properties of KBF:Ln (Ln = Pr, Nd, Sm, Dy) nanocrystalline particles (a–d).

## 4. Conclusions

In conclusion, we have successfully prepared pure and doped KBF nanocrystalline particles via a rapid water-based precipitation route. With varying the ratios of F/Bi and reaction temperatures, the morphology of samples can be tuned from cubic shape to irregular polyhedron. The as-prepared KBF:Ln (Ln = Pr, Nd, Sm, Dy, Eu, and Tb) nanophosphors exhibit intense characteristic photoluminescence from the electron transitions of the doped trivalent lanthanides. Meanwhile, the morphology and photoluminescence performance of these KBF:Ln nanophosphors remain almost unchanged even when immersed in water for 30 days, indicating the obtained KBF:Ln samples have excellent chemical stability. The merit of multicolor luminescence in these KBF:Ln nanocrystalline particles holds great promise when applied to a wide range of important fields, such as optoelectronic devices and optical thermometry.

**Supplementary Materials:** The following supporting information can be downloaded at: https://www.mdpi.com/article/10.3390/cryst12070963/s1, Figure S1: Emission spectra of the KBF:Ln (Ln = Eu, Tb) samples doped with different $Ln^{3+}$ concentrations; Figure S2: temperature-dependent emission spectra of the KBF:25%Eu (a) and KBF:30%Tb (b) samples; Figure S3: Excitation line of $BaSO_4$ and emission spectrum of KBF:25%Eu (a) and KBF:30%Tb (b) samples. Inset shows the magnification of the emission spectrum. Table S1: The crystallite size of pure KBF synthesized at different reaction temperatures. Table S2: The crystallite size of the lanthanide doped KBF nanocrystalline particles.

**Author Contributions:** Conceptualization, W.W., S.M., D.C. and Y.L.; methodology, W.W., S.M. and D.C.; software, S.M. and D.C.; validation, W.W. and Y.L.; data curation, S.M. and D.C.; writing—original draft preparation, W.W., S.M. and D.C.; writing—review and editing, Y.L. All authors have read and agreed to the published version of the manuscript.

**Funding:** This work was financially supported by the National Natural Science Foundation of China (Grant No. 51902184), Key Research and the Development Program of Shandong Province (Major Scientific and Technological Innovation Project) (Grant No. 2021CXGC011101), the Natural Science Foundation of Shandong Province (Grant Nos. ZR2019BEM028 and ZR2020ME028) and the "Qi-Lu Young Scholar Fund" (Grant No. 31370088963167) from Shandong University.

**Institutional Review Board Statement:** Not applicable.

**Informed Consent Statement:** Not applicable.

**Data Availability Statement:** All supporting and actual data are presented in the manuscript.

**Conflicts of Interest:** The authors declare they have no known competing financial interests or personal relationships that could appear to influence the work reported in this paper.

## References

1. Terraschke, H.; Wickleder, C. UV, blue, green, yellow, red, and small: Newest developments on $Eu^{2+}$-doped nanophosphors. *Chem. Rev.* **2015**, *115*, 11352–11378. [CrossRef] [PubMed]
2. Algar, W.R.; Massey, M.; Rees, K.; Higgins, R.; Krause, K.D.; Darwish, G.H.; Peveler, W.J.; Xiao, Z.; Tsai, H.Y.; Gupta, R.; et al. Photoluminescent nanoparticles for chemical and biological analysis and imaging. *Chem. Rev.* **2021**, *121*, 9243–9358. [CrossRef] [PubMed]
3. Ren, W.; Lin, G.; Clarke, C.; Zhou, J.; Jin, D. Optical nanomaterials and enabling technologies for high-security-level anticounterfeiting. *Adv. Mater.* **2020**, *32*, 1901430. [CrossRef] [PubMed]
4. Kim, H.; Beack, S.; Han, S.; Shin, M.; Lee, T.; Park, Y.; Kim, K.S.; Yetisen, A.K.; Yun, S.H.; Kwon, W.; et al. Multifunctional photonic nanomaterials for diagnostic, therapeutic, and theranostic applications. *Adv. Mater.* **2018**, *30*, 1701460. [CrossRef] [PubMed]
5. Chen, D.; Zhang, L.; Liang, Y.; Wang, W.; Yan, S.; Bi, J.; Sun, K. Yolk–shell structured $Bi_2SiO_5$:$Yb^{3+}$,$Ln^{3+}$ (Ln = Er, Ho, Tm) upconversion nanophosphors for optical thermometry and solid-state lighting. *CrystEngComm* **2020**, *22*, 4438. [CrossRef]
6. Wang, F.; Liu, X. Multicolor tuning of lanthanide-doped nanoparticles by single wavelength excitation. *Acc. Chem. Res.* **2014**, *47*, 1378–1385. [CrossRef] [PubMed]
7. Zeng, Z.; Huang, B.; Wang, X.; Lu, L.; Lu, Q.; Sun, M.; Wu, T.; Ma, T.; Xu, J.; Xu, Y.; et al. Multimodal luminescent $Yb^{3+}$/$Er^{3+}$/$Bi^{3+}$-doped perovskite single crystals for X-ray detection and anti-counterfeiting. *Adv. Mater.* **2020**, *32*, 2004506. [CrossRef] [PubMed]
8. Xie, X.; Li, Z.; Zhang, Y.; Guo, S.; Pendharkar, A.I.; Lu, M.; Huang, L.; Huang, W.; Han, G. Emerging ≈800 nm excited lanthanide-doped upconversion nanoparticles. *Small* **2017**, *13*, 1602843. [CrossRef] [PubMed]
9. Huang, H.; Chen, J.; Liu, Y.; Lin, J.; Wang, S.; Huang, F.; Chen, D. Lanthanide-doped core@multishell nanoarchitectures: Multimodal excitable upconverting/downshifting luminescence and high-level anti-counterfeiting. *Small* **2020**, *16*, 2000708. [CrossRef]
10. Vetrone, F.; Naccache, R.; Zamarrón, A.; de la Fuente, A.J.; Sanz-Rodríguez, F.; Maestro, L.M.; Rodriguez, E.M.; Jaque, D.; Solé, J.G.; Capobianco, J.A. Temperature sensing using fluorescent nanothermometers. *ACS Nano* **2010**, *4*, 3254–3258. [CrossRef]
11. Zhuang, Y.; Chen, D.; Chen, W.; Zhang, W.; Su, X.; Deng, R.; An, Z.; Chen, H.; Xie, R.J. X-ray-charged bright persistent luminescence in $NaYF_4$:$Ln^{3+}$@$NaYF_4$ nanoparticles for multidimensional optical information storage. *Light Sci. Appl.* **2021**, *10*, 132. [CrossRef] [PubMed]
12. Mai, H.X.; Zhang, Y.W.; Si, R.; Yan, Z.G.; Sun, L.D.; You, L.P.; Yan, C.H. High-quality sodium rare-earth fluoride nanocrystals: Controlled synthesis and optical properties. *J. Am. Chem. Soc.* **2006**, *128*, 6426–6436. [CrossRef] [PubMed]
13. Gnanasammandhan, M.K.; Idris, N.M.; Bansal, A.; Huang, K.; Zhang, Y. Near-IR photoactivation using mesoporous silica-coated $NaYF_4$:Yb,Er/Tm upconversion nanoparticles. *Nat. Protoc.* **2016**, *11*, 688–713. [CrossRef]
14. Yi, G.S.; Chow, G.M. Synthesis of hexagonal-phase $NaYF_4$:Yb,Er and $NaYF_4$:Yb,Tm nanocrystals with efficient up-conversion fluorescence. *Adv. Funct. Mater.* **2006**, *16*, 2324–2329. [CrossRef]

15. Liu, D.; Xu, X.; Du, Y.; Qin, X.; Zhang, Y.; Ma, C.; Wen, S.; Ren, W.; Goldys, E.M.; Piper, J.A.; et al. Three-dimensional controlled growth of monodisperse sub-50 nm heterogeneous nanocrystals. *Nat. Commun.* **2016**, *7*, 10254. [CrossRef] [PubMed]
16. Mohan, R. Green bismuth. *Nat. Chem.* **2010**, *2*, 336. [CrossRef] [PubMed]
17. Swart, H.C.; Kroon, R.E. Ultraviolet and visible luminescence from bismuth doped materials. *Opt. Mater. X* **2019**, *2*, 100025. [CrossRef]
18. Zhou, Z.; Wang, X.; Yi, X.; Ming, H.; Ma, Z.; Peng, M. Rechargeable and sunlight-activated $Sr_3Y_2Ge_3O_{12}$:$Bi^{3+}$ UV–Visible-NIR persistent luminescence material for night-vision signage and optical information storage. *Chem. Eng. J.* **2021**, *421*, 127820. [CrossRef]
19. Zeng, Z.; Sun, M.; Zhang, S.; Zhang, H.; Shi, X.; Ye, S.; Huang, B.; Du, Y.; Yan, C.H. Rare-earth-based perovskite $Cs_2AgScCl_6$:Bi for strong full visible spectrum emission. *Adv. Funct. Mater.* **2022**, 2204780. [CrossRef]
20. Lei, P.; An, R.; Yao, S.; Wang, Q.; Dong, L.; Xu, X.; Du, K.; Feng, J.; Zhang, H. Ultrafast synthesis of novel hexagonal phase $NaBiF_4$ upconversion nanoparticles at room temperature. *Adv. Mater.* **2017**, *29*, 1700505. [CrossRef]
21. Back, M.; Ueda, J.; Ambrosi, E.; Cassandro, L.; Cristofori, D.; Ottini, R.; Riello, P.; Sponchia, G.; Asami, K.; Tanabe, S.; et al. Lanthanide-doped bismuth-based fluoride nanocrystalline particles: Formation, spectroscopic investigation, and chemical stability. *Chem. Mater.* **2019**, *31*, 8504–8514. [CrossRef]
22. Saraf, R.; Shivakumara, C.; Behera, S.; Dhananjaya, N.; Nagabhushana, H. Synthesis of $Eu^{3+}$-activated BiOF and BiOBr phosphors: Photoluminescence, Judd–Ofelt analysis and photocatalytic properties. *RSC Adv.* **2015**, *5*, 9241–9254. [CrossRef]
23. Wei, D.; Huang, Y.; Seo, H.J. $Eu^{3+}$-doped $Bi_7O_5F_{11}$ microplates with simultaneous luminescence and improved photocatalysis. *APL Mater.* **2020**, *8*, 081109. [CrossRef]
24. Du, P.; Huang, X.; Yu, J.S. Facile synthesis of bifunctional $Eu^{3+}$-activated $NaBiF_4$ red-emitting nanoparticles for simultaneous white light-emitting diodes and field emission displays. *Chem. Eng. J.* **2018**, *337*, 91–100. [CrossRef]
25. Sarkar, S.; Dash, A.; Mahalingam, V. Strong stokes and upconversion luminescence from ultrasmall $Ln^{3+}$-doped $BiF_3$ (Ln = $Eu^{3+}$, $Yb^{3+}/Er^{3+}$) nanoparticles confined in a polymer matrix. *Chem. Asian J.* **2014**, *9*, 447–451. [CrossRef]
26. Zhao, S.; Tian, R.; Shao, B.; Feng, Y.; Yuan, S.; Dong, L.; Zhang, L.; Wang, Z.; You, H. One-pot synthesis of $Ln^{3+}$-doped porous $BiF_3$@PAA nanospheres for temperature sensing and pH-responsive drug delivery guided by CT imaging. *Nanoscale* **2020**, *12*, 695–702. [CrossRef]
27. An, R.; Lei, P.; Zhang, P.; Xu, X.; Feng, J.; Zhang, H. Near-infrared optical and X-ray computed tomography dual-modal imaging probe based on novel lanthanide-doped $K_{0.3}Bi_{0.7}F_{2.4}$ upconversion nanoparticles. *Nanoscale* **2018**, *10*, 1394–1402. [CrossRef] [PubMed]
28. Gao, X.; Song, F.; Ju, D.; Zhou, A.; Khan, A.; Chen, Z.; Sang, X.; Feng, M.; Liu, L. Room-temperature ultrafast synthesis, morphology and upconversion luminescence of $K_{0.3}Bi_{0.7}F_{2.4}$:$Yb^{3+}/Er^{3+}$ nanoparticles for temperature-sensing application. *CrystEngComm* **2020**, *22*, 7066–7074. [CrossRef]
29. Gao, X.; Song, F.; Khan, A.; Chen, Z.; Ju, D.; Sang, X. Room temperature synthesis, Judd Ofelt analysis and photoluminescence properties of down-conversion $K_{0.3}Bi_{0.7}F_{2.4}$:$Eu^{3+}$ orange red phosphors. *J. Lumin.* **2021**, *230*, 117707. [CrossRef]
30. Du, P.; Wan, X.; Luo, L.; Li, W.; Li, L. Thermally stable $Tb^{3+}/Eu^{3+}$-codoped $K_{0.3}Bi_{0.7}F_{2.4}$ nanoparticles with multicolor luminescence for white-light-emitting diodes. *ACS Appl. Nano Mater.* **2021**, *4*, 7062–7071. [CrossRef]
31. Chen, D.; Bi, J.; Wang, W.; Wang, X.; Zhang, Y.; Liang, Y. Rapid aqueous-phase synthesis of highly stable $K_{0.3}Bi_{0.7}F_{2.4}$ upconversion nanocrystalline particles at low temperature. *Inorg. Chem. Front.* **2021**, *8*, 1039–1048. [CrossRef]
32. Chen, D.; Miao, S.; Liang, Y.; Wang, W.; Yan, S.; Bi, J.; Sun, K. Controlled synthesis and photoluminescence properties of $Bi_2SiO_5$:$Eu^{3+}$ core-shell nanospheres with an intense $^5D_0 \rightarrow {}^7F_4$ transition. *Opt. Mater. Express* **2021**, *11*, 355–370. [CrossRef]
33. Wang, Y.; Chen, D.; Zhuang, Y.; Chen, W.; Long, H.; Chen, H.; Xie, R.J. $NaMgF_3$:$Tb^{3+}$@$NaMgF_3$ nanoparticles containing deep traps for optical information storage. *Adv. Opt. Mater.* **2021**, *9*, 2100624. [CrossRef]
34. Zhang, Y.; Miao, S.; Liang, Y.; Liang, C.; Chen, D.; Shan, Z.; Sun, K.; Wang, X.J. Blue LED-pumped intense short-wave infrared luminescence based on $Cr^{3+}$-$Yb^{3+}$-codoped phosphors. *Light Sci. & Appl.* **2022**, *11*, 136.
35. Yuan, C.; Li, R.; Liu, Y.; Zhang, L.; Zhang, Y.; Leniec, G.; Sun, P.; Liu, Z.; Luo, Z.; Dong, R.; et al. Efficient and broadband $LiGaP_2O_7$:$Cr^{3+}$ phosphors for smart near-infrared light-emitting diodes. *Laser Photonics Rev.* **2021**, *15*, 2100227. [CrossRef]
36. Liang, Y.; Liu, F.; Chen, Y.; Wang, X.; Sun, K.; Pan, Z. Red/near-infrared/short-wave infrared multi-band persistent luminescence in $Pr^{3+}$-doped persistent phosphors. *Dalton Trans.* **2017**, *46*, 11149–11153. [CrossRef]
37. Wang, B.; Lin, H.; Xu, J.; Chen, H.; Lin, Z.; Huang, F.; Wang, Y. Design, preparation, and characterization of a novel red long-persistent perovskite phosphor: $Ca_3Ti_2O_7$:$Pr^{3+}$. *Inorg. Chem.* **2015**, *54*, 11299–11306. [CrossRef]
38. Chen, D.; Liang, Y.; Miao, S.; Bi, J.; Sun, K. $Nd^{3+}$-doped $Bi_2SiO_5$ nanospheres for stable ratiometric optical thermometry in the first biological window. *J. Lumin.* **2021**, *234*, 117967. [CrossRef]
39. Li, Y.C.; Chang, Y.H.; Lin, Y.F.; Chang, Y.S.; Lin, Y.J. Synthesis and luminescent properties of $Ln^{3+}$ ($Eu^{3+}$, $Sm^{3+}$, $Dy^{3+}$)-doped lanthanum aluminum germanate $LaAlGe_2O_7$ phosphors. *J. Alloys Compd.* **2007**, *439*, 367–375. [CrossRef]
40. Tian, T.; Feng, H.; Zhang, Y.; Zhou, D.; Shen, H.; Wang, H.; Xu, J. Crystal growth and luminescence properties of $Dy^{3+}$ and $Ge^{4+}$ co-doped $Bi_4Si_3O_{12}$ single crystals for high power warm white LED. *Crystals* **2017**, *7*, 249. [CrossRef]

Article

# Infrared Photoluminescence of Nd-Doped Sesquioxide and Fluoride Nanocrystals: A Comparative Study

Fulvia Gennari [1], Milica Sekulić [2], Tanja Barudžija [3], Željka Antić [2], Miroslav D. Dramićanin [2] and Alessandra Toncelli [1,4,5,*]

[1] Dipartimento di Fisica, Università di Pisa, Largo B. Pontecorvo 3, 6127 Pisa, Italy; fulvia.gennari@phd.unipi.it
[2] Centre of Excellence for Photoconversion, Vinča Institute of Nuclear Sciences—National Institute of the Republic of Serbia, University of Belgrade, 11001 Belgrade, Serbia; msekulic@vinca.rs (M.S.); zeljkaa@gmail.com (Ž.A.); dramican@vinca.rs (M.D.D.)
[3] Department of Theoretical Physics and Condensed Matter Physics, Vinča Institute of Nuclear Sciences—National Institute of the Republic of Serbia, University of Belgrade, 11001 Belgrade, Serbia; tbarudzija@vinca.rs
[4] Istituto Nanoscienze CNR, Piazza San Silvestro 12, 56127 Pisa, Italy
[5] Istituto Nazionale di Fisica Nucleare-Sezione di Pisa, Largo B. Pontecorvo 3, 56127 Pisa, Italy
* Correspondence: alessandra.toncelli@unipi.it; Tel.: +39-050-2214-556

**Abstract:** Lanthanide ions possess various emission channels in the near-infrared region that are well known in bulk crystals but are far less studied in samples with nanometric size. In this work, we present the infrared spectroscopic characterization of various Nd-doped fluoride and sesquioxide nanocrystals, namely Nd:Y$_2$O$_3$, Nd:Lu$_2$O$_3$, Nd:Sc$_2$O$_3$, Nd:YF$_3$, and Nd:LuF$_3$. Emissions from the three main emission bands in the near-infrared region have been observed and the emission cross-sections have been calculated. Moreover, another decay channel at around 2 μm has been observed and ascribed to the $^4F_{3/2} \rightarrow {}^4I_{15/2}$ transition. The lifetime of the $^4F_{3/2}$ level has been measured under LED pumping. Emission cross-sections for the various compounds are calculated in the 1 μm, 900 nm, and 1.3 μm regions and are of the order of $10^{-20}$ cm$^2$ in agreement with the literature results. Those in the 2 μm region are of the order of $10^{-21}$ cm$^2$.

**Keywords:** nanoparticles; infrared spectroscopy; Nd-luminescence

## 1. Introduction

Lanthanide-doped nanocrystals are widely studied systems for their visible emission features thanks to their unparalleled advantages over other types of materials such as their excellent thermomechanical properties and chemical stability, the large Stokes shift and sharp emission lines, and their long emission lifetimes. In particular, upconverting nanocrystals have received great attention for many different applications in the biomedical field, such as biomedical imaging, drug delivery, and photodynamic therapy [1–3], as well as for thermometric measurements [4,5] and for security applications [6], just to name a few.

Lanthanide ions also possess many efficient near-infrared emission transitions that have been exploited for laser emission in bulk crystals [7]. The possibility to exploit the infrared emission of lanthanide-activated nanocrystals can determine a paradigm shift for some applications and can also open the way to a lot of new types of applications, for example, for deep tissue imaging [8], image-guided surgery [9], and forensic science [10]. For example, nanocrystals with infrared emission have added values for biomedical applications such as the reduction of tissue absorption, light scattering, and autofluorescence. Among the various proposed materials, lanthanide nanocrystals with their intriguing emission properties are among the most promising materials. Moreover, Nd shows some very intense emissions in various infrared regions at around 900 nm, 1064 nm, and 1300 nm. All these emissions come from the decay from the $^4F_{3/2}$ to the lower-lying $^4I_{9/2}$, $^4I_{11/2}$, and

$^4I_{13/2}$ and have been widely exploited even for laser emission, but Nd ions also possess a weaker emission band at around 2 μm that has rarely been observed even in bulk crystals.

Sesquioxides are an important class of oxide crystals that possess good thermal and physical properties, have relatively low phonon energy compared with other oxides, and can be grown to good quality [11]. Unfortunately, the high temperature required for the growth of this class of materials as single crystals (around 2400 °C) makes this process quite demanding [12]. For this reason, the same compositions have been produced in fiber, ceramic, or nanopowder form. $Y_2O_3$ is probably the most widely studied sesquioxide when doped with Nd as bulk crystal [13], single crystal fiber [14], ceramic [15–17], and nanocrystals [18,19], but also other isomorphs such as $Lu_2O_3$ [20,21] and $Sc_2O_3$ [22,23] have shown very interesting emission properties when doped with Nd. In general, the focus of the spectroscopic investigations of these materials is limited to the visible absorption bands and to the main emission channel at around 1 micron, for which there is some inconsistency among the published values of the stimulated emission cross-section, especially when estimated with different techniques. Moreover, Nd also possesses other interesting emission channels at around 900 nm and 1300 nm from which even laser emission has been obtained [14], but very few reports of the emission cross-sections in these regions can be found in the literature. Last but not least, the emission at around 2 μm has never been reported to the best of the authors' knowledge.

Fluoride crystals are considered the preferred choice for emissions in the near-infrared, thanks to their good thermomechanical properties combined with low-phonon energy values, but the bulk crystal growth of this class of materials is complicated due to the high purity needed both for the starting chemicals and for the growth atmosphere.

Synthesis of these materials in nanometric form is accomplished by a polymer complex solution technique (oxides) and a low-temperature, solid-state method (fluorides) to study the infrared emission properties of these materials.

## 2. Materials and Methods

For syntheses of materials, the following chemicals were used: $Y_2O_3$ (Alfa Aeser, 99.99%), $Sc_2O_3$ (Alfa Aeser, 99.99%), $Lu_2O_3$ (Alfa Aeser, 99.99%), $Nd_2O_3$ (Alfa Aeser, 99.9%), polyethylene glycol (molecular weight 200, Alfa Aeser), nitric acid ($HNO_3$, Macron, 65%), and ammonium hydrogen difluoride ($NH_4HF_2$, Sigma−Aldrich, 98.5%). Nd-doped sesquioxide nanocrystals were prepared by the polymer complex solution method as previously described [24,25]. In brief, the stoichiometric ratio of oxide precursors was dissolved in a hot nitric acid at 130 °C until reaching the completely transparent solution. Then, the polyethylene glycol was added to the solution at a mass ratio of 1:1 to the mass of oxides. The solution was stirred at 80 °C until the nitrate gasses dissipated and a clear gel was formed. The gel was pre-sintered for 2 h at 800 °C in a ceramic crucible to produce a voluminous white powder, which was subsequently formed into pellets and calcined for 24 h at 1100 °C. Nd-doped fluorides were prepared by a low-temperature, solid-state synthesis accompanied by fluorination, as previously described [26]. In brief, the appropriate amounts of oxides were mixed with $NH_4HF_2$, thoroughly ground in an agate mortar to ensure homogeneity, and then heated in two steps, in the air at 170 °C for 20 h and in the reducing atmosphere (Ar−10% $H_2$) at 500 °C for 3 h.

The structure of the obtained nanomaterials was checked by X-ray powder diffraction (XRD) using the Rigaku SmartLab device (measurement settings: Cu-K$\alpha_{1,2}$ radiation, $\lambda$ = 0.1540 nm, ambient temperature, 2θ range 10–90°, measurement step 0.02°, and counting time 1 min/°). Scanning electron images were acquired by a field emission TESCAN MIRA3 microscope. Diffuse spectral reflectance measurements were performed on the FEI TECNAI G2 X-TWIN microscope. Measurements of diffuse reflection spectra were performed on a Thermo Evolution 600 spectrometer equipped with an integrating sphere and using the $BaSO_4$ spectrum as a white standard.

For infrared emission measurements, the sample was pumped by an 808 nm diode laser with about 400 mW output power. The emitted luminescence was collected by a parabolic mirror and was sent to an FTIR spectrometer (Magna860, Nicodom Ltd., Praha, Czech Republic) equipped with an MCT cooled detector. The resolution of the emission measurements was set to 1 cm$^{-1}$. All the spectra were corrected for the spectral response of the system using a blackbody source. Lifetimes of excited states were acquired after LED pumping at around 520 nm. The emission was collected by a lens, filtered by suitable filters to cut spurious pump light, and then sent to a fiber-coupled Si detector (OE-200-UV, Femto, Berlin, Germany). The amplification factor of the detector was $10^9$ in high-speed mode, so that the response time of the system was 17 µs.

## 3. Results

XRD patterns shown in Figure 1a confirm that the crystal structures of prepared sesquioxides are cubic bixbyite, space group Ia-3, and for prepared fluoride nanocrystals, it is orthorhombic, space group P$_{nma}$. No reflections belonging to impurity phases were observed. The average particle sizes of sesquioxides are around 350 nm (Figure 1b) and around 500 nm in fluorides (Figure 1c).

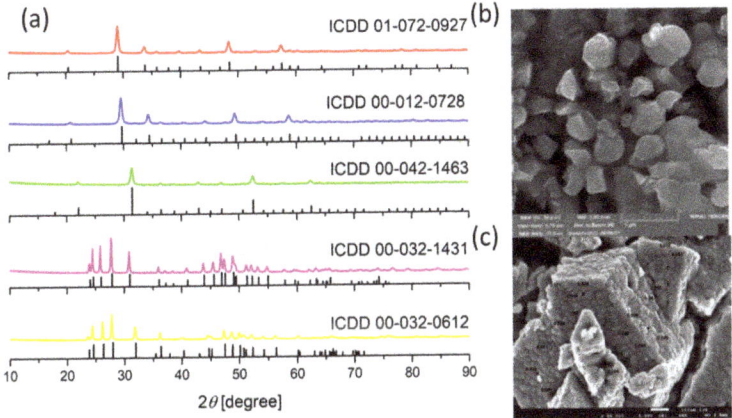

**Figure 1.** (a) XRD patterns of 3%Nd:Y$_2$O$_3$, 3%Nd:Lu$_2$O$_3$, 3%Nd:Sc$_2$O$_3$, 5%Nd:YF$_3$, and 5%Nd:LuF$_3$ (top to bottom) with respective ICDD data; (b) SEM image of 3%Nd:Y$_2$O$_3$, (c) SEM image of 5%Nd:YF$_3$.

### 3.1. Visible and Near-Infrared Spectroscopy

Diffuse reflection spectra of Nd-doped sesquioxides and fluorides are shown in Figure 2a,b, respectively. Measurements reveal typical absorptions of trivalent Nd located in low-energy phonon hosts, among which the strongest absorption around 800 nm is due to electronic transitions to $^4F_{5/2}$ and $^2H_{9/2}$ from the ground state.

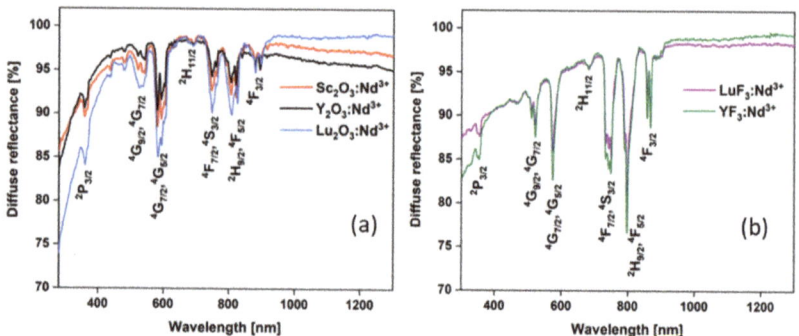

**Figure 2.** Diffuse reflection spectra of (a) 3%Nd:Y$_2$O$_3$, 3%Nd:Lu$_2$O$_3$, and 3%Nd:Sc$_2$O$_3$, and (b) 5%Nd:YF$_3$ and 5%Nd:LuF$_3$.

### 3.2. Sesquioxides

All sesquioxide samples show four emission bands that are composed of a series of well-separated peaks, as expected by the strong crystal field of these crystal matrixes [27]. The first band extends from 11,500 cm$^{-1}$ to 10,000 cm$^{-1}$ and corresponds to the $^4F_{3/2} \to {}^4I_{9/2}$ transition, the second extends from 9600 cm$^{-1}$ to 8600 cm$^{-1}$ and corresponds to the $^4F_{3/2} \to {}^4I_{11/2}$ transition, and the third extends from 7800 cm$^{-1}$ to 6700 cm$^{-1}$ and corresponds to the $^4F_{3/2} \to {}^4I_{13/2}$ transition. The peak position agrees with the energy level position reported in the literature [27]. Spectra are very similar among the various compositions. In fact, we can notice a strong similarity in the shape of these emission spectra, with only a small shift of the emission features and small differences in the relative emission intensity among the three compounds. This is not unexpected since Y$_2$O$_3$, Sc$_2$O$_3$, and Lu$_2$O$_3$ are isomorphs. When going from Y$_2$O$_3$ to Lu$_2$O$_3$ and to Sc$_2$O$_3$, the emission features experience a tendency to redshift that is more pronounced for the longest wavelength emission peaks within each band. This can be ascribed to the increasing crystal field strength in the three compounds [20]. The strongest peaks of the first band are located at 10,560 cm$^{-1}$ (947 nm) in Y$_2$O$_3$, 10,240 cm$^{-1}$ (977 nm) in Lu$_2$O$_3$, and 10,350 cm$^{-1}$ (966 nm) in Sc$_2$O$_3$. As usual for Nd-doped compounds, the strongest emission band is the one located at around 1 micron with maxima at 9265 cm$^{-1}$ (1079 nm) for Y$_2$O$_3$, 9253 cm$^{-1}$ (1081 nm) for Lu$_2$O$_3$, and 9237 cm$^{-1}$ (1083 nm) for Sc$_2$O$_3$. The maxima of the 1.3 μm band are located at 7363 cm$^{-1}$ (1358 nm) for Y$_2$O$_3$, 7352 cm$^{-1}$ (1360 nm) for Lu$_2$O$_3$, and 7311 cm$^{-1}$ (1368 nm) for Sc$_2$O$_3$. Moreover, in all cases, we were able to observe a fourth emission band in the 2 μm region that extends from about 4500 cm$^{-1}$ to about 6000 cm$^{-1}$. This band is usually considered very weak, and the emission has rarely been reported in the literature, even in bulk crystals. As for the other bands, also in this region, the shapes of the spectra look very similar for the three compounds with a tendency to red-shifting when passing from Y$_2$O$_3$ to Lu$_2$O$_3$ and to Sc$_2$O$_3$. The highest peaks are located at 4800 cm$^{-1}$ (2083 nm) for Y$_2$O$_3$, 4760 cm$^{-1}$ (2101 nm) for Lu$_2$O$_3$, and 4632 cm$^{-1}$ (2159 nm) for Sc$_2$O$_3$.

From the emission spectra, we calculated the emission cross-section of the $^4F_{3/2} \to {}^4I_i$ (i = 9/2, 11/2, 13/2, 15/2) emission bands with the following equation [28]:

$$\sigma_{em}(\nu) = \frac{c^2 I(\nu)}{8\pi\tau n^2 h\nu^3 \int \frac{I(\nu)}{h\nu} d\nu} \tag{1}$$

where $c$ is the speed of light in vacuum, $h$ is Planck's constant, $I(\nu)$ is the fluorescence signal, and $n$ and $\tau$ are the crystal refractive indexes at 1 μm wavelength and the radiative lifetime, respectively, both taken from the literature as reported in Table 1 for the various compounds. For LuF$_3$, we could not find proper references to published values; therefore, we used the values of the isomorph compound YF$_3$. In Equation (1), the integral is over the whole emission region of the $^4F_{3/2}$ decay channels, including the 2 μm emission band.

It is worth mentioning that we performed all the calculations in the frequency domain using Equation (1), because the experimental data were acquired with an FTIR that works at fixed wavenumber intervals, instead of using the equivalent expression in wavelength, as reported in Equation (14) of ref [28] that must be used when working with grating spectrometers.

**Table 1.** Parameters used for cross-section calculation.

| Compound | n | Ref. | τ (µs) | Ref. |
|---|---|---|---|---|
| Nd:$Y_2O_3$ | 1.90 | [29] | 318 | [18] |
| Nd:$Lu_2O_3$ | 1.91 | [30] | 300 | [31] |
| Nd:$Sc_2O_3$ | 1.97 | [32] | 230 | [33] |
| Nd:$YF_3$ | 1.45 | [34] | 783 | [35] |

Figure 3a–c show the emission cross-sections of all the Nd-doped sesquioxides in the 11,500 $cm^{-1}$–6500 $cm^{-1}$ region measured at 1%Nd doping level for oxides and 5%Nd doping level for fluorides because these were the samples with the highest emission intensities. In this region, we can distinguish the three main emission bands. The shape and peak position of the various bands qualitatively agree with published results, when available, and the cross-section peak intensities we obtained are compared with the literature results in Tables 2–6. It is evident that large discrepancies are present among the literature results, especially for the most studied of these compounds, such as Nd:$Y_2O_3$, where many different estimates are present. Our results compare well with the variation interval of published values. In all cases, the highest emission cross-section is that of the $^4F_{3/2} \rightarrow {}^4I_{11/2}$ transition, and our calculations for this band are in good agreement with published results. The emission cross-section of the other decay channels is not always known in the literature, and when present, our results compare well with published values.

It may be worth noting that these results are similar or slightly lower than the emission cross-section of well-known laser crystals. For example, the maximum emission cross-section of YLF is about $2 \times 10^{-20}$, $18 \times 10^{-20}$, and $3 \times 10^{-20}$ $cm^2$ for the $^4F_{3/2} \rightarrow {}^4I_{9/2}$, $^4F_{3/2} \rightarrow {}^4I_{11/2}$, and $^4F_{3/2} \rightarrow {}^4I_{13/2}$ transitions, respectively [36].

**Table 2.** Emission cross-sections of Nd:$Y_2O_3$.

| | Decay Channel | | $\sigma_{em}$ ($10^{-20}$ $cm^2$) | | | | | |
|---|---|---|---|---|---|---|---|---|
| | $^4F_{3/2} \rightarrow$ | This work | [14] | [18] | [16] | [15] | [17] | [13] |
| Nd:$Y_2O_3$ | $^4I_{9/2}$ | 2.4 | - | - | - | - | 4.89 | 1.8 |
| | $^4I_{11/2}$ | 7.3 | 6.9 | 1.73 | 7.24 | 5.13 | 6.35 | 6.8 |
| | $^4I_{13/2}$ | 1.5 | 5.5 | - | - | - | 0.92 | - |
| | $^4I_{15/2}$ | 0.07 | - | - | - | - | - | - |

**Table 3.** Emission cross-sections of Nd:$Lu_2O_3$.

| | Decay Channel | | $\sigma_{em}$ ($10^{-20}$ $cm^2$) | | | |
|---|---|---|---|---|---|---|
| | $^4F_{3/2} \rightarrow$ | This work | [20] | [21] | [27] | [37] |
| Nd:$Lu_2O_3$ | $^4I_{9/2}$ | 2.4 | - | - | 1.9 | - |
| | $^4I_{11/2}$ | 5.9 | 8.49 | 6.5 | 5.0 | 6.5 |
| | $^4I_{13/2}$ | 1.3 | - | - | 3.1 | - |
| | $^4I_{15/2}$ | 0.04 | - | - | - | - |

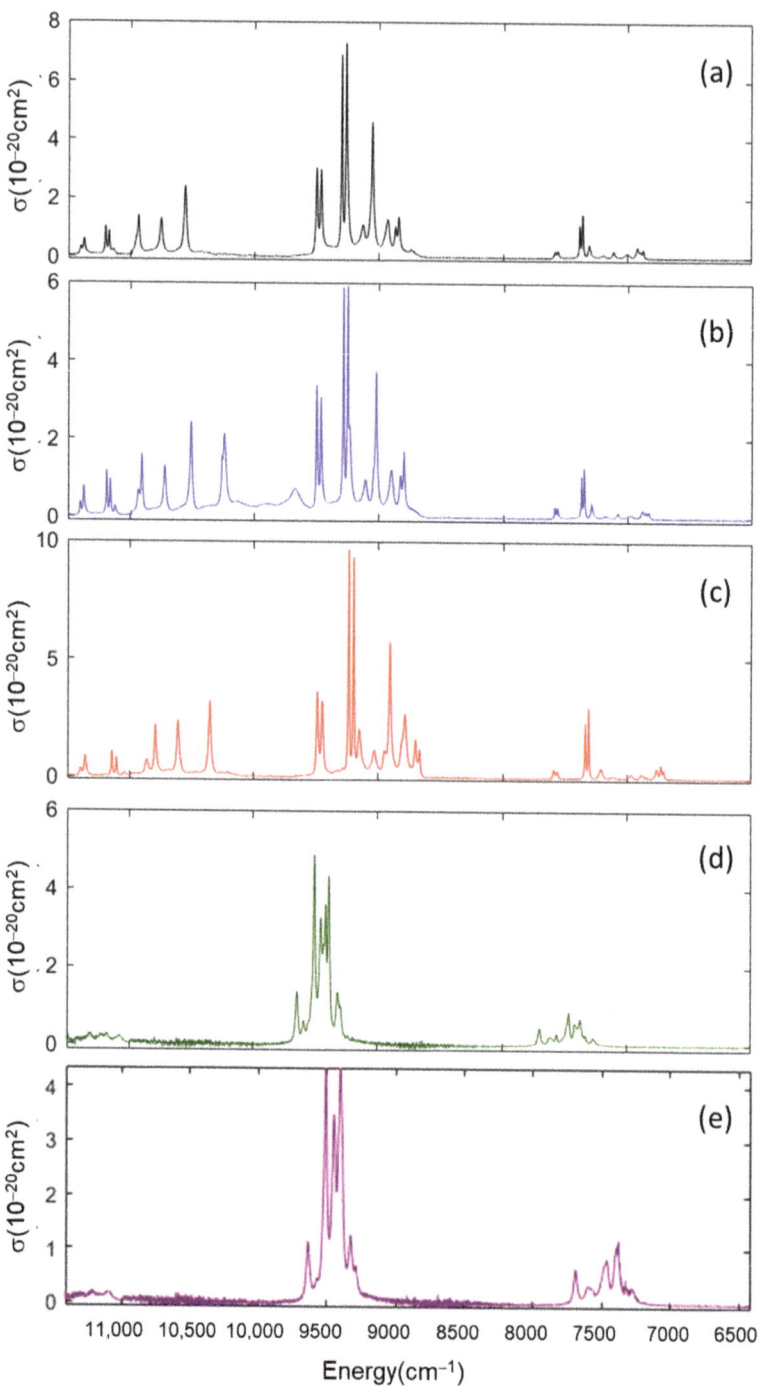

**Figure 3.** Emission cross-section of the various samples in the 6000–12,000 cm$^{-1}$ range: (**a**) 1%Nd:Y$_2$O$_3$; (**b**) 1%Nd:Lu$_2$O$_3$; (**c**) 1%Nd:Sc$_2$O$_3$; (**d**) 5%Nd:YF$_3$; (**e**) 5%Nd:LuF$_3$.

**Table 4.** Emission cross-sections of Nd:Sc$_2$O$_3$.

|  | Decay Channel | $\sigma_{em}$ (10$^{-20}$ cm$^2$) | |
| --- | --- | --- | --- |
|  | $^4F_{3/2} \to$ | This work | [33] |
| Nd:Sc$_2$O$_3$ | $^4I_{9/2}$ | 3.2 | - |
|  | $^4I_{11/2}$ | 9.7 | 9.5 |
|  | $^4I_{13/2}$ | 3 | - |
|  | $^4I_{15/2}$ | 0.09 | - |

**Table 5.** Emission cross-sections of Nd:YF$_3$.

|  | Decay Channel | $\sigma_{em}$ (10$^{-20}$ cm$^2$) | |
| --- | --- | --- | --- |
|  | $^4F_{3/2} \to$ | This work | [35] * |
| Nd:YF$_3$ | $^4I_{9/2}$ | 0.3 | 0.51 |
|  | $^4I_{11/2}$ | 4.9 | 0.74 |
|  | $^4I_{13/2}$ | 0.9 | 0.4 |
|  | $^4I_{15/2}$ | 0.06 | 0.032 |

* calculated.

**Table 6.** Emission cross-sections of Nd:LuF$_3$.

|  | Decay Channel | $\sigma_{em}$ (10$^{-20}$ cm$^2$) |
| --- | --- | --- |
|  | $^4F_{3/2} \to$ | This work |
| Nd:LuF$_3$ | $^4I_{9/2}$ | 0.2 |
|  | $^4I_{11/2}$ | 4.7 |
|  | $^4I_{13/2}$ | 1.1 |
|  | $^4I_{15/2}$ | 0.1 |

The stimulated emission cross-section for the $^4F_{3/2} \to {^4I_{15/2}}$ decay is depicted in Figure 4a–c for all investigated compounds. This transition appears as a series of separated groups of peaks of increasing intensity. The highest emission cross-section is observed at around 2.1 µm in all compounds.

We also measured the $^4F_{3/2}$ decay time under LED pumping on 3% and 1% doped samples. The decay profile is always exponential and lifetime values measured on 1% doped samples are reported in Tables 7–9 and compared with the literature values on low concentration samples, whenever available. On higher doped samples, concentration quenching effects make the lifetime shorter than the radiative value; we measured 217 µs, 211 µs, and 324 µs in 3%Nd-doped Y$_2$O$_3$, Lu$_2$O$_3$, and Sc$_2$O$_3$, respectively. The product of quantum efficiency and the dopant concentration can be considered as a figure of merit of the material [17]. In the case of 3%Nd:Y$_2$O$_3$, for example, considering a radiative lifetime of 354 µs, this value is 1.8, about 2.7 times higher than that obtained by Kumar and co-workers for the same doping level [17] from which laser emission has been obtained. The values obtained for the other compounds at 3% doping level are 1.8 for Lu$_2$O$_3$ and 2.8 for Sc$_2$O$_3$.

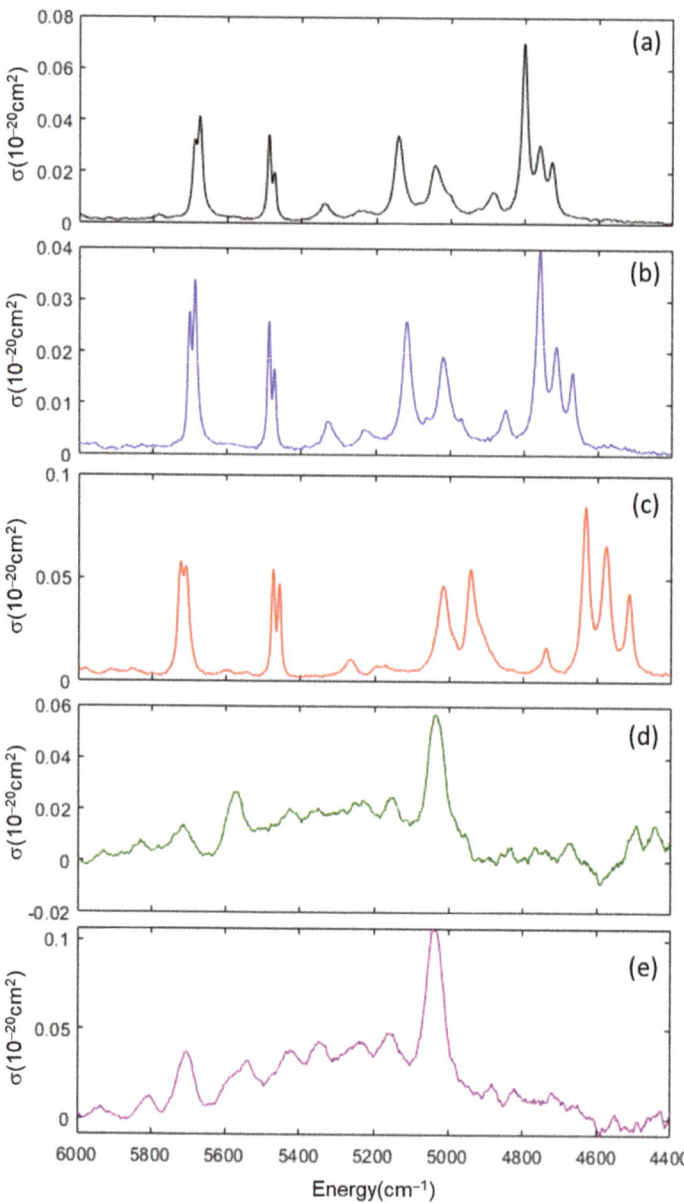

**Figure 4.** Emission cross-section of the various samples in the 4400–6000 cm$^{-1}$ range: (**a**) 1%Nd:Y$_2$O$_3$; (**b**) 1%Nd:Lu$_2$O$_3$; (**c**) 1%Nd:Sc$_2$O$_3$; (**d**) 5%Nd:YF$_3$; (**e**) 5%Nd:LuF$_3$.

**Table 7.** Decay time of 1%Nd:Y$_2$O$_3$.

|  |  | τ (μs) | | | | | | |
| --- | --- | --- | --- | --- | --- | --- | --- | --- |
|  |  | This work | [13] | [18] | [16] | [15] | [14] | [17] |
| Nd:Y$_2$O$_3$ |  | 320 | 300 |  | 321 | 232 | 340 | 315 |
|  | Radiative |  | 378 | 318 | 322 |  |  | 354 |

**Table 8.** Decay time of 1%Nd:Lu$_2$O$_3$.

|  |  | τ (µs) | | | |
|---|---|---|---|---|---|
|  |  | This work | [20] | [31] | [38] |
| Nd:Lu$_2$O$_3$ |  | 420 | 286 | 300 |  |
|  | Radiative |  | 344 |  | 165 |

**Table 9.** Decay time of 1%Nd:Sc$_2$O$_3$.

|  |  | τ (µs) | | | |
|---|---|---|---|---|---|
|  |  | This work | [33] | [39] | [40] |
| Nd:Sc$_2$O$_3$ |  | 335 | 180 | 224 | 260 |
|  | Radiative |  | 344 |  |  |

For Sc$_2$O$_3$, we investigated the dependence of the emission intensity and of the lifetime as a function of the doping level from 0.5% to 7%. The results are shown in Figure 5. As expected, both the emission intensity and the lifetime decrease with the concentration. The low-doping level value of the lifetime is slightly lower, but consistent with the theoretical radiative lifetime reported in Table 1, but the high concentration values are typically much longer than those measured in Y$_2$O$_3$ with similar doping levels. These results indicate that concentration quenching in Sc$_2$O$_3$ is not very strong and confirm the high quality of our samples.

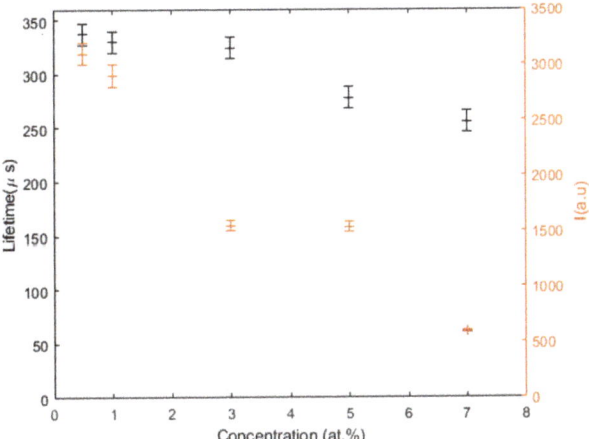

**Figure 5.** Lifetime of Nd:Sc$_2$O$_3$ (black, left axis) and emission intensity (right, red axis) as a function of the doping level.

### 3.3. Fluorides

We also acquired the emission spectra from 5%Nd:YF$_3$ and 5%Nd:LuF$_3$ samples and calculated the emission cross-section with Equation (1), as for sesquioxides. Results are shown in Figure 3d,e for the 11,500 cm$^{-1}$–6500 cm$^{-1}$ region and in Figure 4d,e for the 6000 cm$^{-1}$–4000 cm$^{-1}$ region. Since the decay time of LuF$_3$ is not known in the literature, the value for YF$_3$ has been used, instead. The emission intensity of fluoride samples is, in general, much weaker than that of sesquioxide samples. This can be ascribed either to the higher Nd doping level of our fluoride samples that can cause concentration quenching effects, or to a worse matching of the emission wavelength of our pump diode that causes lower absorption. In all cases, the emission is dominated by the 1-micron band. The emission cross-sections of the two compounds have similar shapes and intensity, as expected from the fact that the two compounds are isomorphs, and are much different from that of sesquioxides. The Stark splitting of the energy levels is in general smaller, and single peaks usually merge into continuous bands. The maximum emission cross-section recorded in the 1 µm region is $5 \times 10^{-20}$ cm$^2$ and $4.7 \times 10^{-20}$ cm$^2$ for YF$_3$ and LuF$_3$, respectively. The emission cross-section in the 2-micron region follows the same features already described: the shape is very similar between FY$_3$ and LuF$_3$ and is composed of an almost featureless band with a few peaks with maximum intensity of about $1 \times 10^{-21}$ cm$^2$.

Emission lifetimes of the $^4$F$_{3/2}$ level have been recorded under LED pumping, and results are reported in Tables 10 and 11 and compared with the literature for YF$_3$. Measured decay times are 170 µs and 120 µs for YF$_3$ and LuF$_3$, respectively. If compared to the radiative lifetime of YF$_3$ of 783 µs determined in [35], we can observe that concentration quenching at this high doping level is strong.

**Table 10.** Decay time of Nd:YF$_3$.

|  |  |  | τ (µs) |  |
| --- | --- | --- | --- | --- |
|  |  | This work |  | [35] |
| Nd:YF$_3$ | 5% | 169 |  |  |
|  | Low C |  |  | 588 |
|  | Radiative |  |  | 783 |

**Table 11.** Decay time of Nd:LuF$_3$.

|  |  | τ (µs) |
| --- | --- | --- |
|  |  | This work |
| Nd:LuF$_3$ | 5% | 119 |

These results show that fluoride materials generally show broader and weaker emission features in all wavelength regions, although fluoride crystals have lower phonon energy. This is probably due to the longer radiative lifetime of fluoride materials, but we cannot rule out interaction with possible quenching centers that are known to severely affect the emission efficiency of lanthanide-doped fluoride materials. The highest emission cross-sections are obtained from Nd:Sc$_2$O$_3$ in all regions.

## 4. Conclusions

We have synthesized and characterized a set of different Nd-doped fluoride and oxide nanocrystals, namely Nd:Y$_2$O$_3$, Nd:Lu$_2$O$_3$, Nd:Sc$_2$O$_3$, Nd:YF$_3$, and Nd:LuF$_3$. Under 808 nm pumping, we observed the three main emission bands in the near-infrared region, and we measured the lifetime of the $^4$F$_{3/2}$ level under LED pumping. In all cases, we were able to detect the weak 2-micron emission from the $^4$F$_{3/2} \rightarrow {}^4$I$_{15/2}$. Using the emission and lifetime data, we calculated the emission cross-sections of the various emission bands for all the compounds. Oxide materials generally showed narrower emissions, higher emission cross-sections, and shorter lifetimes. The results are in good agreement with the literature data, whenever available.

**Author Contributions:** Conceptualization, A.T. and M.D.D.; methodology, A.T., Ž.A. and M.D.D.; investigation, F.G., M.S. and T.B.; data curation, F.G.; writing—original draft preparation, F.G.; writing—review and editing, A.T. and M.D.D.; supervision, A.T. and M.D.D.; funding acquisition, M.D.D. All authors have read and agreed to the published version of the manuscript.

**Funding:** Authors from Serbia acknowledge funding from the Ministry of Education, Science and Technological Development of the Republic of Serbia.

**Institutional Review Board Statement:** Not applicable.

**Informed Consent Statement:** Not applicable.

**Data Availability Statement:** All data will be available on request.

**Conflicts of Interest:** The authors declare no conflict of interest.

# References

1. Bouzigues, C.; Gacoin, T.; Alexandrou, A. Biological applications of rare-earth based nanoparticles. *ACS Nano* **2011**, *5*, 8488–8505. [CrossRef]
2. Lim, X. The nanolight revolution is coming. *Nature* **2016**, *531*, 26–28. [CrossRef] [PubMed]
3. Yang, M.; Liang, Y.; Gui, Q.; Zhao, B.; Jin, D.; Lin, M.; Yan, L.; You, H.; Dai, L.; Liu, Y. Multifunctional luminescent nanomaterials from NaLa(MoO$_4$)$_2$:Eu$^{3+}$/Tb$^{3+}$ with tunable decay lifetimes, emission colors and enhanced cell viability. *Sci. Rep.* **2015**, *5*, 11844. [CrossRef] [PubMed]
4. Jaque, D.; Vetrone, F. Luminescence nanothermometry. *Nanoscale* **2012**, *4*, 4301. [CrossRef]
5. Dramićanin, M.D. Sensing temperature via downshifting emissions of lanthanide-doped metal oxides and salts. A review. *Methods Appl. Fluoresc.* **2016**, *4*, 042001. [CrossRef]
6. Blumenthal, T.; Meruga, J.; Stanley May, P.; Kellar, J.; Cross, W.; Ankireddy, K.; Vunnam, S.; Luu, Q.N. Patterned direct-write and screen-printing of NIR-to-visible upconverting inks for security applications. *Nanotechnology* **2012**, *23*, 185305. [CrossRef]
7. Cornacchia, F.; Toncelli, A.; Tonelli, M. 2-µm lasers with fluoride crystals: Research and development. *Prog. Quantum Electron.* **2009**, *33*, 61–109. [CrossRef]
8. Zhao, J.; Zhong, D.; Zhou, S. NIR-I-to-NIR-II fluorescent nanomaterials for biomedical imaging and cancer therapy. *J. Mater. Chem. B* **2018**, *6*, 349–365. [CrossRef]
9. Qu, Z.; Shen, J.; Li, Q.; Xu, F.; Wang, F.; Zhang, X.; Fan, C. Near-IR emissive rare-earth nanocrystals for guided surgery. *Theranostics* **2020**, *10*, 2631–2644. [CrossRef] [PubMed]
10. Gee, W.J. Recent trends concerning upconversion nanoparticles and near-IR emissive lanthanide materials in the context of forensic applications. *Aust. J. Chem.* **2019**, *72*, 164. [CrossRef]
11. Denker, B.; Shklovsky, E. (Eds.) *Handbook of Solid-State Lasers: Materials, Systems and Applications*; Woodhead Publishing Series in Electronic and Optical Materials; Elsevier: Amsterdam, The Netherlands, 2013; ISBN 9780857092724.
12. Toncelli, A.; Xu, J.; Tredicucci, A.; Heuer, A.M.; Kränkel, C. Mid-infrared spectroscopic characterization of Pr$^{3+}$:Lu$_2$O$_3$. *Opt. Mater. Express* **2019**, *9*, 4464. [CrossRef]
13. Walsh, B.M.; McMahon, J.M.; Edwards, W.C.; Barnes, N.P.; Equall, R.W.; Hutcheson, R.L. Spectroscopic characterization of Nd:Y$_2$O$_3$: Application toward a differential absorption lidar system for remote sensing of ozone. *J. Opt. Soc. Am. B* **2002**, *19*, 2893. [CrossRef]
14. Stone, J.; Burrus, C.A. Nd:Y$_2$O$_3$ single-crystal fiber laser: Room-temperature cw operation at 1.07- and 1.35-µm wavelength. *J. Appl. Phys.* **1978**, *49*, 2281. [CrossRef]
15. Zhang, L.; Huang, Z.; Pan, W. High transparency Nd:Y$_2$O$_3$ ceramics prepared with La$_2$O$_3$ and ZrO$_2$ additives. *J. Am. Ceram. Soc.* **2015**, *98*, 824–828. [CrossRef]
16. Hou, X.; Zhou, S.; Jia, T.; Lin, H.; Teng, H. Effect of Nd concentration on structural and optical properties of Nd:Y$_2$O$_3$ transparent ceramic. *J. Lumin.* **2011**, *131*, 1953–1958. [CrossRef]
17. Kumar, G.A.; Lu, J.; Kaminskii, A.A.; Ueda, K.-I.; Yagi, H.; Yanagitani, T. Spectroscopic and stimulated emission characteristics of Nd$^{3+}$ in transparent Y$_2$O$_3$ ceramics. *IEEE J. Quantum Electron.* **2006**, *42*, 643–650. [CrossRef]
18. Cui, X.; Lu, J.; Gao, C.; Hou, C.; Wei, W.; Peng, B. Luminescence properties of Nd$^{3+}$-doped Y$_2$O$_3$ nanocrystals in organic media. *Appl. Phys. A* **2011**, *103*, 27–32. [CrossRef]
19. Belli Dell'Amico, D.; Biagini, G.; Bongiovanni, G.; Chiaberge, S.; di Giacomo, A.; Labella, L.; Marchetti, F.; Marra, G.; Mura, A.; Quochi, F.; et al. A convenient preparation of nano-powders of Y$_2$O$_3$, Y$_3$Al$_5$O$_{12}$ and Nd:Y$_3$Al$_5$O$_{12}$ and study of the photoluminescent emission properties of the neodymium doped oxide. *Inorg. Chim. Acta* **2018**, *470*, 149–157. [CrossRef]
20. Hao, L.; Wu, K.; Cong, H.; Yu, H.; Zhang, H.; Wang, Z.; Wang, J. Spectroscopy and laser performance of Nd:Lu$_2$O$_3$ crystal. *Opt. Express* **2011**, *19*, 17774. [CrossRef] [PubMed]
21. Liu, Z.; Toci, G.; Pirri, A.; Patrizi, B.; Feng, Y.; Chen, X.; Hu, D.; Tian, F.; Wu, L.; Vannini, M.; et al. Fabrication and optical property of Nd:Lu$_2$O$_3$ transparent ceramics for solid-state laser applications. *J. Inorg. Mater.* **2021**, *36*, 210. [CrossRef]
22. Wang, Y.; Lu, B.; Sun, X.; Sun, T.; Xu, H. Synthesis of nanocrystalline Sc$_2$O$_3$ powder and fabrication of transparent Sc$_2$O$_3$ ceramics. *Adv. Appl. Ceram.* **2011**, *110*, 95–98. [CrossRef]
23. Ubaldini, A.; Carnasciali, M.M. Raman characterisation of powder of cubic RE$_2$O$_3$ (RE = Nd, Gd, Dy, Tm, and Lu), Sc$_2$O$_3$ and Y$_2$O$_3$. *J. Alloys Compd.* **2008**, *454*, 374–378. [CrossRef]
24. Lojpur, V.M.; Ahrenkiel, P.S.; Dramićanin, M.D. Color-tunable up-conversion emission in Y$_2$O$_3$:Yb$^{3+}$, Er$^{3+}$ nanocrystals prepared by polymer complex solution method. *Nanoscale Res. Lett.* **2013**, *8*, 131. [CrossRef]
25. Krsmanović, R.; Antić, Ž.; Bártová, B.; Dramićanin, M.D. Characterization of rare-earth doped Lu$_2$O$_3$ nanopowders prepared with polymer complex solution synthesis. *J. Alloys Compd.* **2010**, *505*, 224–228. [CrossRef]
26. Ćirić, A.; Aleksić, J.; Barudžija, T.; Antić, Ž.; Đorđević, V.; Medić, M.; Periša, J.; Zeković, I.; Mitrić, M.; Dramićanin, M.D. Comparison of three ratiometric temperature readings from the Er$^{3+}$ up-conversion emission. *Nanomaterials* **2020**, *10*, 627. [CrossRef] [PubMed]

27. Chang, N.C. Energy levels and crystal-field splittings of Nd$^{3+}$ in yttrium oxide. *J. Chem. Phys.* **1966**, *44*, 4044–4050. [CrossRef]
28. Aull, B.; Jenssen, H. Vibronic interactions in Nd:YAG resulting in nonreciprocity of absorption and stimulated emission cross sections. *IEEE J. Quantum Electron.* **1982**, *18*, 925–930. [CrossRef]
29. Nigara, Y. Measurement of the optical constants of yttrium oxide. *Jpn. J. Appl. Phys.* **1968**, *7*, 404–408. [CrossRef]
30. Medenbach, O.; Dettmar, D.; Shannon, R.D.; Fischer, R.X.; Yen, W.M. Refractive index and optical dispersion of rare earth oxides using a small-prism technique. *J. Opt. A Pure Appl. Opt.* **2001**, *3*, 174–177. [CrossRef]
31. Von Brunn, P.; Heuer, A.M.; Fornasiero, L.; Huber, G.; Kränkel, C. Efficient laser operation of Nd$^{3+}$:Lu$_2$O$_3$ at various wavelengths between 917 nm and 1463 nm. *Laser Phys.* **2016**, *26*, 084003. [CrossRef]
32. Belosludtsev, A.; Juškevičius, K.; Ceizaris, L.; Samuilovas, R.; Stanionytė, S.; Jasulaitienė, V.; Kičas, S. Correlation between stoichiometry and properties of scandium oxide films prepared by reactive magnetron sputtering. *Appl. Surf. Sci.* **2018**, *427*, 312–318. [CrossRef]
33. Kuzminykh, Y.; Kahn, A.; Huber, G. Nd$^{3+}$ doped Sc$_2$O$_3$ waveguiding film produced by pulsed laser deposition. *Opt. Mater.* **2006**, *28*, 883–887. [CrossRef]
34. Barnes, N.P.; Gettemy, D.J. Temperature variation of the refractive indices of yttrium lithium fluoride. *J. Opt. Soc. Am.* **1980**, *70*, 1244–1247. [CrossRef]
35. Tan, M.C.; Kumar, G.A.; Riman, R.E.; Brik, M.G.; Brown, E.; Hommerich, U. Synthesis and optical properties of infrared-emitting YF$_3$:Nd nanocrystals. *J. Appl. Phys.* **2009**, *106*, 063118. [CrossRef]
36. Turri, G.; Webster, S.; Bass, M.; Toncelli, A. Temperature-dependent stimulated emission cross-section in Nd$^{3+}$:YLF crystal. *Materials* **2021**, *14*, 431. [CrossRef]
37. Zhou, D.; Shi, Y.; Xie, J.; Ren, Y.; Yun, P. Fabrication and luminescent properties of Nd$^{3+}$-doped Lu$_2$O$_3$ transparent ceramics by pressureless sintering. *J. Am. Ceram. Soc.* **2009**, *92*, 2182–2187. [CrossRef]
38. Zhou, D.; Cheng, Y.; Ren, Y.Y.; Shi, Y.; Xie, J.J. Fabrication and spectroscopic properties of Nd:Lu$_2$O$_3$ transparent ceramics for laser media. In *Ceramic Materials and Components for Energy and Environmental Applications*; Jiang, D., Zeng, Y., Singh, M., Heinrich, J., Eds.; Ceramic Transactions Series; John Wiley & Sons, Inc.: Hoboken, NJ, USA, 2010; pp. 605–610. ISBN 978-0-470-64084-5.
39. Fornasiero, L.; Mix, E.; Peters, V.; Heumann, E.; Petermann, K.; Huber, G. *Advanced Solid-State Lasers*; OSA Trend in Optics and Photonics Series; Optical Society of America: Washington, DC, USA, 1999; Volume 26, p. 249.
40. Zverev, G.M.; Kolodnyi, G.Y.; Smirnov, A.I. Optical spectra of Nd$^{3+}$ in single crystals of scandium and yttrium oxides. *Opt. Spectrosc.* **1967**, *23*, 325–327.

Article

# Optical Properties of Yttria-Stabilized Zirconia Single-Crystals Doped with Terbium Oxide

Yazhao Wang, Zhonghua Zhu, Shengdi Ta, Zeyu Cheng, Peng Zhang, Ninghan Zeng, Bernard Albert Goodman, Shoulei Xu * and Wen Deng *

School of Physical Science and Technology, Guangxi University, 100 East Daxue Road, Nanning 530004, China; 2007301142@st.gxu.edu.cn (Y.W.); 2107301199@st.gxu.edu.cn (Z.Z.); 2007301125@st.gxu.edu.cn (S.T.); 2007301021@st.gxu.edu.cn (Z.C.); 2007301183@st.gxu.edu.cn (P.Z.); 2107301006@st.gxu.edu.cn (N.Z.); bernard_a_goodman@outlook.com (B.A.G.)
* Correspondence: xsl@gxu.edu.cn (S.X.); wdeng@gxu.edu.cn (W.D.)

**Abstract:** A series of yttria-stabilized zirconia single-crystals doped with 0.000–0.250 mol% $Tb_4O_7$ was prepared by the optical floating-zone method. As shown by XRD and Raman spectroscopy, all of the crystals had a cubic-phase structure. These were initially orange–yellow in color, which is indicative of the presence of $Tb^{4+}$ ions, but they then became colorless after being annealed in a $H_2/Ar$ atmosphere as a result of the reduction of $Tb^{4+}$ to $Tb^{3+}$. The absorption spectra of the unannealed samples show both the $4f^8 \rightarrow 4f^75d^1$ transition of $Tb^{3+}$ ions and the $Tb^{4+}$ charge-transfer band. In addition, the transmittance of the crystals was increased by annealing. Under irradiation with 300 nm of light, all of the single-crystal samples showed seven emission peaks in the visible region, corresponding to the decay from the $^5D_{3,4}$ excited state of $Tb^{3+}$ to the $^7F_J$ (J = 6–0) states. The most intense emission was at 544 nm, which corresponds to the typical strong green emission from the $^5D_4 \rightarrow ^7F_5$ transition in $Tb^{3+}$ ions.

**Keywords:** yttria-stabilized zirconia; $Tb_4O_7$-doped; annealed in $H_2/Ar$ atmosphere; optical properties; optical floating-zone method

## 1. Introduction

In recent years, the preparation and properties of optical materials containing rare earth ions have attracted wide-spread attention [1,2]. As a result of their partially filled 4f electronic configurations, rare earth ions have unique luminescence properties, including sharp emission peaks, long lifetimes, high quantum yields, and flexible tunability [3], and various optical materials, such as phosphors [4–6], glass ceramics [7–9], and optical crystals [10–13], are based on their excellent optical properties [2]. Among the rare earths, terbium (Tb) is an important element for green light-emitting devices, although its chemistry is complicated by its redox properties, which are manifested as $Tb^{3+}$ ions in the reduced state and $Tb^{4+}$ ions in the oxidized state. $Tb^{3+}$ has a $4f^8$ electronic configuration, whereas $Tb^{4+}$ is $4f^7$. Furthermore, an electron in $Tb^{3+}$ can be excited to the 5d-energy level ($4f^8 \rightarrow 4f^75d^1$) and this can then undergo non-radiative relaxation from the $^5D_{3,4}$ excited state and finally transition to the $^7F_J$ (J = 6–0) states with strong green emission near 545 nm [14]. The energy-level difference between the $^5D_4$ state and the $Tb^{3+}$ ground state is about 15,000 $cm^{-1}$, which makes multi-phonon relaxation negligible. Therefore, emissions from the $^5D_4$ state usually have high quantum efficiency and good thermal stability [15], and the luminescence of $Tb^{3+}$ ions in phosphors and lasers has been shown to have applications in medical imaging [8,16–18].

Terbium oxide in the form $Tb_4O_7$ contains both $Tb^{3+}$ and $Tb^{4+}$ ions. The standard redox potential (relative to the hydrogen electrode) of $Tb^{3+}/Tb^{4+}$ is + 3.1 V [19,20], which is far greater than that of the most easily reduced rare earth ions, such as $Eu^{2+}/Eu^{3+}$ (−0.35 V) [19]. Therefore, $Tb^{4+}$ is relatively stable and optical materials doped with $Tb^{4+}$

appear yellow as a result of $Tb^{4+}$ absorption [21]. Although $Tb^{3+}$ and $Tb^{4+}$ can coexist in the oxide, $Tb^{4+}$ ions do not emit visible light after excitation; thus, its presence should not affect the green emission of $Tb^{3+}$. However, investigating the optics and other physical and chemical properties of systems containing a combination of $Tb^{3+}$ and $Tb^{4+}$ should increase our understanding of their luminescence properties and help develop new applications of such rare earth-doped luminescence materials [21].

The absorption and emission intensity, light efficiency, luminescence lifetime, and quenching process of rare earth ion-based luminescent materials are strongly influenced by the host material [22]. Zirconia ($ZrO_2$) is a wide-bandgap semiconductor with a high melting point, high dielectric constant, good corrosion resistance, and low phonon energy [23], which is conducive to the observation of the luminescence of rare earth ions [24]. Besides its low phonon energy, a high refractive index and photochemical stability make $ZrO_2$ an ideal optical host material [25]. However, it is difficult to produce large high-quality zirconia crystals because of large volume changes, which accompany transitions between its three-phase structures (monoclinic (m-$ZrO_2$) below 1443 K, tetragonal (t-$ZrO_2$) between 1443 K and 2643 K, and cubic (c-$ZrO_2$) above 2643 K). This commonly leads to the cracking of the crystals that are cooling from melts, which is not conducive to industrial production [26]. However, the high-temperature c-$ZrO_2$ form, which is most suitable for many types of technological applications, can be stabilized at room temperature by the addition of an appropriate amount of an oxide stabilizer, such as $Y_2O_3$ [27,28], and several studies have shown that the addition of 8.0 mol% $Y_2O_3$ to $ZrO_2$ (abbreviated to 8YSZ) effectively stabilizes the cubic-phase structure [29–31]. As a result, crystals with good optical transparency, a high refractive index, and stable photo-thermo-chemical properties can be produced [29,32,33]. Previous studies of Tb-doped zirconia have concentrated mainly on phosphors and nanomaterials [34], although Soares et al. successfully produced YSZ: $Tb^{3+}$ fibers with a tetragonal crystal structure by the laser floating-zone method [35]. However, the luminescence properties of Tb-doped YSZ single-crystals in the cubic phase have not been reported previously.

In this paper, crystals were grown by the optical floating-zone method (OFZ) based on $ZrO_2$ stabilized with 8 mol% $Y_2O_3$, which was partially replaced with various amounts of $Tb_4O_7$. Its structure was characterized by X-ray diffractometry (XRD) and Raman spectroscopy, whilst the Tb oxidation states were determined by X-ray photoelectron spectroscopy (XPS). Optical properties were investigated using a combination of ultraviolet-visible (UV-Vis) absorption and transmission, photoluminescence excitation (PLE), photoluminescence emission (PL), and fluorescence-lifetime determination. These results could be explained by the physical mechanisms of luminescence and it lays the foundation for the wider application of green-laser and medical-imaging areas.

## 2. Experimental

### 2.1. Preparation of Crystals

$ZrO_2$, $Y_2O_3$, and $Tb_4O_7$ powders (Aladdin, Shanghai, China) with 99.99% purity were weighed with a precision balance scale in the stoichiometric ratio of $(ZrO_2)_{92}(Y_2O_3)_{8-x}(Tb_4O_7)_x$ ($x$ = 0.000, 0.050, 0.075, 0.100, 0.150, 0.200, and 0.250). The molar concentrations of the oxides used for the preparation of each sample are shown in Table 1. The weighed powders were suspended in ethanol and stirred on a magnetic stirrer for 24 h before drying in a constant-temperature oven at 85 °C for 24 h. The dried sample was added to cylindrical rubber molds, vacuum sealed, and set in a 50 MPa isostatic press for 20 min. The molds were then removed and the mixed powder bars were sintered at 1500 °C for 10 h to produce polycrystalline ceramic samples. These samples were used as the feed-and-seed rods for the crystal production in an optical suspension-zone furnace (FZ-T-12000-X-VII-VPO-GU-PC, Crystal Systems Co., Yamanashi, Japan). The optical floating zone could heat materials with high melting points and produce crystal samples with a high degree of purity. After production, some crystal samples were annealed at 1500 °C in a reducing atmosphere (7% $H_2$ + 93% Ar). After preparation, single-crystal rods were sliced and polished to produce disks with a thickness of 1 mm for optical measurements and fractions were ground into powders for structure determination. Figure 1 shows a photo of the $(ZrO_2)_{92}(Y_2O_3)_{8-x}(Tb_4O_7)_x$ ($x$ = 0.200) crystals before and after annealing in 7% $H_2$ + 93% Ar. The original single-crystal sample was orange–yellow, with good uniformity and high transparency, but after annealing, it became colorless.

**Table 1.** Composition of $(ZrO_2)_{92}(Y_2O_3)_{8-x}(Tb_4O_7)_x$.

| Samples | Composition (mol%) | | |
| --- | --- | --- | --- |
| | $ZrO_2$ | $Y_2O_3$ | $Tb_4O_7$ |
| $(ZrO_2)_{92}(Y_2O_3)_{8.00}$ | 92.00 | 8.000 | 0.000 |
| $(ZrO_2)_{92}(Y_2O_3)_{7.950}(Tb_4O_7)_{0.050}$ | 92.00 | 7.950 | 0.050 |
| $(ZrO_2)_{92}(Y_2O_3)_{7.925}(Tb_4O_7)_{0.075}$ | 92.00 | 7.925 | 0.075 |
| $(ZrO_2)_{92}(Y_2O_3)_{7.900}(Tb_4O_7)_{0.100}$ | 92.00 | 7.900 | 0.100 |
| $(ZrO_2)_{92}(Y_2O_3)_{7.850}(Tb_4O_7)_{0.150}$ | 92.00 | 7.850 | 0.150 |
| $(ZrO_2)_{92}(Y_2O_3)_{7.800}(Tb_4O_7)_{0.200}$ | 92.00 | 7.800 | 0.200 |
| $(ZrO_2)_{92}(Y_2O_3)_{7.750}(Tb_4O_7)_{0.250}$ | 92.00 | 7.750 | 0.250 |

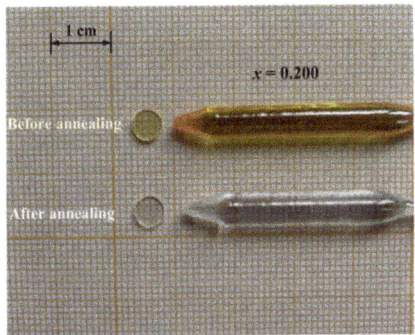

**Figure 1.** Photo showing the $(ZrO_2)_{92}(Y_2O_3)_{7.800}(Tb_4O_7)_{0.200}$ crystal rods and disks before and after annealing at 1500 °C in a $H_2$/Ar atmosphere.

### 2.2. Phase and Structure Characterization

The phase and structure characterization of the samples was performed by X-ray diffractometry (XRD) using a DX-2700 X-ray powder diffractometer (Dandong Hao Yuan Company, Dandong City, Liaoning Province) with Cu K alpha ($\lambda$ = 0.15406 nm) as the radiation source and Raman spectroscopy with a Finder One laser micro-Raman spectrometer, in which excitation was at 532 nm. X-ray photoelectron spectroscopy (ESCALAB 250XI+, Thermo Fisher Scientific Company, Shanghai, China) with a monochromatic Al-target X-ray source was used to characterize the elemental composition and oxidation states of the samples. The absorption and transmission spectra of the crystal samples were measured in the range of 200–800 nm using a UV-Vis spectrophotometer (UV-2700, Shimadzu, Kyoto, Japan) and both excitation and emission spectra were measured at room temperature with a ZLF-325 photoluminescence spectrometer (Zolix Instruments Co., Ltd., Beijing, China). The fluorescence-lifetime spectroscopy measurements were performed with an Edinburgh FLS1000 transient fluorescence spectrometer.

## 3. Results and Discussion

### 3.1. Crystal-Phase Structure Analysis

The XRD patterns from the powders of the $(ZrO_2)_{92}(Y_2O_3)_{8-x}(Tb_4O_7)_x$ crystal samples before annealing all consist of six diffraction peaks at ~30.12°, 34.94°, 50.18°, 59.64°, 60.56°, and 73.68° 2θ, which correspond to the (111), (200), (220), (311), (222), and (400) planes, respectively, of the c-$ZrO_2$ structure (PDF 04-006-5589). There was no evidence of diffraction peaks from the ZrO$_2$ monoclinic and tetragonal phases, and the phase structure was independent of the Tb$_4$O$_7$-doping concentration (in the range of $x$ = 0.000–0.250). The annealed crystal samples show similar results (e.g., Figure 2b). As shown in Table 2, the unit cell parameters and volume were essentially constant in the unannealed crystals, but they increased in the annealed crystals. In 8YSZ, a considerable number of anion vacancies were introduced into the ZrO$_2$ structure to maintain the charge balance and computer simulations have shown that such vacancies tend to avoid direct association with Y$^{3+}$ ions [36]. Consequently, there is a tendency for Zr$^{4+}$ to adopt a seven-coordinate configuration, whilst Y$^{3+}$ is coordinated to eight oxygen atoms. Considering that the radii of Tb$^{4+}$ and Zr$^{4+}$ are similar (0.76 and 0.78 Å, respectively, for the seven-coordination configuration), as are Tb$^{3+}$ and Y$^{3+}$ (1.04 and 1.019 Å, respectively, for the eight-coordination configuration) [37], we expect that Tb$^{4+}$ will substitute for Zr$^{4+}$ and Tb$^{3+}$ for Y$^{3+}$ in the unannealed crystals. Such substitutions would have little or no effect on the cell dimensions, as observed by XRD. A reduction of Tb$^{4+}$ to Tb$^{3+}$ because of the annealing results in an increase in the fraction of trivalent ions in the crystals and this must be accompanied by a decrease in the total anion charge. This could be achieved by a loss of oxygen from the structure or protonation of an appropriate number of structural oxygen atoms. However, the c-ZrO$_2$ structure accommodates a considerable range of Y$^{3+}$ ions; thus, there should be no problem in accommodating the reduced Tb. Nevertheless, the larger size of Tb$^{3+}$ compared to Tb$^{4+}$ results in a lattice expansion, as shown in Table 2.

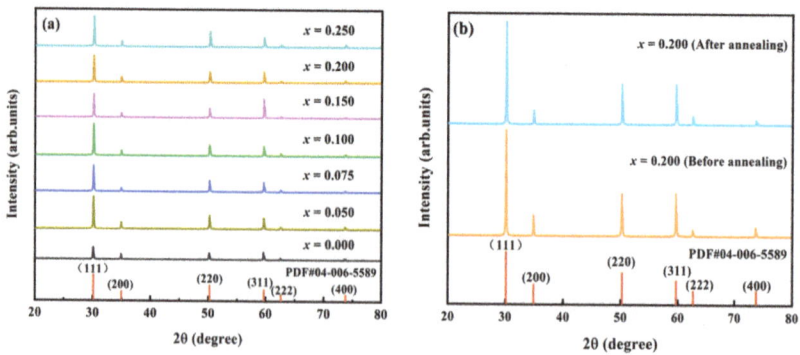

**Figure 2.** (a) XRD patterns of (ZrO$_2$)$_{92}$(Y$_2$O$_3$)$_{8-x}$(Tb$_4$O$_7$)$_x$ ($x$ = 0.000–0.250) crystal powders before annealing. (b) XRD pattern of (ZrO$_2$)$_{92}$(Y$_2$O$_3$)$_{7.800}$(Tb$_4$O$_7$)$_{0.200}$ crystal powders before and after annealing at 1500 °C in a H$_2$/Ar atmosphere.

**Table 2.** The lattice parameter of (ZrO$_2$)$_{92}$(Y$_2$O$_3$)$_{8-x}$(Tb$_4$O$_7$)$_x$ crystals.

| Samples | Lattice Parameter (nm) | Cell Volume (nm$^3$) | Before or after Annealing |
|---|---|---|---|
| (ZrO$_2$)$_{92}$(Y$_2$O$_3$)$_{8.00}$ | 0.5139 | 0.1357 | Before annealing |
| (ZrO$_2$)$_{92}$(Y$_2$O$_3$)$_{7.950}$(Tb$_4$O$_7$)$_{0.050}$ | 0.5137 | 0.1355 | Before annealing |
| (ZrO$_2$)$_{92}$(Y$_2$O$_3$)$_{7.925}$(Tb$_4$O$_7$)$_{0.075}$ | 0.5138 | 0.1357 | Before annealing |
| (ZrO$_2$)$_{92}$(Y$_2$O$_3$)$_{7.900}$(Tb$_4$O$_7$)$_{0.100}$ | 0.5138 | 0.1356 | Before annealing |
| (ZrO$_2$)$_{92}$(Y$_2$O$_3$)$_{7.850}$(Tb$_4$O$_7$)$_{0.150}$ | 0.5138 | 0.1357 | Before annealing |
| (ZrO$_2$)$_{92}$(Y$_2$O$_3$)$_{7.800}$(Tb$_4$O$_7$)$_{0.200}$ | 0.5138 | 0.1357 | Before annealing |
| (ZrO$_2$)$_{92}$(Y$_2$O$_3$)$_{7.750}$(Tb$_4$O$_7$)$_{0.250}$ | 0.5137 | 0.1356 | Before annealing |
| (ZrO$_2$)$_{92}$(Y$_2$O$_3$)$_{7.800}$(Tb$_4$O$_7$)$_{0.200}$ | 0.5155 | 0.1370 | After annealing |

Raman spectroscopy is a sensitive method for distinguishing the phase structure of zirconia crystals because m-ZrO$_2$ has 18 Raman vibration modes (9A$_g$ + 9B$_g$) [29,38] and t-ZrO$_2$ has 6 strong Raman vibration modes (A$_{1g}$ + 2B$_{1g}$ + 3E$_g$) [17], whereas c-ZrO$_2$ has only 1 (F$_{2g}$) [39]. The results in Figure S1 (see Supplementary Materials) show that the

Raman spectra of all the $(ZrO_2)_{92}(Y_2O_3)_{8-x}(Tb_4O_7)_x$ ($x$ = 0.050–0.250) crystal disks before annealing have a single strong vibration at 620 cm$^{-1}$ under excitation with a laser light of 532 nm. Furthermore, the spectra recorded before and after annealing (Figure 3) were similar and, thus, all are consistent with the c-ZrO$_2$ structure.

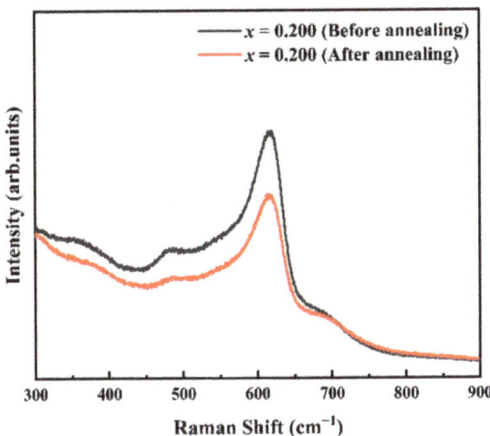

**Figure 3.** Raman spectra of $(ZrO_2)_{92}(Y_2O_3)_{7.800}(Tb_4O_7)_{0.200}$ crystals before and after annealing at 1500 °C in a H$_2$/Ar atmosphere.

### 3.2. X-ray Photoelectron Spectroscopy (XPS)

X-ray photoelectron spectroscopy (XPS) was used to investigate the oxidation states of the elements in the $(ZrO_2)_{92}(Y_2O_3)_{7.800}(Tb_4O_7)_{0.200}$ single-crystals before and after annealing. The survey spectrum (Figure 4) showed clear signals from the matrix elements Zr, Y, and O, and a weak signal from Tb. The expansion of the Tb 3$d$ energy-level spectra of the crystalline samples before and after annealing are shown in Figure 5. The Tb 3$d$ core-level spectrum contained peaks from the Tb $3d_{3/2}$ and Tb $3d_{5/2}$ spin-orbit states. In the unannealed sample, the peaks of Tb $3d_{3/2}$ and Tb $3d_{5/2}$ were at about 1277.1 and 1241.8.4 eV, respectively, and shifted to about 1276.8 and 1240.9 eV in the annealed sample. These are similar to the values of about 1277 and 1241 eV reported by Zhang et al. for the Tb$^{3+}$ $3d_{3/2}$ and Tb$^{3+}$ $3d_{5/2}$ peaks, respectively [40]. However, the low Tb content in the present work resulted in insufficient signal-to-noise values to allow for a conclusive assignment to the environments for Tb$^{3+}$ and Tb$^{4+}$. The XPS spectrum of the O 1$s$ core-energy level (Figure 6) showed two peaks at 531.8 and 530.2 eV, corresponding to O$^{2-}$ in the unannealed sample [41]. The positions of these peaks were similar to that in the annealed sample (531.6 and 529.9 eV) and there was also a small change in their relative intensities. We suggest that this is the result of a decrease in O$^{2-}$ ions bound to two metal ions in the 4+ oxidation state (Zr$^{4+}$ + some Tb$^{4+}$) and an increase in those bound to one 4+ and one 3+ ion as a result of the reduction of Tb$^{4+}$ to Tb$^{3+}$. The XPS spectra of the matrix elements Zr and Y for the before-and-after annealing samples showed that Zr and Y were essentially unaffected by annealing [36,42] (see Supplementary Materials).

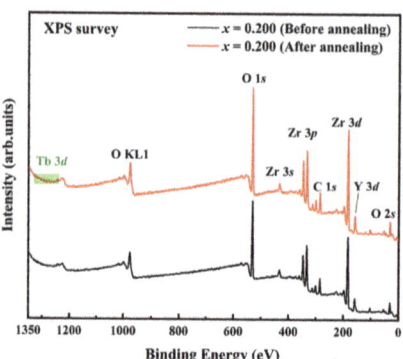

**Figure 4.** XPS survey spectrum of $(ZrO_2)_{92}(Y_2O_3)_{7.800}(Tb_4O_7)_{0.200}$ single-crystals before and after annealing in a $H_2/Ar$ atmosphere.

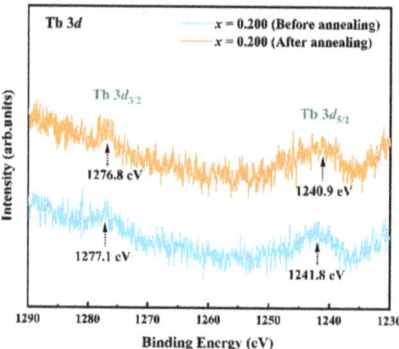

**Figure 5.** XPS Tb 3d core-energy-level spectrum.

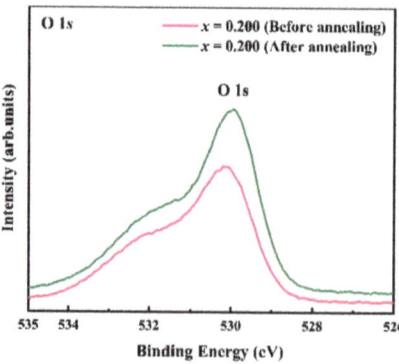

**Figure 6.** XPS O 1s core-energy-level spectrum.

### 3.3. Absorption and Transmission Spectrum

The absorption spectra of the $(ZrO_2)_{92}(Y_2O_3)_{8-x}(Tb_4O_7)_x$ ($x$ = 0.000–0.250) crystal disks in the range of 220–800 nm are shown in Figure 7. In the absence of Tb, there was a single intense absorption at 243 nm, corresponding to the direct absorption in the wide-bandgap semiconductor $ZrO_2$ [32,43]. This transition was also present in the $Tb_4O_7$-doped samples, which show a total of three absorption peaks at 245, 300, and 370 nm. The peak at 300 nm corresponds to the $4f^8 \rightarrow 4f^7 5d^1$ transition in $Tb^{3+}$ ions and the broad peak at

370 nm may originate from $Tb^{4+}$ [20]. As the Tb doping concentration increased, the $Tb^{3+}$ absorption peak underwent a red shift similar to that reported for Tb-doped $Bi_2MoO_6$ [44]; thus, this demonstrates that the presence of Tb affects the light absorption in the $ZrO_2$ host.

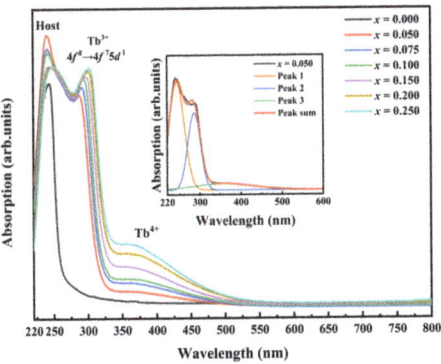

**Figure 7.** The absorption spectrum of $(ZrO_2)_{92}(Y_2O_3)_{8-x}(Tb_4O_7)_x$ ($x$ = 0.000–0.250) crystal discs before annealing (the inset is the result of Gaussian peak fitting of the absorption spectrum of the $(ZrO_2)_{92}(Y_2O_3)_{7.950}(Tb_4O_7)_{0.050}$ crystal).

Transmission spectra of the $(ZrO_2)_{92}(Y_2O_3)_{7.800}(Tb_4O_7)_{0.200}$ crystal disks were also measured before and after annealing, and were shown in Figure 8. The annealed crystals showed only two transmission peaks at 246 nm and 302 nm, and thus supports the assignment of the peak at 370 nm in the unannealed sample to $Tb^{4+}$ ions. The large width of this peak (from 330 to 550 nm), which is much larger than that of the $Tb^{3+}$ $4f^8 \rightarrow 4f^75d^1$ transition (280–335 nm), suggests that it may correspond to a charge-transfer transition in the $Tb^{4+}$ ions [45]. The transmittance of the annealed crystals in the range of 550–800 nm was about 86%, whereas it was about 78% in the unannealed sample; thus, this shows that the transmittance of the crystals increased after annealing.

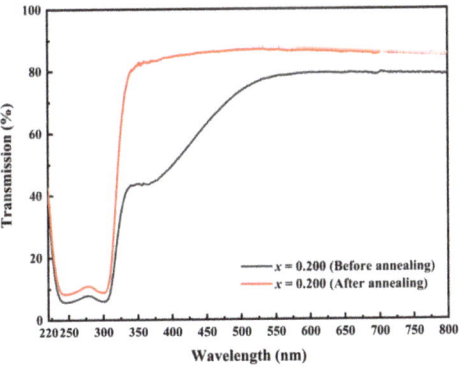

**Figure 8.** Transmission spectra of $(ZrO_2)_{92}(Y_2O_3)_{7.800}(Tb_4O_7)_{0.200}$ crystals before and after annealing at 1500 °C in a $H_2$/Ar atmosphere.

### 3.4. Photoluminescence Excitation (PLE) Spectra

The PLE spectra of the $(ZrO_2)_{92}(Y_2O_3)_{8-x}(Tb_4O_7)_x$ ($x$ = 0.050–0.250) crystals monitored at 544 nm are shown in Figure 9. There were three excitation peaks in the range of 200–500 nm at 300, 377, and 484 nm. The excitation peak centered at 300 nm (230–360 nm) is broad and corresponds to the spin-allowed $4f^8 \rightarrow 4f^75d^1$ transition of $Tb^{3+}$ ions [46]. In contrast, the weak excitation peaks at 377 nm and 484 nm are sharp, which correspond to the forbidden electric-dipole transitions $^7F_6 \rightarrow {}^5D_3$ and $^7F_6 \rightarrow {}^5D_4$ in the $Tb^{3+}$ 4f configuration [47,48].

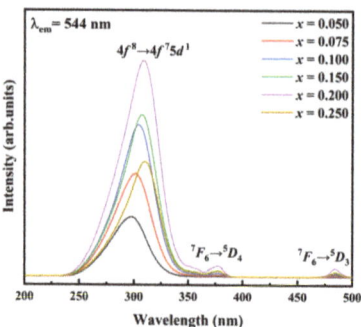

**Figure 9.** Excitation spectra of $(ZrO_2)_{92}(Y_2O_3)_{8-x}(Tb_4O_7)_x$ ($x$ = 0.050–0.250) crystals before annealing and monitored at 544 nm.

The PLE spectra of the $(ZrO_2)_{92}(Y_2O_3)_{7.800}(Tb_4O_7)_{0.200}$ crystals shown in Figure 10 remained unchanged during the annealing and demonstrated that there was no $Tb^{4+}$-excitation band. The excitation peak at 300 nm was red-shifted with the increasing $Tb_4O_7$ concentration, which probably reflects the changes in the energy of the lowest 5d-level of $Tb^{3+}$ in the YSZ host. This has been shown to differ between different host materials and may be influenced by two independent factors: (1) the nephelauxetic effect, which corresponds to a shortening of the metal-ligand distance as a result of a decrease in the coordination number and results in a shift in the center of the mass of the 5d electrons from the free-ion energy level; and (2) a crystal field effect, which causes the 5d manifold to split into various sub-levels [49]. Considering that both the XRD and Raman results show that the phase structure of the crystals does not change, we presume that the red-shift phenomenon in these samples was caused by the nephelauxetic effect [50]. In Figure 9, the intensity of the excitation spectrum of the $(ZrO_2)_{92}(Y_2O_3)_{8-x}(Tb_4O_7)_x$ crystal sample first increases with the concentration of $Tb_4O_7$, reaches a maximum at $x$ = 0.200, and then decreases, which may be the result of the cross-relaxation between two adjacent $Tb^{3+}$ ions [51].

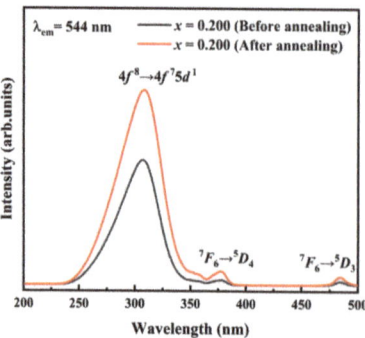

**Figure 10.** Excitation spectra of $(ZrO_2)_{92}(Y_2O_3)_{7.800}(Tb_4O_7)_{0.200}$ crystals before and after annealing at 1500 °C in a $H_2/Ar$ atmosphere monitored at 544 nm.

### 3.5. Photoluminescence Emission (PL) Spectra

The PL spectra of the unannealed $(ZrO_2)_{92}(Y_2O_3)_{8-x}(Tb_4O_7)_x$ crystal disks following irradiation at 300 nm at room temperature are shown in Figure 11. A bright green emission was observed and consisted of a total of seven peaks, all of which originate from 4f→4f transitions in $Tb^{3+}$ ions. The weak peaks at 385, 422, and 441 nm, which are shown in the expanded form in Figure 12, are blue and correspond to the $Tb^{3+}\ ^5D_3 \rightarrow ^7F_J$ (J = 6,5,4) transitions, whereas the peaks at 489, 544, 585, 621, and 677 nm are green and correspond to the $Tb^{3+}\ ^5D_4 \rightarrow ^7F_J$ (J = 6,5,4,3,1) transitions. The strongest emission peak at 544 nm arose from the $^5D_4 \rightarrow ^7F_5$ transition and is the typical strong-green emission from $Tb^{3+}$ ions. No additional emission peaks that could originate from $Tb^{4+}$ were observed in the unannealed samples (e.g., Figure 12), indicating that $Tb^{4+}$ did not emit in the visible-light region, but the intensity of the spectrum of the annealed sample was appreciably higher because of the higher concentration of $Tb^{3+}$. With the increasing $Tb_4O_7$ concentration, the intensity of the blue emission ($^5D_3 \rightarrow ^7F_6$) first increased and then decreased, with the highest intensity at $x = 0.100$ (Figure 12), whereas the green emission ($^5D_4 \rightarrow ^7F_5$) first increased to reach a maximum intensity at $x = 0.200$ and then decreased with higher $x$ (Figure 11). These different trends are the result of the cross-relaxation (CR) between neighboring $Tb^{3+}$ ions. Considering that the excited states of two $Tb^{3+}$ ions resonate with the ground state, the CR process can be expressed as:

$$Tb^{3+}\ (^5D_3) + Tb^{3+}\ (^7F_0) = Tb^{3+}\ (^5D_4) + Tb^{3+}\ (^7F_6). \tag{1}$$

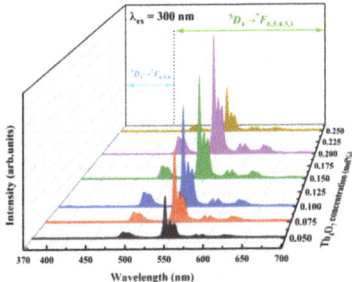

**Figure 11.** Emission spectra of $(ZrO_2)_{92}(Y_2O_3)_{8-x}(Tb_4O_7)_x$ ($x = 0.050$–$0.250$) crystals before annealing.

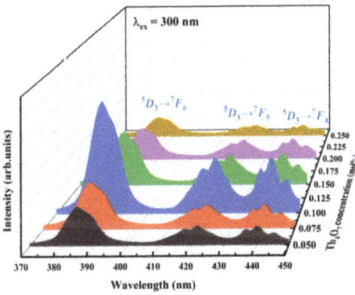

**Figure 12.** Expansion of the emission spectra in the range 370–450 nm of $(ZrO_2)_{92}(Y_2O_3)_{8-x}(Tb_4O_7)_x$ ($x = 0.050$–$0.250$) crystals before annealing.

The energy difference between the $^5D_3$ and $^5D_4$ states is about 5600 cm$^{-1}$, whilst that between the $^7F_0$ and $^7F_6$ states is about 5800 cm$^{-1}$, with the result that there is little difference in the total energy on each side of the above equation. Therefore, when the concentration of $Tb^{3+}$ is high, energy can be transferred from the $^5D_3$ state of one $Tb^{3+}$ ion to the $^5D_4$ state of a neighboring $Tb^{3+}$ ion, and this results in a quenching of the emission from $Tb^{3+}$ ions at the $^5D_3$ high-energy level and an increased emission from the lower $^5D_4$-energy level.

The emission spectra of the $(ZrO_2)_{92}(Y_2O_3)_{7.800}(Tb_4O_7)_{0.200}$ crystals before and after annealing in $H_2$ (Figure 13) confirm that all of the emissions in the visible region arise from $Tb^{3+}$ $4f \rightarrow 4f$ transitions and indicate that their positions are unaffected by the presence of $Tb^{4+}$. Thus, the $Tb^{3+}$ emission is independent of $Tb^{4+}$ in $Tb^{3+}$ and $Tb^{4+}$ co-doped systems. However, the emission intensity of the annealed crystals was stronger than that of the unannealed crystals probably because of the higher $Tb^{3+}$ concentration as a result of the $Tb^{4+}$ reduction. The various excitation and emission transitions for $Tb^{3+}$ ions are summarized in Figure 14.

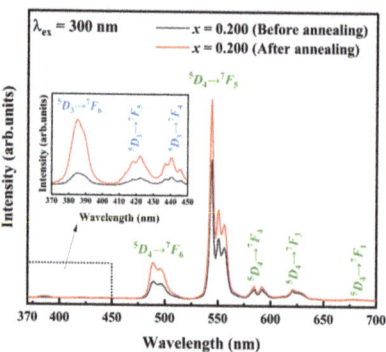

**Figure 13.** Emission spectrum of $(ZrO_2)_{92}(Y_2O_3)_{7.800}(Tb_4O_7)_{0.200}$ crystals before and after annealing at 1500 °C in a $H_2/Ar$ atmosphere.

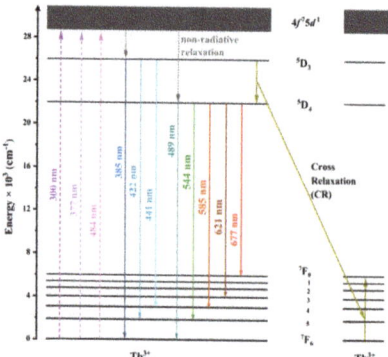

**Figure 14.** Energy-level-transition diagram of $Tb^{3+}$ ions in the yttria-stabilized zirconia single-crystal.

### 3.6. Fluorescence Decay

The fluorescence-decay lifetime is an important parameter for understanding energy-transfer mechanisms and the curve for the fluorescence decay of the $(ZrO_2)_{92}(Y_2O_3)_{7.800}(Tb_4O_7)_{0.200}$ crystals before annealing is shown in Figure 15 for the 544 nm ($^5D_4 \rightarrow {}^7F_5$, green) peak produced by the excitation at 300 nm.

The decay curve fits to a double exponential function, i.e.,

$$I(t) = A_1 \exp\left(\frac{-t}{\tau_1}\right) + A_2 \exp\left(\frac{-t}{\tau_2}\right) \qquad (2)$$

where $I(t)$ is the PL intensity, $\tau_1$ and $\tau_2$ represent the fast and slow components of the luminescence lifetime, respectively, and $A_1$ and $A_2$ are fitting parameters. Additionally, the average decay time $\tau$ of the sample is defined as:

$$\tau = \frac{A_1\tau_1^2 + A_2\tau_2^2}{A_1\tau_1 + A_2\tau_2} \qquad (3)$$

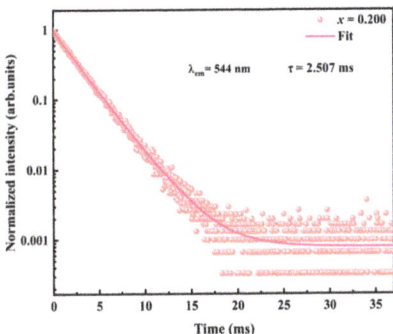

**Figure 15.** Fluorescence-decay curve for the $Tb^{3+}$ $^5D_4 \to ^7F_5$ transition in the $(ZrO_2)_{92}(Y_2O_3)_{7.800}(Tb_4O_7)_{0.200}$ single-crystals before annealing.

The values for these parameters obtained by computer fitting are $A_1 = 0.49369$, $A_2 = 0.50033$, $\tau_1 = 2.079$ ms, and $\tau_2 = 2.817$ ms. From these results, $\tau$ is calculated to be 2.507 ms, which is longer than the values reported for that of BSGdCaTb glass (2.31 ms) [52] or $Tb^{3+}$-doped germanium borosilicate (GBS) magneto-optical glass (2.317 ms) [53]. Furthermore, the double exponential function indicates that the $Tb^{3+}$ ions may be located in two different environments in the yttria-stabilized zirconia crystal.

## 4. Conclusions

Yttria-stabilized zirconia crystals doped with terbium oxide ($Tb_4O_7$) were prepared by the optical floating-zone method and shown to be in the cubic form without cracks or inclusions. The crystals initially contained Tb in both the $Tb^{3+}$ and $Tb^{4+}$ forms but the $Tb^{4+}$ was reduced to $Tb^{3+}$ by annealing crystals at 1500 °C in an $Ar/H_2$ atmosphere, although the $Tb^{4+}$ ions were reduced to $Tb^{3+}$ through annealing. The XPS results showed that Zr, Y, and O were essentially unaffected because the change in the relative intensities of the peaks in the O spectra were small. The absorption spectra of the crystals showed three peaks at 240, 300, and 370 nm, corresponding to the $Tb^{3+}$ $4f^8 \to 4f^75d^1$ transition and the $Tb^{4+}$ charge-transfer band. This spectrum also showed a red shift with increasing Tb, indicating that Tb-doping affects the light absorption properties of the YSZ host. The transmission spectra showed that the crystals had higher transmittance after annealing, although both the excitation and emission spectra correspond to transitions within the 4f and 5d configurations of $Tb^{3+}$ ($4f^8 \to 4f^75d^1$ and $4f \to 4f$ configurations). No excitation or emission peaks were observed from $Tb^{4+}$ and its presence did not influence the positions or shapes of the $Tb^{3+}$ peaks. The PL spectra also showed emissions from the $Tb^{3+}$ $^5D_3$ and $^5D_4$ states, which exhibited different relationships with the level of Tb-doping as a result of the cross-relaxation between neighboring $Tb^{3+}$ ions. The strongest emission peak from the crystals was located at 544 nm and corresponds to the $Tb^{3+}$ $^5D_4 \to ^7F_5$ transition, which is the typical strong green-light emission of $Tb^{3+}$ ions. The intensity of this emission increased initially with an increasing $Tb_4O_7$ concentration, reached a maximum at $x = 0.200$, and then decreased at a higher doping level. Overall, these results suggest that $(ZrO_2)_{92}(Y_2O_3)_{8-x}(Tb_4O_7)_x$ single-crystals may have potential applications as green lasers and for use in medical imaging.

**Supplementary Materials:** The following supporting information can be downloaded at: https://www.mdpi.com/article/10.3390/cryst12081081/s1, Figure S1: Raman spectra of $(ZrO_2)_{92}(Y_2O_3)_{8-x}(Tb_4O_7)_x$ ($x = 0.050$–0.250) crystals before annealing; Figure S2: XPS Zr 3d core-energy-level spectrum of $(ZrO_2)_{92}(Y_2O_3)_{7.800}(Tb_4O_7)_{0.200}$ single-crystals before and after annealing at 1500 °C in a $H_2/Ar$ atmosphere; Figure S3: XPS Y 3d core-energy-level spectrum of $(ZrO_2)_{92}(Y_2O_3)_{7.800}(Tb_4O_7)_{0.200}$ single-crystals before and after annealing at 1500 °C in a $H_2/Ar$ atmosphere.

**Author Contributions:** Conceptualization, Y.W. and S.X.; methodology, Y.W. and Z.Z.; software, Y.W., S.T. and Z.C.; validation, Y.W., S.X., and W.D.; formal analysis, Y.W., Z.C. and P.Z.; investigation, Y.W., N.Z. and P.Z.; resources, W.D.; data curation, Y.W., Z.Z. and S.T.; writing—original draft preparation, Y.W.; writing—review and editing, Y.W., B.A.G. and W.D.; visualization, B.A.G., W.D. and S.X.; supervision, S.X. and W.D.; project administration, W.D.; funding acquisition, W.D. All authors have read and agreed to the published version of the manuscript.

**Funding:** This work was financially supported by the Key Research and Development Plan Project of Guangxi, China, under grant number Guike AB18281007, and by the National Natural Science Foundations of China under grant numbers 11975004 and 12175047.

**Data Availability Statement:** Not applicable.

**Conflicts of Interest:** The authors declare no conflict of interest.

## References

1. Zhang, M.; Zhai, X.; Sun, M.; Ma, T.; Huang, Y.; Huang, B.; Du, Y.; Yan, C. When rare earth meets carbon nanodots: Mechanisms, applications and outlook. *Chem. Soc. Rev.* **2020**, *49*, 9220–9248. [CrossRef] [PubMed]
2. Zhang, C.; Lin, J. Defect-related luminescent materials: Synthesis, emission properties and applications. *Chem. Soc. Rev.* **2012**, *41*, 7938–7961. [CrossRef] [PubMed]
3. Chen, P.; Han, W.; Zhao, M.; Su, J.; Li, Z.; Li, D.; Pi, L.; Zhou, X.; Zhai, T. Recent Advances in 2D Rare Earth Materials. *Adv. Funct. Mater.* **2020**, *31*, 1616–3028. [CrossRef]
4. Brüninghoff, R.; Engelsen, D.; Fern, G.R.; Ireland, T.G.; Dhillon, R.; Silver, J. Nanosized $(Y_{1-x}Gd_x)_2O_2S$: $Tb^{3+}$ particles: Synthesis, photoluminescence, cathodoluminescence studies and a model for energy transfer in establishing the roles of $Tb^{3+}$ and $Gd^{3+}$. *RSC Adv.* **2016**, *6*, 42561–42571. [CrossRef]
5. Guan, H.; Sheng, Y.; Xu, C.; Dai, Y.; Xie, X.; Zou, H. Energy transfer and tunable multicolor emission and paramagnetic properties of $GdF_3$: $Dy^{3+}$, $Tb^{3+}$, $Eu^{3+}$ phosphors. *Phys. Chem. Chem. Phys.* **2016**, *18*, 19807–19819. [CrossRef]
6. Jung, J.-Y. Luminescent Color-Adjustable Europium and Terbium Co-Doped Strontium Molybdate Phosphors Synthesized at Room Temperature Applied to Flexible Composite for LED Filter. *Crystals* **2022**, *12*, 552. [CrossRef]
7. Santos, S.N.C.; Paula, K.T.; Almeida, J.M.P.; Hernandes, A.C.; Mendonça, C.R. Effect of $Tb^{3+}/Yb^{3+}$ in the nonlinear refractive spectrum of CaLiBO glasses. *J. Non-Cryst. Solids* **2019**, *524*, 00223093. [CrossRef]
8. Quang, V.X.; Van Do, P.; Ca, N.X.; Thanh, L.D.; Tuyen, V.P.; Tan, P.M.; Hoa, V.X.; Hien, N.T. Role of modifier ion radius in luminescence enhancement from $^5D_4$ level of $Tb^{3+}$ ion doped alkali-alumino-telluroborate glasses. *J. Lumin.* **2020**, *221*, 00222313. [CrossRef]
9. Han, S.; Tao, Y.; Du, Y.; Yan, S.; Chen, Y.; Chen, D. Luminescence Behavior of $GdVO_4$: Tb Nanocrystals in Silica Glass-Ceramics. *Crystals* **2020**, *10*, 396. [CrossRef]
10. Wu, M.-y.; Qu, P.-f.; Wang, S.-y.; Guo, Z.; Cai, D.-f.; Li, B.-b. Investigation of multi-segmented Nd:YAG/NdYVO$_4$ crystals and their laser performance end-pumped by a fiber coupled diode laser. *Optik* **2019**, *179*, 367–372. [CrossRef]
11. Sriwongsa, K.; Limkitjaroenporn, P.; Hongtong, W.; Chaiphaksa, W.; Kaewkhao, J.; Kim, H.J. Non-Proportionality Electron Response and Energy Resolution of LaBr$_3$:Ce and LuYAP:Ce Scintillating Crystals. *J. Korean Phys. Soc.* **2019**, *75*, 672–677. [CrossRef]
12. Kawaguchi, N.; Kimura, H.; Akatsuka, M.; Okada, G.; Kawano, N.; Fukuda, K.; Yanagida, T. Scintillation Characteristics of Pr:CaF$_2$ Crystals for Charged-particle Detection. *Sens. Mater.* **2018**, *30*, 0914–4935. [CrossRef]
13. Yang, Y.; Xu, S.; Li, S.; Wu, W.; Pan, Y.; Wang, D.; Hong, X.; Cheng, Z.; Deng, W. Luminescence Properties of Ho$_2$O$_3$-Doped Y$_2$O$_3$ Stabilized ZrO$_2$ Single Crystals. *Crystals* **2022**, *12*, 415. [CrossRef]
14. Yan, X.; Fern, G.R.; Withnall, R.; Silver, J. Effects of the host lattice and doping concentration on the colour of Tb$^{3+}$ cation emission in Y$_2$O$_2$S: Tb$^{3+}$ and Gd$_2$O$_2$S: Tb$^{3+}$ nanometer sized phosphor particles. *Nanoscale* **2013**, *5*, 8640–8646. [CrossRef]
15. Wu, D.; Xiao, W.; Zhang, L.; Zhang, X.; Hao, Z.; Pan, G.-H.; Luo, Y.; Zhang, J. Simultaneously tuning the emission color and improving thermal stability via energy transfer in apatite-type phosphors. *J. Mater. Chem. C* **2017**, *5*, 11910–11919. [CrossRef]
16. Xu, J.; Xu, X.-D.; Hou, W.-T.; Shi, Z.-L.; Zhao, H.-Y.; Xue, Y.-Y.; Shi, J.-J.; Liu, B.; Li, N. Research Progress of Rare-earth Doped Laser Crystals in Visible Region. *J. Inorg. Mater.* **2019**, *34*, 573–589.
17. Lovisa, L.X.; Gomes, E.O.; Gracia, L.; Santiago, A.A.G.; Li, M.S.; Andrés, J.; Longo, E.; Bomio, M.R.D.; Motta, F.V. Integrated experimental and theoretical study on the phase transition and photoluminescent properties of ZrO$_2$: $x$ Tb$^{3+}$ ($x = 1, 2, 4$ and 8 mol %). *Mater. Res. Bull.* **2022**, *145*, 00255408. [CrossRef]
18. Metz, P.W.; Marzahl, D.-T.; Majid, A.; Kränkel, C.; Huber, G. Efficient continuous wave laser operation of Tb$^{3+}$-doped fluoride crystals in the green and yellow spectral regions. *Laser Photonics Rev.* **2016**, *10*, 335–344. [CrossRef]
19. Kaszewski, J.; Witkowski, B.S.; Wachnicki, Ł.; Przybylińska, H.; Kozankiewicz, B.; Mijowska, E.; Godlewski, M. Reduction of Tb$^{4+}$ ions in luminescent Y$_2$O$_3$: Tb nanorods prepared by microwave hydrothermal method. *J. Rare Earths* **2016**, *34*, 774–781. [CrossRef]

20. Gompa, T.P.; Ramanathan, A.; Rice, N.T.; La Pierre, H.S. The chemical and physical properties of tetravalent lanthanides: Pr, Nd, Tb, and Dy. *Dalton Trans.* **2020**, *49*, 15945–15987. [CrossRef]
21. López-Pacheco, G.; Padilla-Rosales, I.; López, R.; González, F. Revisiting the Charge Transfer State in Tetravalent Lanthanide Doped Oxides: Up to Date Phenomenological Description. *ECS J. Solid State Sci. Technol.* **2021**, *10*, 2162–8777. [CrossRef]
22. Shyichuk, A.; Meinrath, G.; Lis, S. Pairs of Ln(III) dopant ions in crystalline solid luminophores: An ab initio computational study. *J. Rare Earths* **2016**, *34*, 820–827. [CrossRef]
23. Kaszewski, J.; Borgstrom, E.; Witkowski, B.S.; Wachnicki, Ł.; Kiełbik, P.; Slonska, A.; Domino, M.A.; Narkiewicz, U.; Gajewski, Z.; Hochepied, J.-F.; et al. Terbium content affects the luminescence properties of $ZrO_2$: Tb nanoparticles for mammary cancer imaging in mice. *Opt. Mater.* **2017**, *74*, 16–26. [CrossRef]
24. Lovisa, L.X.; Araújo, V.D.; Tranquilin, R.L.; Longo, E.; Li, M.S.; Paskocimas, C.A.; Bomio, M.R.D.; Motta, F.V. White photoluminescence emission from $ZrO_2$ co-doped with $Eu^{3+}$, $Tb^{3+}$ and $Tm^{3+}$. *J. Alloys Compd.* **2016**, *674*, 245–251. [CrossRef]
25. Meetei, S.D.; Singh, S.D.; Sudarsan, V. Polyol synthesis and characterizations of cubic $ZrO_2$: $Eu^{3+}$ nanocrystals. *J. Alloys Compd.* **2012**, *514*, 174–178. [CrossRef]
26. Huang, H.-J.; Wang, M.-C. The phase formation and stability of tetragonal $ZrO_2$ prepared in a silica bath. *Ceram. Int.* **2013**, *39*, 1729–1739. [CrossRef]
27. Reddy, C.V.; Reddy, I.N.; Shim, J.; Kim, D.; Yoo, K. Synthesis and structural, optical, photocatalytic, and electrochemical properties of undoped and yttrium-doped tetragonal $ZrO_2$ nanoparticles. *Ceram. Int.* **2018**, *44*, 12329–12339. [CrossRef]
28. Vasanthavel, S.; Kannan, S. Structural investigations on the tetragonal to cubic phase transformations in zirconia induced by progressive yttrium additions. *J. Phys. Chem. Solids* **2018**, *112*, 100–105. [CrossRef]
29. Xu, S.; Tan, X.; Liu, F.; Zhang, L.; Huang, Y.; Goodman, B.A.; Deng, W. Growth and optical properties of thulia-doped cubic yttria stabilized zirconia single crystals. *Ceram. Int.* **2019**, *45*, 15974–15979. [CrossRef]
30. Wang, X.; Tan, X.; Xu, S.; Liu, F.; Goodman, B.A.; Deng, W. Preparation and up-conversion luminescence of Er-doped yttria stabilized zirconia single crystals. *J. Lumin.* **2020**, *219*, 00222313. [CrossRef]
31. Wang, D.; Wu, W.; Tan, X.; Goodman, B.A.; Xu, S.; Deng, W. Upconversion Visible Light Emission in Yb/Pr Co-Doped Yttria-Stabilized Zirconia (YSZ) Single Crystals. *Crystals* **2021**, *11*, 1328. [CrossRef]
32. Li, S.; Xu, S.; Wang, X.; Wang, D.; Goodman, B.A.; Hong, X.; Deng, W. Optical properties of gadolinia-doped cubic yttria stabilized zirconia single crystals. *Ceram. Int.* **2021**, *47*, 3346–3353. [CrossRef]
33. Hong, X.; Xu, S.; Wang, X.; Wang, D.; Li, S.; Goodman, B.A.; Deng, W. Growth, structure and optical spectroscopic properties of dysprosia-doped cubic yttria stabilized zirconia (YSZ) single crystals. *J. Lumin.* **2021**, *231*, 00222313. [CrossRef]
34. Vidya, Y.S.; Gurushantha, K.; Nagabhushana, H.; Sharma, S.C.; Anantharaju, K.S.; Shivakumara, C.; Suresh, D.; Nagaswarupa, H.P.; Prashantha, S.C.; Anilkumar, M.R. Phase transformation of $ZrO_2$: $Tb^{3+}$ nanophosphor: Color tunable photoluminescence and photocatalytic activities. *J. Alloys Compd.* **2015**, *622*, 86–96. [CrossRef]
35. Soares, M.R.N.; Nico, C.; Rodrigues, J.; Peres, M.; Soares, M.J.; Fernandes, A.J.S.; Costa, F.M.; Monteiro, T. Bright room-temperature green luminescence from YSZ: $Tb^{3+}$. *Mater. Lett.* **2011**, *65*, 1979–1981. [CrossRef]
36. Devanathan, R.; Weber, W.; Singhal, S.; Gale, J. Computer simulation of defects and oxygen transport in yttria-stabilized zirconia. *Solid State Ion.* **2006**, *177*, 1251–1258. [CrossRef]
37. Shannon, R.D. Revised effective ionic radii and systematic studies of interatomic distances in halides and chalcogenides. *Acta Crystallogr. Sect. A* **1976**, *32*, 751–767. [CrossRef]
38. Phillippi, C.M.; Mazdiyasni, K.S. Infrared and Raman Spectra of Zirconia Polymorphs. *J. Am. Ceram. Soc.* **1971**, *54*, 254–258. [CrossRef]
39. Tan, X.; Xu, S.; Zhang, L.; Liu, F.; Goodman, B.A.; Deng, W. Preparation and optical properties of $Ho^{3+}$-doped YSZ single crystals. *Appl. Phys. A* **2018**, *124*, 1–7. [CrossRef]
40. Zhang, S.; Li, Y.; Lv, Y.; Fan, L.; Hu, Y.; He, M. A full-color emitting phosphor $Ca_9Ce(PO_4)_7$: $Mn^{2+}$, $Tb^{3+}$: Efficient energy transfer, stable thermal stability and high quantum efficiency. *Chem. Eng. J.* **2017**, *322*, 314–327. [CrossRef]
41. Ullah, B.; Lei, W.; Cao, Q.-S.; Zou, Z.-Y.; Lan, X.-K.; Wang, X.-H.; Lu, W.-Z.; Chen, X.M. Structure and Microwave Dielectric Behavior of A-Site-Doped $Sr_{(1-1.5x)}Ce_xTiO_3$ Ceramics System. *J. Am. Ceram. Soc.* **2016**, *99*, 3286–3292. [CrossRef]
42. Velu, S.; Suzuki, K.; Gopinath, C.S.; Yoshida, H.; Hattori, T. XPS, XANES and EXAFS investigations of $CuO/ZnO/Al_2O_3/ZrO_2$ mixed oxide catalysts. *Phys. Chem. Chem. Phys.* **2002**, *4*, 1990–1999. [CrossRef]
43. Pan, G.-H.; Zhang, L.; Wu, H.; Qu, X.; Wu, H.; Hao, Z.; Zhang, L.; Zhang, X.; Zhang, J. On the luminescence of $Ti^{4+}$ and $Eu^{3+}$ in monoclinic $ZrO_2$: High performance optical thermometry derived from energy transfer. *J. Mater. Chem. C* **2020**, *8*, 4518–4533. [CrossRef]
44. Li, H.; Li, W.; Gu, S.; Wang, F.; Zhou, H. In-built $Tb^{4+}/Tb^{3+}$ redox centers in terbium-doped bismuth molybdate nanograss for enhanced photocatalytic activity. *Catal. Sci. Technol.* **2016**, *6*, 3510–3519. [CrossRef]
45. Dorenbos, P. Systematic behaviour in trivalent lanthanide charge transfer energies. *J. Phys.-Condens Matter* **2003**, *15*, 8417–8434. [CrossRef]
46. Zhang, P.; Su, Y.; Teng, F.; He, Y.; Zhao, C.; Zhang, G.; Xie, E. Luminescent enhancement in $ZrO_2$: $Tb^{3+}$, $Gd^{3+}$ nanoparticles by active-shell modification. *CrystEngComm* **2014**, *16*, 1378–1383. [CrossRef]
47. Jarucha, N.; Wantana, N.; Kaewkhao, J.; Sareein, T. Studying the properties of $Gd_2O_3$–$WO_3$–CaO–$SiO_2$–$B_2O_3$ glasses doped with $Tb^{3+}$. *Semicond. Phys. Quantum Electron. Optoelectron.* **2020**, *23*, 276–281. [CrossRef]

48. Sztolberg, D.; Brzostowski, B.; Dereń, P.J. Spectroscopic properties of LaAlO$_3$ single-crystal doped with Tb$^{3+}$ ions. *Opt. Mater.* **2018**, *78*, 292–294. [CrossRef]
49. Wang, X.; Wang, Y. Synthesis, Structure, and Photoluminescence Properties of Ce$^{3+}$-Doped Ca$_2$YZr$_2$Al$_3$O$_{12}$: A Novel Garnet Phosphor for White LEDs. *J. Phys. Chem. C* **2015**, *119*, 16208–16214. [CrossRef]
50. Zheng, T.; Luo, L.; Du, P.; Lis, S.; Rodríguez-Mendoza, U.R.; Lavín, V.; Martín, I.R.; Runowski, M. Pressure-triggered enormous redshift and enhanced emission in Ca$_2$Gd$_8$Si$_6$O$_{26}$: Ce$^{3+}$ phosphors: Ultrasensitive, thermally-stable and ultrafast response pressure monitoring. *Chem. Eng. J.* **2022**, *443*, 13858947. [CrossRef]
51. Boruc, Z.; Fetlinski, B.; Kaczkan, M.; Turczynski, S.; Pawlak, D.; Malinowski, M. Temperature and concentration quenching of Tb$^{3+}$ emissions in Y$_4$Al$_2$O$_9$ crystals. *J. Alloys Compd.* **2012**, *532*, 92–97. [CrossRef]
52. Kesavulu, C.R.; Kim, H.J.; Lee, S.W.; Kaewkhao, J.; Kaewnuam, E.; Wantana, N. Luminescence properties and energy transfer from Gd$^{3+}$ to Tb$^{3+}$ ions in gadolinium calcium silicoborate glasses for green laser application. *J. Alloys Compd.* **2017**, *704*, 557–564. [CrossRef]
53. Mo, Z.X.; Guo, H.W.; Liu, P.; Shen, Y.D.; Gao, D.N. Luminescence properties of magneto-optical glasses containing Tb$^{3+}$ ions. *J. Alloys Compd.* **2016**, *658*, 967–972. [CrossRef]

Article

# Saturation Spectroscopic Studies on $Yb^{3+}$ and $Er^{3+}$ Ions in $Li_6Y(BO_3)_3$ Single Crystals

Gábor Mandula *, Zsolt Kis, Krisztián Lengyel, László Kovács and Éva Tichy-Rács

Wigner Research Centre for Physics, Konkoly-Thege Miklós út 29-33, H-1121 Budapest, Hungary
* Correspondence: mandula.gabor@wigner.hu

**Abstract:** The results of a series of pump–probe spectral hole-burning experiments are presented on $Yb^{3+}$- or $Er^{3+}$-doped $Li_6Y(BO_3)_3$ (LYB) single crystals in the temperature range of 2–14 K and 9–28 K, respectively. The spectral hole has a complex structure for $Yb^{3+}$ with superposed narrow and broad bands, while a single absorption hole has been observed for $Er^{3+}$. Population relaxation times ($T_1$) at about 850 ± 60 µs and 1010 ± 50 µs and dipole relaxation times ($T_2$) with values of 1100 ± 120 ns and 14.2 ± 0.3 ns have been obtained for the two components measured for the $Yb^{3+}$:$^2F_{7/2}$—$^2F_{5/2}$ transition. $T_1$ = 402 ± 8 µs and $T_2$ = 11.9 ± 0.2 ns values have been found for the $Er^{3+}$:$^4I_{15/2}$—$^4I_{11/2}$ excitation. The spectral diffusion rate at about 1 and 5 MHz/ms has been determined for the narrow and broad spectral line in $Yb^{3+}$-doped crystal, respectively. The temperature dependence of the spectral hole halfwidth has also been investigated.

**Keywords:** saturation spectroscopy; LYB; ytterbium; erbium; spectral hole

---

**Citation:** Mandula, G.; Kis, Z.; Lengyel, K.; Kovács, L.; Tichy-Rács, É. Saturation Spectroscopic Studies on $Yb^{3+}$ and $Er^{3+}$ Ions in $Li_6Y(BO_3)_3$ Single Crystals. *Crystals* **2022**, *12*, 1151. https://doi.org/10.3390/cryst12081151

**Academic Editors:** Alessandra Toncelli and Željka Antić

Received: 27 July 2022
Accepted: 13 August 2022
Published: 16 August 2022

**Publisher's Note:** MDPI stays neutral with regard to jurisdictional claims in published maps and institutional affiliations.

**Copyright:** © 2022 by the authors. Licensee MDPI, Basel, Switzerland. This article is an open access article distributed under the terms and conditions of the Creative Commons Attribution (CC BY) license (https://creativecommons.org/licenses/by/4.0/).

## 1. Introduction

Coherent quantum optical processes in rare-earth (RE) ion-doped dielectric crystals are of great interest in the quest for physical systems in order to realize quantum information processing devices. The successful realization of coherent quantum control procedures raises serious demands of atomic systems, such as a sufficiently long population, $T_1$ and dipole relaxation, $T_2$ times of the involved quantum states. Among several choices, e.g., magnetic- or laser-trapped ultracold single atoms or atomic clouds or coherent spin dynamics in quantum dots, a potential way to localize atomic systems is doping a single crystal with RE ions. The primary motivation of choosing $Er^{3+}$ dopant ion is that its $^4I_{13/2}$—$^4I_{15/2}$ transition is resonant with the frequency of the telecommunication laser field with a wavelength of about ~1.5 µm. A number of oxide crystals have been tested as a host of erbium to obtain a long coherence time in this wavelength range [1]. Moreover, the coherence time depends not only on the host material but on the concentration of the dopant, the temperature of the sample and the applied magnetic field as well [2–5]. Erbium-doped crystals have been used in several coherent quantum optical applications, such as the control of dispersion and group velocity [6], diode laser frequency stabilization [7], interference detection of spontaneous emission of light from two solid state atomic ensembles [8], electromagnetically induced transparency [9], ultraslow light propagation via coherent population oscillation [10], and quantum memories [11–15].

For practical applications, there are other relevant wavelength intervals beside the telecommunication domain. One of them is the near infrared range between 980–1100 nm. The ytterbium-doped crystal and fiber lasers radiate in the wavelength range between 1030–1080 nm, which can be pumped with ~980 nm light. There are several optical devices optimized to this wavelength domain [16,17].

$Yb^{3+}$ and $Er^{3+}$ dopant ions are perspective candidates for coherent quantum optical applications because the transitions $^2F_{5/2}$—$^2F_{7/2}$ and $^4I_{11/2}$—$^4I_{15/2}$, respectively, lay in the wavelength range 960–980 nm. Our goal is to investigate the use of the LYB crystal as a host for these ions according to the previous properties. Additionally, the solution of the problem about the interaction between the RE ions and the borate lattice could show us new research directions to develop borate crystals with targeted compositions.

Our aim was to study the ytterbium and erbium ions as dopants in a novel host crystal: lithium yttrium borate $Li_6Y(BO_3)_3$ (LYB). This member of the borate crystal family can be grown by both the Bridgman and the Czochralski methods [18–20], which have a monoclinic structure with the $P2_1/c$ space group. The advantage of LYB crystal is that rare-earth dopants can incorporate into Y site with $C_1$ symmetry in the lattice without any distortion because of the charge neutral substitution and the similar ion radius. LYB:$Yb^{3+}$ and LYB:$Er^{3+}$ have already been investigated by optical spectroscopic methods [21–25] to determine the energy terms of the dopant ions. In addition, LYB:$Yb^{3+}$ has been used for short pulse laser applications [26] and as a diode-pumped laser with Er co-doping [27].

A detailed description of the spectral hole burning procedure carried out by using sequential pump and probe pulses derived from a single laser beam passing through lithium niobate crystals can be found, e.g., in our previous papers [28,29]. The RE ions can be considered as effective two-level systems, where the laser-induced transition between the ground and excited states can be considered as a dipole transition. Thus, we suppose a set of effective two-state atoms interacting with a long and intensive pump and then a short and sufficiently weaker read-out probe pulse with a much longer time delay than the dipole relaxation time. In addition, the probe pulse needs to be much shorter than the population relaxation time but much longer than the dipole relaxation time. If these requirements are met, then by scanning the frequency of the probe field around the pumping frequency the attenuation of the probe field determines a Lorentzian curve, which is called spectral hole. The lifetime of the spectral hole was found to be single exponential with a characteristic time constant $T_1$. The halfwidth of the spectral hole depends on the intensity of the pumping pulse, thus the value of $T_2$ comes out only in the low intensity limit, which can be achieved by the Z-scan method described in the next section.

## 2. Materials and Methods

Lithium yttrium borate single crystals doped with 0.1 mol% $Yb^{3+}$, 0.05 mol% or 1.0 mol% $Er^{3+}$ were grown by the Czochralski method [20]. LYB has a monoclinic structure of the $P2_1/c$ space group. The dielectric x-axis coincides with the crystallographic b-axis <010>, while the dielectric z-axis is perpendicular to the optical plane (-102). Although the crystal is biaxial, it behaves as uniaxial positive at $\lambda = 1064$ nm [21]. $Yb^{3+}$-doped sample was prepared with incident plane (010) and a thickness of 0.6 mm, while the $Er^{3+}$-doped sample was prepared with incident plane (-102) and a thickness of 2 mm. The high resolution absorption spectra of $Er^{3+}$- or $Yb^{3+}$-doped LYB crystals have been measured at 9 K by a Bruker IFS 66v/S or a Bruker IFS 120 FTIR spectrometer, respectively. In case of the $Er^{3+}$ dopant, the 1 mol% concentration sample was more suitable for spectroscopy than the one containing fewer $Er^{3+}$ ions, while for saturation spectroscopy, lower dopant concentration is required. The absorption spectra between 10,250 and 10,450 $cm^{-1}$ wavenumbers are shown in Figure 1. In the case of the $Yb^{3+}$-doped crystal, a single absorption peak was found at 10,283.5 $cm^{-1}$ with a halfwidth of 0.18 $cm^{-1}$, corresponding to the transition from the lowest Stark level of $^2F_{7/2}$ multiplet to the lowest crystal field level of $^2F_{5/2}$ multiplet of the $Yb^{3+}$ ions in LYB at $T = 9$ K. The other two crystal field components of the $^2F_{5/2}$ excited state lie out of this range and are not suitable for our laser spectroscopic measurements. However, all possible six absorption bands of the $^4I_{15/2}$—$^4I_{11/2}$ transition of $Er^{3+}$ can be identified in this wavenumber range. The positions and halfwidths of the investigated absorption lines are listed in Table 1.

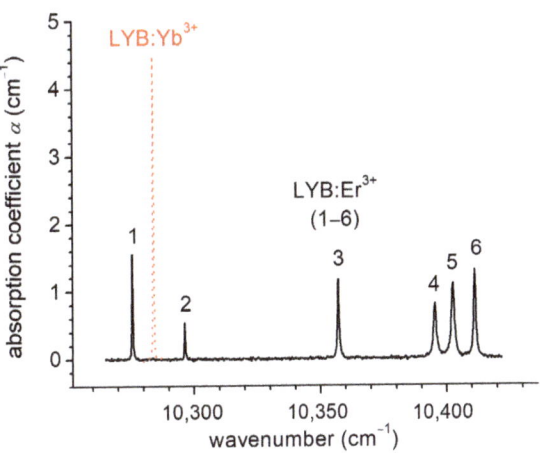

**Figure 1.** Absorption spectra of LYB:Er$^{3+}$, $c_{Er}$ = 1 mol% (black, with numbered peaks) and LYB: Yb$^{3+}$, $c_{Yb}$ = 0.1 mol% (red, dotted) between 10,250 and 10,450 cm$^{-1}$, measured at 9 K.

**Table 1.** Positions and halfwidths of the absorption lines in the spectrum of LYB:Er$^{3+}$ and LYB:Yb$^{3+}$ crystals at $T$ = 9 K shown in Figure 1, and Stark energies of the crystal field splittings are determined by L. Kovács et al. [25] and J. Sablayrolles et al. [22].

|  |  |  | Wavenumber, Stark Energy (cm$^{-1}$) | Halfwidth (cm$^{-1}$) | Reference |
|---|---|---|---|---|---|
| Er$^{3+}$ | $^4I_{15/2}$ | 1 | 0 |  | [25] |
|  |  | 2 | 47 |  | [25] |
|  |  | 3 | 76 |  | [25] |
|  |  | 4 | 113 |  | [25] |
|  |  | 5 | 368 |  | [25] |
|  |  | 6 | 466 |  | [25] |
|  |  | 7 | 497 |  | [25] |
|  |  | 8 | 524 |  | [25] |
|  | $^4I_{11/2}$ | 1 | 10,275.34 | 0.14 | present work |
|  |  | 2 | 10,295.97 | 0.15 | present work |
|  |  | 3 | 10,356.69 | 0.51 | present work |
|  |  | 4 | 10,394.97 | 1.17 | present work |
|  |  | 5 | 10,402.08 | 1.13 | present work |
|  |  | 6 | 10,410.85 | 0.77 | present work |
| Yb$^{3+}$ | $^2F_{7/2}$ | 1 | 0 |  | [22] |
|  |  | 2 | 367 |  | [22] |
|  |  | 3 | 507 |  | [22] |
|  |  | 4 | 676 |  | [22] |
|  | $^2F_{5/2}$ | 1 | 10,283.58 | 0.18 | present work |
|  |  | 2 | 10,475 | ≈50 | present work |
|  |  | 3 | 10,810 | ≈80 | present work |

The measurement setup used for the spectral hole-burning experiments can be seen in Figure 2. The light source was a stabilized external cavity diode laser (Sacher Lasertechnik,

Manually Tunable Littrow Laser System–Lynx S3, tunable between 930 and 985 nm) with current, temperature, mechanical and piezo wavelength-tuning capability. For the narrow spectral hole measurements, a Toptica Photonics DL Pro grating stabilized tunable single-mode diode laser was used. This laser has a typical linewidth of 25 kHz with 5 µs integration time and less than 100 kHz in a millisecond time scale.

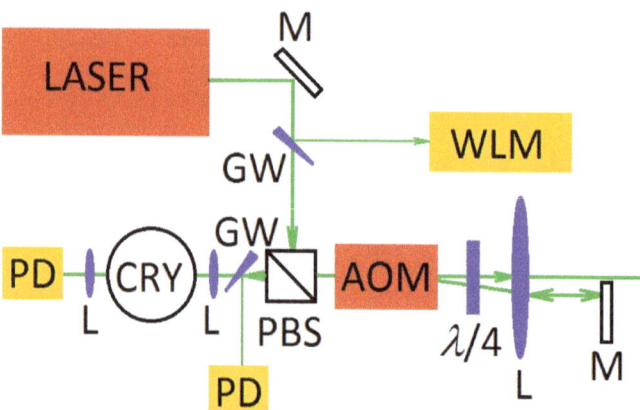

**Figure 2.** Experimental setup for spectral hole-burning measurements at 9 K. The red box labelled as LASER involves an external cavity diode laser, a beam profiler, a Faraday isolator, a space filtering beam expander, and a lambda half-wave plate to prepare vertical polarization. M = mirror, GW = glass wedge, WLM = wavelength meter, PBS = polarizing beam splitter cube, AOM = acousto-optical modulator, λ/4 = lambda quarter-wave plate, L = lens, CRY = cryostat with crystal sample, and PD = avalanche photodiode with focusing lens system. For the lower temperature measurements, polarization maintaining a single mode optical fiber was used instead of the beam profiler and space filtering beam expander.

The wavelength of the laser beam was measured real time with a High Finesse WS-6 laser wavelength meter (WLM). The further details of the setup and the measurement method are described in our earlier paper [28]. The measurements presented in the following section were carried out by using an acousto-optic modulator (AOM) for laser frequency modulation, except for the temperature dependence of the broad spectral hole and all experiments on $Er^{3+}$-doped sample, where laser frequency was modulated by the piezo modulation capability of the laser. The samples were mounted in a closed-cycle helium cryostat and cooled down between 9–30 K temperatures for the laser spectroscopic measurements. For the measurement of the narrow spectral hole of the LYB:$Yb^{3+}$ sample, a liquid helium cryostat was used with an additional helium pump in order to cool down the sample to near 2 K.

To determine the low-intensity limit of the spectral hole halfwidth, a Z-scan measurement setup was established. The pumping laser intensity was varied from 1.5 to 190 W/cm$^2$ for $Yb^{3+}$- and from 9 to 700 W/cm$^2$ for the $Er^{3+}$-doped LYB, while the confocal lens pair in front and behind the cryostat was moved together from the starting position to shift the focal plane passing through the sample. All intensity data given here are the peak value calculated for a two-dimensional Gaussian beam profile. The probing pulse with an intensity of about one tenth of the pumping pulse was delayed at about 10 and 30 µs for $Yb^{3+}$ and $Er^{3+}$ ions, respectively. Assuming a Gaussian laser beam profile, a theoretical function can be fitted to the halfwidth (FWHM) of the spectral hole as the function of the sample position to determine the homogeneous linewidth in the low-intensity limit:

$$\sigma^2(z) = \frac{f_2^2}{2} + \frac{kf_2}{2\left[1+\left(\frac{z-z_0}{z_R}\right)^2\right]} + \sqrt{\frac{f_2^4}{4} + \frac{kf_2^3}{2\left[1+\left(\frac{z-z_0}{z_R}\right)^2\right]}}, \quad (1)$$

where $\sigma$ is the spectral hole halfwidth, $f_2 = 4/T_2$, $z_R$ is the Rayleigh length, and $k$ depends on the material parameters, $T_1$, optical power, and beam waist:

$$k = \frac{4T_1|d_{eg}|^2 \mu_0 cP}{\hbar^2 n \pi w_0^2}. \quad (2)$$

Here, $d_{eg}$ is the dipole momentum associated with the transition, $\mu_0$ is the permeability of the vacuum, $c$ is the velocity of the light, $P$ is the power of the focused beam, $\hbar$ is the reduced Planck constant, $n$ is the effective refractive index of the sample, and $w_0$ is the beam waist.

### 3. Results

For the LYB:Yb$^{3+}$ sample a special double spectral hole structure consisting of a narrow and a broad spectral hole (see Figure 3) was observed at about 10,283.58 cm$^{-1}$, similarly to our measurement on LiNbO$_3$:Yb$^{3+}$ [28]. A potential reason for such a double spectral hole system can be the instantaneous spectral diffusion (ISD) [30]. In the LYB:Er$^{3+}$ crystal, a single absorption hole was generated at the $^4I_{15/2}$–$^4I_{11/2}$ electronic transition at about 10,275.36 cm$^{-1}$ (Figure 4). For the higher wavenumber spectral lines in LYB:Er$^{3+}$ (see absorption spectra in Figure 1), we found only weak spectral hole burning effect with a very broad halfwidth (>1400 MHz).

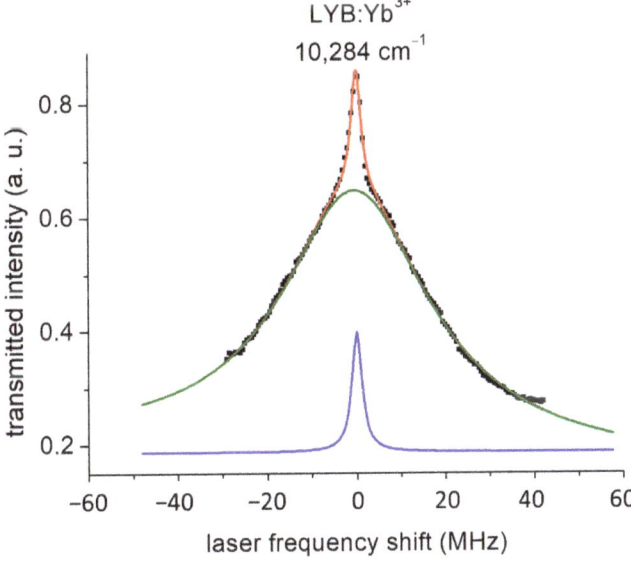

**Figure 3.** Double peak system of the spectral hole in LYB:Yb$^{3+}$. A peak of 30–50 MHz width is superposed with a narrow peak of 2–3 MHz width. Black squares indicate the measured points, olive and blue lines indicate the fitted broad and narrow Lorentzian curves, and the red line shows the sum of two Lorentzians.

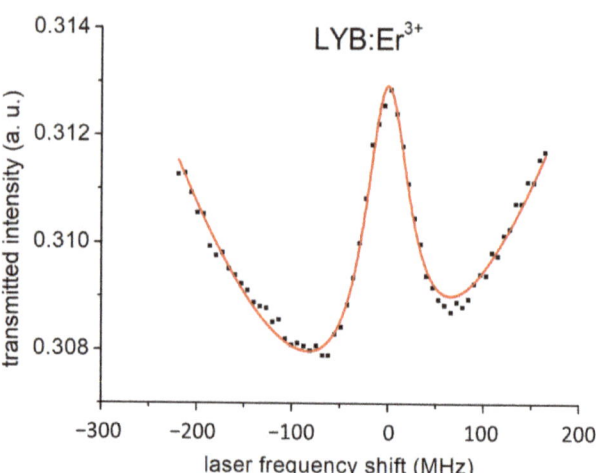

**Figure 4.** Spectral hole burned in LYB:Er$^{3+}$. The transmittance curve can be fitted by the superposition of the inhomogeneously broadened spectral line and a pure Lorentzian spectral hole.

Using the Z-scan technique, homogeneous linewidths of about 0.29 ± 0.03 and 22.5 ± 0.5 MHz, corresponding to the $T_2$ dipole relaxation times of about 1100 ± 120 and 14.2 ± 0.3 ns, were determined in LYB:Yb$^{3+}$ crystal for the narrow and broad spectral hole components, respectively (for the narrow one see Figure 5). In Figure 5, the Rabi frequency $\Omega_w$ of the pump pulse in the middle of the beam, calculated from the fitted parameters, is indicated on the right scale. For Er$^{3+}$-doped LYB crystal a dipole relaxation time of about 11.9 ± 0.2 ns (corresponding to a homogeneous linewidth of about 26.7 ± 0.4 MHz) was determined by weighted averaging of the results of two measurements in different polarization directions. Although there is a strong polarization dependence in the absorption and the spectral hole depth, any significant polarization dependence was obtained neither in spectral hole halfwidth nor in relaxation times.

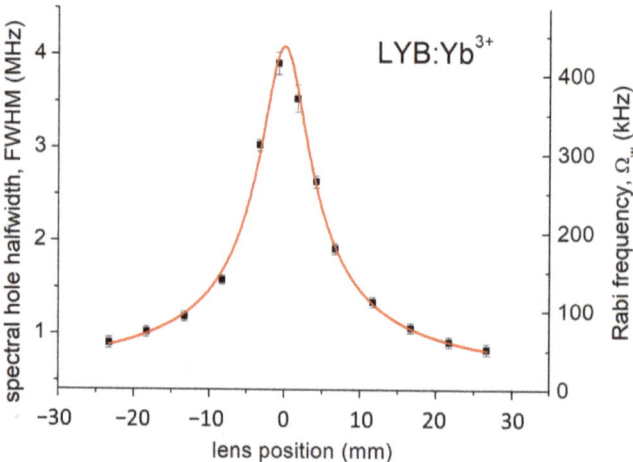

**Figure 5.** The width of the narrow spectral hole as a function of distance between the focal plane of the lens and the sample (Z-scan measurement) measured in LYB:Yb$^{3+}$ at $T$ = 2.2 K. The fit of Equation (1) is shown as a solid curve (left scale). The right scale approximately shows the Rabi frequency calculated from the fitted parameters.

The determination of population relaxation time ($T_1$) in LYB:Yb$^{3+}$ and LYB:Er$^{3+}$ crystals is shown in Figure 6, where the points represent the depths of the spectral holes as a function of delay time. In case of Yb$^{3+}$ dopant, the linear behavior of the measured data in logarithmic scale shows a single exponential decay process both for the narrow ($T_1 = 850 \pm 60$ μs) and broad ($T_1 = 1010 \pm 50$ μs) spectral holes. In addition, the line broadening as a function of delay time, i.e., the spectral diffusion rates at about 1 and 5 MHz/ms, have also been determined for the narrow and broad spectral lines during these measurements, respectively, at a temperature of 9 K and an intensity of 40 mW/cm$^2$ in the center of the beam at position $z = -17$ mm (see Figure 7). For this reason, the initial width is larger than its minimal value. The measurement of the population relaxation in the LYB:Er$^{3+}$ crystal exhibits a single exponential decrease with a time constant of $402 \pm 8$ μs, as a weighted average of results for polarizations parallel to the dielectric axes $x$ and $y$, since there is no reason to expect polarization-dependence for $T_1$.

In Figure 8, the temperature dependence of the narrow spectral hole width measured in the LYB:Yb$^{3+}$ crystal is shown, where the solid (red) line is a fitted curve using the direct one-phonon coupling model [31–33]:

$$\Gamma(T) = \Gamma_0 + a \frac{\exp(\frac{\hbar \omega_{ph}}{k_B T})}{\left[\exp(\frac{\hbar \omega_{ph}}{k_B T}) - 1\right]^2} \tag{3}$$

where $\Gamma(T)$ is the halfwidth (i.e., full width at half maximum FWHM) of the spectral hole at a temperature $T$; $\Gamma_0 = \Gamma(T)_{T \to 0}$; $a = 2(\delta w)^2 / \gamma$; $\delta w$ is the coupling constant; $\gamma$ is the bandwidth; and $\omega_{ph}$ is the frequency of the phonon band.

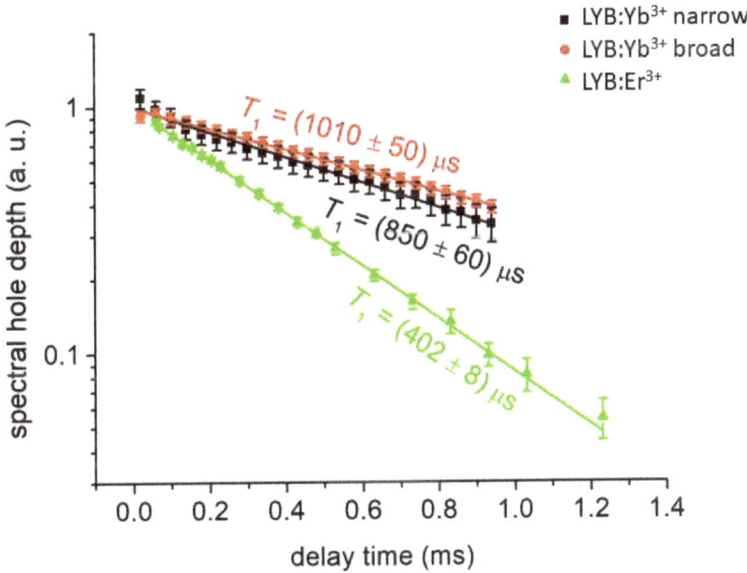

Figure 6. The depth of the spectral holes as a function of delay time between the pump and probe pulses for the Yb$^{3+}$- and Er$^{3+}$-doped LYB crystals.

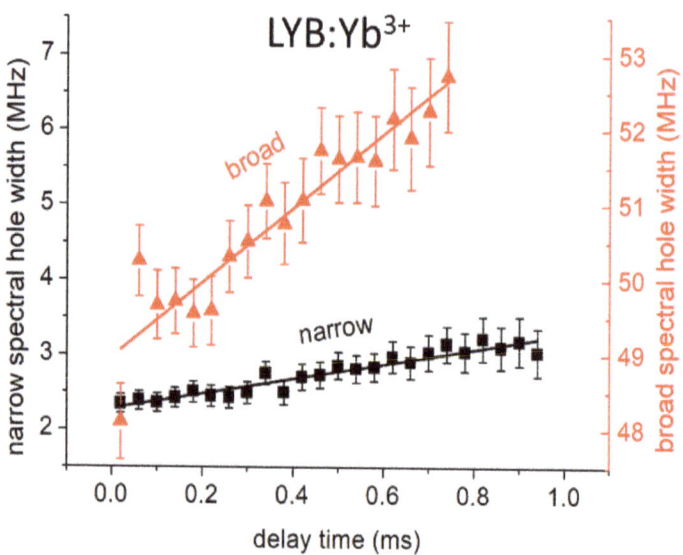

**Figure 7.** The impact of the spectral diffusion on the width of the spectral hole components in LYB:Yb$^{3+}$ at $T = 9$ K.

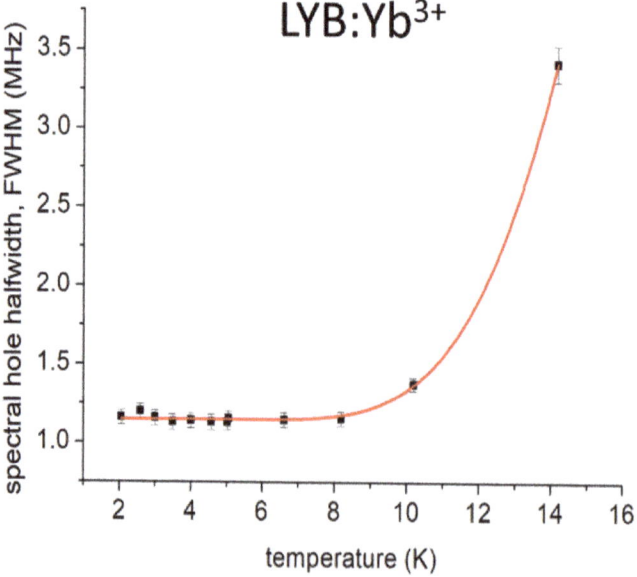

**Figure 8.** The width of the narrow spectral hole as a function of temperature measured in LYB:Yb$^{3+}$. The dots with error bars are the measured values, whereas the solid curve is the fitted exponential function, which originates from the direct one-phonon relaxation model. The measurement was taken at $z = -25$ mm position of the Z-scan setup, with 500 µs pumping time. For this reason, the halfwidth values and their zero-temperature limit are larger than its zero-intensity limit.

Because of the very different halfwidths, the broad and narrow spectral holes were measured by using the piezo and AOM scanning methods, respectively. As a consequence, the measurements of the broad and narrow spectral holes were carried out with a pumping intensity of 2.6 W/cm² and 9 W/cm² and with a delay time of 200 and 20 µs, respectively. Similar temperature dependence was obtained for the LYB:Er$^{3+}$ crystal. One can see that lowering the temperature of the sample below 9 K does not cause further significant decrease of the spectral hole halfwidth. The one-phonon coupling model describes the temperature dependence well for both crystals. The indicated $\omega_0$ parameter, characterizing the phonons playing a role during the relaxation, indicates similar relaxation processes through the lattice vibrations in LYB:Er$^{3+}$ and for the narrow spectral hole in LYB:Yb$^{3+}$ crystals.

For comparison, the $T_1$ and $T_2$ parameters measured on Yb$^{3+}$- and Er$^{3+}$-doped LiNbO$_3$ [28,29] and Yb$^{3+}$-doped Y$_2$SiO$_5$ crystals [34] are collected in Table 2.

**Table 2.** Results of spectral hole-burning experiments on LYB:Yb$^{3+}$ and LYB:Er$^{3+}$, compared with those of Yb$^{3+}$- and Er$^{3+}$-doped stoichiometric (sLN) and congruent lithium niobate (cLN), and with Y$_2$SiO$_5$:Yb$^{3+}$. L is the length of transmitted light path through the sample, i.e., the thickness of the sample; $T_1$ and $T_2$ are the population relaxation and dipole relaxation time constants of the transition, respectively; and $\omega_0$ is the characteristic phonon energy for the direct one-phonon coupling process in spatial frequency units.

| Material | Yb/Er Conc (mol%) | α (cm$^{-1}$) | L (mm) | $T_1$ (µs) | $T_2$ (ns) | $\omega_0$ (cm$^{-1}$) |
|---|---|---|---|---|---|---|
| LYB:Yb$^{3+}$ narrow | 0.1 * | 9.6 | 0.60 | 850 ± 60 | 1100 ± 120 | 58 ± 6 |
| broad | | | | 1010 ± 50 | 14.2 ± 0.3 | 145 ± 8 |
| sLN:Yb$^{3+}$ [28] narrow | 0.09 | 26 | 0.57 | 266 ± 17 | 134 ± 7 | 143 ± 20 |
| broad | | | | 438 ± 29 | 18.2 ± 0.5 | 54 ± 9 |
| cLN:Yb$^{3+}$ [28] narrow | 0.18 | 11.7 | 0.48 | 386 ± 34 | 240 ± 20 | 65 ± 8 |
| broad | | | | 420 ± 20 | 16 ± 1 | 89 ± 7 |
| Y$_2$SiO$_5$:Yb$^{3+}$ [34] | 0.005 | | | 4920 ± 10 | 103,000 ± 10,000 | |
| LYB:Er$^{3+}$ | 0.05 * | 3.1 | 2.00 | 402 ± 8 | 11.9 ± 0.2 | 58 ± 3 |
| sLN:Er$^{3+}$ [29] | 0.1 | 3.4 | 1.85 | 270 ± 140 | 6.84 ± 0.13 | 44 ± 3 |
| cLN:Er$^{3+}$ [29] | 0.1 | 0.44 | 1.89 | 160 ± 80 | 5.83 ± 0.11 | 37 ± 4 |

* in melt.

## 4. Discussion

In the case of the Yb$^{3+}$ ion, the population relaxation time ($T_1$) of both absorption holes is at least twice larger in LYB than in LiNbO$_3$. Additionally, though the dipole relaxation time of the broad hole is essentially similar in all cases, that of the narrow hole is much larger in the case of LYB. Similarly, the dipole and population relaxation times of the Er$^{3+}$ electronic transition in LYB crystal are almost double those measured in LiNbO$_3$. The increase of the population relaxation time can be explained through the higher level of defects in the lattice of LiNbO$_3$ due to the incorporation of RE$^{3+}$ ions as compared to the LYB crystal, where the incorporation into Y$^{3+}$ sites can be established without charge compensation and lattice distortion. The difference between the dipole relaxation times may originate from the different magnetic environment of the incorporated RE ions. In the LiNbO$_3$ crystal, the fluctuating magnetic field of the Nb$^{5+}$ ions may influence the electronic state of the RE ions in a much stronger manner than any of the ions in the LYB crystal, resulting in significantly shorter relaxation time.

In addition, the temperature dependence of the spectral hole parameters can also give information about the RE ion-lattice coupling, e.g., phonon coupling parameters. Although the phonon band structure of the LiNbO$_3$ and LYB crystals differs significantly, the similar $\omega_0$ coupling frequencies may show that some of the low frequency vibrational modes, probably localized ones, play an important role in the relaxation process in both crystals. In case of Er$^{3+}$, the transition between the 2nd and 3rd Stark level of the $^4I_{11/2}$ excited state corresponds to 60.7 cm$^{-1}$ [26] (see Table 1.), which is very near to the fitted phonon frequency at 58 cm$^{-1}$ and may confirm that the temperature dependence of the spectral hole width can be explained by the direct one-phonon coupling relaxation process. On the other hand, in case of Yb$^{3+}$, a coupling phonon frequency of $58 \pm 6$ and $145 \pm 8$ cm$^{-1}$ for the narrow and the broad spectral hole components was obtained, respectively, although both are quite far from even the smallest crystal field splitting value of 194 cm$^{-1}$ of the $^2F_{5/2}$ excited state [22]. It has to be mentioned, however, that low energy phonons around 130–160 cm$^{-1}$ have been detected in the Raman spectra of Ce-doped LYB crystals [35], which is in good agreement with the value of $\omega_0$ obtained for the broad spectral hole component.

Among the other phonon-coupling relaxation processes, a possible one is the Raman relaxation. Its contribution to the spectral hole broadening is the following:

$$\Delta \Gamma_R = \alpha \left(\frac{T}{T_D}\right)^7 \int_0^{x_0} \frac{x^6 exp(x)}{(exp(x)-1)^2} dx, \qquad (4)$$

where $\alpha$ is the electron–phonon coupling parameter for the Raman process, $T_D = \hbar \omega_D / k_B$ is the effective Debye temperature, $\omega_D$ is the Debye cut-off frequency, $k_B$ is the Boltzmann-constant, and $x_0 = T_D/T$ [32]. The integral has been calculated numerically, and we found that the single phonon relaxation formula in Equation (3), with appropriate change of the parameter $\omega_{ph} \to \omega_D$, is almost identical with Equation (4) in the range of $T/T_D = 0$–0.2. The parameter fitting of this equation describes the temperature dependence of the spectral hole broadening both for the narrow and the broad components quite well, and results in a value for $T_D$ equal to $111 \pm 6$ K and $131 \pm 3$ K, respectively. Since the numerical value of the expression in Equation (4) is practically the same as that given by the formula for the direct one-phonon model in Equation (3), the contribution of the direct one-phonon and the Raman relaxation cannot be separated only by fitting the temperature dependence of the spectral hole broadening. The temperature dependence of the spectral hole width may consist of contributions of more than one effect with similar temperature dependence. Orbach and multiphonon processes do not likely give significant contribution in relaxation for trivalent lanthanide ions [33].

## 5. Conclusions

A pump–probe type saturation spectroscopic experiment has been successfully employed to measure the relaxation parameters of the Yb$^{3+}$: $^2F_{7/2}$—$^2F_{5/2}$ and Er$^{3+}$: $^4I_{15/2}$—$^4I_{11/2}$ transitions in LYB single crystals in the temperature range of 2–14 and 9–28 K, respectively. The dipole and population relaxation times were found to be at about twice those measured in stoichiometric and congruent lithium niobate single crystals, except for the dipole relaxation time of the narrow component in Yb$^{3+}$, where the increase is much larger. The small number of defects due to the neutral charge substitution of Y for the RE ions in LYB crystals explains the longer population relaxation time as compared to lithium niobate. The difference in the dipole relaxation times can be understood by the fluctuating magnetic field, which seems to be stronger in LiNbO$_3$ due to the high nuclear magnetic moment of the Nb$^{5+}$ ion. Although LYB as a host material seems to be better than LiNbO$_3$, the coherence time of excitation of the Yb$^{3+}$ dopant is still much shorter than in Y$_2$SiO$_5$, and it is likely that there is a similar situation with the Er$^{3+}$ dopant as well as for the $^4I_{11/2}$–$^4I_{15/2}$ transition.

**Author Contributions:** Conceptualization, Z.K.; methodology, Z.K.; software, G.M.; validation, G.M. and Z.K.; formal analysis, G.M.; investigation, G.M., K.L. and É.T.-R.; resources, Z.K. and É.T.-R.; data curation, G.M.; writing—original draft preparation, G.M.; writing—review and editing, Z.K., K.L., L.K. and É.T.-R.; visualization, G.M.; supervision, Z.K. and L.K.; project administration, Z.K.; funding acquisition, Z.K. All authors have read and agreed to the published version of the manuscript.

**Funding:** This research was funded by "Innovációs és Technológiai Minisztérium" (the Ministry of Innovation and Technology) and the "Nemzeti Kutatási Fejlesztési és Innovációs Hivatal" (National Research, Development and Innovation Office) within the Quantum Information National Laboratory of Hungary.

**Institutional Review Board Statement:** Not applicable.

**Informed Consent Statement:** Not applicable.

**Data Availability Statement:** The data presented in this study are available on request from the corresponding author.

**Acknowledgments:** The authors are grateful to Gábor Corradi for useful discussions and to Ivett Hajdara for participating in the early experiments.

**Conflicts of Interest:** The authors declare no conflict of interest.

## References

1. Sun, Y.; Thiel, C.W.; Cone, R.L.; Equall, R.W.; Hutcheson, R.L. Recent progress in developing new rare earth materials for hole burning and coherent transient applications. *J. Lumin.* **2002**, *98*, 281–287. [CrossRef]
2. Bottger, T.; Thiel, C.W.; Sun, Y.; Cone, R.L. Optical decoherence and spectral diffusion at 1.5 µm in $Er^{3+}$:$Y_2SiO_5$ versus magnetic field, temperature, and $Er^{3+}$ concentration. *Phys. Rev. B* **2006**, *73*, 075101. [CrossRef]
3. Bottger, T.; Sun, Y.; Thiel, C.W.; Cone, R.L. Spectroscopy and dynamics of $Er^{3+}$:$Y_2SiO_5$ at 1.5 µm. *Phys. Rev. B* **2006**, *74*, 075107. [CrossRef]
4. Guillot-Noel, O.; Vezin, H.; Goldner, P.; Beaudoux, F.; Vincent, J.; Lejay, J.; Lorgere, I. Direct observation of rare-earth-host interactions in Er:$Y_2SiO_5$. *Phys. Rev. B* **2007**, *76*, 180408. [CrossRef]
5. Thiel, C.W.; Macfarlane, R.M.; Bottger, T.; Sun, Y.; Cone, R.L.; Babbitt, W.R. Optical decoherence and persistent spectral hole burning in $Er^{3+}$:$LiNbO_3$. *J. Lumin.* **2010**, *130*, 1603–1609. [CrossRef]
6. Wang, G.; Xue, Y.; Wu, J.-H.; Gao, J.-Y. Phase dependences of optical dispersion and group velocity in an $Er^{3+}$-doped yttrium aluminium garnet crystal. *J. Phys. B At. Mol. Opt. Phys.* **2006**, *39*, 4409–4417. [CrossRef]
7. Bottger, T.; Sun, Y.; Pryde, G.J.; Reinemer, G.; Cone, R.L. Diode laser frequency stabilization to transient spectral holes and spectral diffusion in $Er^{3+}$:$Y_2SiO_5$ at 1536 nm. *J. Lumin.* **2001**, *9495*, 565–568. [CrossRef]
8. Afzelius, M.; Staudt, M.U.; de Riedmatten, H.; Simon, C.; Hastings-Simon, S.R.; Ricken, R.; Suche, H.; Sohler, W.; Gisin, N. Interference of spontaneous emission of light from two solid-state atomic ensembles. *New J. Phys.* **2007**, *9*, 413. [CrossRef]
9. Xu, H.; Dai, Z.; Jiang, Z. Effect of concentration of the $Er^{3+}$ ion on electromagnetically induced transparency in $Er^{3+}$:YAG crystal. *Phys. Lett. A* **2002**, *294*, 19–25. [CrossRef]
10. Baldit, E.; Bencheikh, K.; Monnier, P.; Levenson, J.A.; Rouget, V. Ultraslow Light Propagation in an Inhomogeneously Broadened Rare-Earth Ion-Doped Crystal. *Phys. Rev. Lett.* **2005**, *95*, 143601. [CrossRef]
11. Staudt, M.U.; Hastings-Simon, S.R.; Nilsson, M.; Afzelius, M.; Scarani, V.; Ricken, R.; Suche, H.; Sohler, W.; Tittel, W.; Gisin, N. Fidelity of an Optical Memory Based on Stimulated Photon Echoes. *Phys. Rev. Lett.* **2007**, *98*, 113601. [CrossRef] [PubMed]
12. Lauritzen, B.; Minar, J.; de Riedmatten, H.; Afzelius, M.; Gisin, N. Approaches for a quantum memory at telecommunication wavelengths. *Phys. Rev. A* **2011**, *83*, 012318. [CrossRef]
13. Thiel, C.W.; Bottger, T.; Cone, R.L. Rare-earth-doped materials for applications in quantum information storage and signal processing. *J. Lumin.* **2011**, *131*, 353–361. [CrossRef]
14. Thiel, C.W.; Sun, Y.; Macfarlane, R.M.; Bottger, T.; Cone, R.L. Rare-earth-doped $LiNbO_3$ and $KTiOPO_4$ (KTP) for waveguide quantum memories. *J. Phys. B At. Mol. Opt. Phys.* **2012**, *45*, 124013. [CrossRef]
15. Dajczgewand, J.; Ahlefeldt, R.; Bottger, T.; Louchet-Chauvet, A.; Le Gouet, J.-L.; Chaneliere, T. Optical memory bandwidth and multiplexing capacity in the erbium telecommunication window. *New J. Phys.* **2015**, *17*, 023031. [CrossRef]
16. Paschotta, R.; Nilsson, J.; Tropper, A.C.; Hanna, D.C. Ytterbium-doped fiber amplifiers. *IEEE J. Quantum Electron.* **1997**, *33*, 1049–1056. [CrossRef]
17. Kränkel, C.; Fagundes-Peters, D.; Fredrich, S.T.; Johannsen, J.; Mond, M.; Huber, G.; Bernhagen, M.; Uecker, R. Continuous wave laser operation of $Yb^{3+}$:$YVO_4$. *Appl. Phys. B* **2004**, *79*, 543–546. [CrossRef]
18. Shekhovtsov, A.N.; Tolmachev, A.V.; Dubovik, M.F.; Dolzhenkova, E.F.; Korshikova, T.I.; Grinyov, B.V.; Baumer, V.N.; Zelenskaya, O.V. Structure and growth of pure and $Ce^{3+}$-doped $Li_6Gd(BO_3)_3$ single crystals. *J. Cryst. Growth* **2002**, *242*, 167–171. [CrossRef]
19. Yang, F.; Pan, S.K.; Ding, D.Z.; Chen, X.F.; Feng, H.; Ren, G.H. Crystal growth and luminescent properties of the Ce-doped $Li_6Lu(BO_3)_3$. *J. Cryst. Growth* **2010**, *312*, 2411–2414. [CrossRef]
20. Peter, A.; Polgar, K.; Toth, M. Synthesis and crystallization of lithium-yttrium orthoborate $Li_6Y(BO_3)_3$ phase. *J. Cryst. Growth* **2012**, *346*, 69–74. [CrossRef]
21. Zhao, Y.W.; Gong, X.H.; Chen, Y.J.; Huang, L.X.; Lin, Y.F.; Zhang, G.; Tan, Q.G.; Luo, Z.D.; Huang, Y.D. Spectroscopic properties of $Er^{3+}$ ions in $Li_6Y(BO_3)_3$ crystal. *Appl. Phys. B* **2007**, *88*, 51–55. [CrossRef]
22. Sablayrolles, J.; Jubera, V.; Guillen, F.; Decourt, R.; Couzi, M.; Chaminade, J.P.; Garcia, A. Infrared and visible spectroscopic studies of the ytterbium doped borate $Li_6Y(BO_3)_3$. *Opt. Commun.* **2007**, *280*, 103–109. [CrossRef]
23. Zhao, Y.; Xie, Y.; Li, Z.; Gong, X.; Chen, Y.; Lin, Y.; Huang, Y. Polarized Spectral Properties of Er-Yb-Codoped $Li_6Y(BO_3)_3$ Single Crystal. In Proceedings of the Symposium on Photonics and Optoelectronics (SOPO), Shanghai, China, 21–23 May 2012; pp. 1–3. [CrossRef]

24. Skvortsov, A.P.; Poletaev, N.K.; Polgar, K.; Peter, A. Absorption spectra of $Er^{3+}$ ions in $Li_6Y(BO_3)_3$ crystals. *Tech. Phys. Lett.* **2013**, *39*, 424–426. [CrossRef]
25. Kovács, L.; Arceiz Casas, S.; Corradi, G.; Tichy-Rács, É.; Kocsor, L.; Lengyel, K.; Ryba-Romanowski, W.; Strzep, A.; Scholle, A.; Greulich-Weber, S. Optical and EPR spectroscopy of $Er^{3+}$ in lithium yttrium borate, $Li_6Y(BO_3)_3$:Er single crystals. *Opt. Mater.* **2017**, *72*, 270–275. [CrossRef]
26. Delaigue, M.; Jubera, V.; Sablayrolles, J.; Chaminade, J.-P.; Garcia, A.; Manek-Honninger, I. Mode-locked and Q-switched laser operation of the Yb-doped $Li_6Y(BO_3)_3$ crystal. *Appl. Phys. B* **2007**, *87*, 693–696. [CrossRef]
27. Zhao, Y.; Lin, Y.F.; Chen, Y.J.; Gong, X.H.; Luo, Z.D.; Huang, Y.D. Spectroscopic properties and diode-pumped 1594 nm laser performance of Er:Yb:$Li_6Y(BO_3)_3$ crystal. *Appl. Phys. B* **2008**, *90*, 461–464. [CrossRef]
28. Kis, Z.; Mandula, G.; Lengyel, K.; Hajdara, I.; Kovacs, L.; Imlau, M. Homogeneous linewidth measurements of $Yb^{3+}$ ions in congruent and stoichiometric lithium niobate crystals. *Opt. Mat.* **2014**, *37*, 845–853. [CrossRef]
29. Mandula, G.; Kis, Z.; Kovacs, L.; Szaller, Z.; Krampf, A. Site-selective measurement of relaxation properties at 980 nm in $Er^{3+}$-doped congruent and stoichiometric lithium niobate crystals. *Appl. Phys. B* **2016**, *122*, 72. [CrossRef]
30. Huang, J.; Zhang, J.M.; Lezama, A.; Mossberg, T.W. Excess dephasing in photon-echo experiments arising from excitation-induced electronic level shifts. *Phys. Rev. Lett.* **1989**, *63*, 78–81. [CrossRef]
31. Yen, W.M.; Scott, W.C.; Schawlow, A.L. Phonon-Induced Relaxation in Excited Optical States of Trivalent Praseodymium in $LaF_3$. *Phys. Rev.* **1964**, *136*, A271–A283. [CrossRef]
32. Henderson, B.; Imbusch, G.F. Chapter 5: Electronic Centres in a Vibrating Crystalline Environment. In *Optical Spectroscopy of Inorganic Solids; Monographs on the Physics and Chemistry of Materials 44*, 1st ed.; Henderson, B., Imbusch, G.F., Eds.; Oxford University Press Inc.: New York, NY, USA, 1989; pp. 183–257. ISBN 9780199298624.
33. Ellens, A.; Andres, H.; Meijerink, A.; Blasse, G. Spectral-line-broadening study of the trivalent lanthanide-ion series. I. Line broadening as a probe of the electron-phonon coupling strength. *Phys. Rev. B* **1997**, *55*, 173–179. [CrossRef]
34. Lim, H.J.; Welinski, S.; Ferrier, A.; Goldner, P.; Morton, J.J.L. Coherent spin dynamics of ytterbium ions in yttrium orthosilicate. *Phys. Rev. B* **2018**, *97*, 064409. [CrossRef]
35. Yuan, D.; Víllora, E.G.; Shimamura, K. Flux-growth of Ce:$LiY_6O_5(BO_3)_3$ single crystals during the Czochralski pulling from nearly congruent Ce:$Li_6Y(BO_3)_3$ and their optical properties. *J. Solid State Chem.* **2021**, *300*, 122286. [CrossRef]

MDPI
St. Alban-Anlage 66
4052 Basel
Switzerland
Tel. +41 61 683 77 34
Fax +41 61 302 89 18
www.mdpi.com

*Crystals* Editorial Office
E-mail: crystals@mdpi.com
www.mdpi.com/journal/crystals

www.ingramcontent.com/pod-product-compliance
Lightning Source LLC
LaVergne TN
LVHW070643100526
838202LV00013B/869